Family Farming and the Worlds to Come

Jean-Michel Sourisseau
Editor

Family Farming and the Worlds to Come

 Springer

Editor
Jean-Michel Sourisseau
CIRAD
Paris, France

ISBN 978-94-017-9357-5   ISBN 978-94-017-9358-2 (eBook)
DOI 10.1007/978-94-017-9358-2
Springer Dordrecht Heidelberg New York London

Library of Congress Control Number: 2014954240

Éditions Quæ, R10, 78026 Versailles cedex, France www.quae.com

© Éditions Quæ, 2015
This work is subject to copyright. All rights are reserved by the Publisher, whether the whole or part of the material is concerned, specifically the rights of translation, reprinting, reuse of illustrations, recitation, broadcasting, reproduction on microfilms or in any other physical way, and transmission or information storage and retrieval, electronic adaptation, computer software, or by similar or dissimilar methodology now known or hereafter developed. Exempted from this legal reservation are brief excerpts in connection with reviews or scholarly analysis or material supplied specifically for the purpose of being entered and executed on a computer system, for exclusive use by the purchaser of the work. Duplication of this publication or parts thereof is permitted only under the provisions of the Copyright Law of the Publisher's location, in its current version, and permission for use must always be obtained from Springer. Permissions for use may be obtained through RightsLink at the Copyright Clearance Center. Violations are liable to prosecution under the respective Copyright Law.
The use of general descriptive names, registered names, trademarks, service marks, etc. in this publication does not imply, even in the absence of a specific statement, that such names are exempt from the relevant protective laws and regulations and therefore free for general use.
While the advice and information in this book are believed to be true and accurate at the date of publication, neither the authors nor the editors nor the publisher can accept any legal responsibility for any errors or omissions that may be made. The publisher makes no warranty, express or implied, with respect to the material contained herein.

Printed on acid-free paper

Springer is part of Springer Science+Business Media (www.springer.com)

Foreword

This book is most welcome in a year dedicated to family farming by the United Nations. It revisits the most common approaches for analyzing and understanding family farming – and emphasizes, rightly so, the multiplicity of forms of family farming that do exist. It is essential to properly define this type of agriculture to better identify its issues and to place it in a larger context and thus assess its contribution to sustainable and equitable development. I thus welcome the initiative taken to make family farming once again central to debates on agricultural development.

Family farming is indeed at the heart of world agriculture. The role it plays in the dynamics of land and agrarian reform is of paramount importance to the transformational projects that many countries of the South intend to undertake in the next 10–20 years. At a time when our territories and lands are being reshaped, rapid and often unpredictable changes require populations to exhibit a high level of resilience to hazards of all kinds. Faced with these challenges, family farming is a rational choice to help maintain the delicate balance of ever dwindling natural resources in order to meet the needs of a growing population. Calibrating our responses to changing daily needs so that we can preserve the legacy that we leave to future generations remains a global challenge. Every continent has to meet this challenge, Africa especially so, given its demographic perspectives and vast agricultural potential.

This publication encapsulates three decades of intellectual investment on a very complex subject. The relative lack of knowledge about the topic can explain – but not excuse – the perpetuation of notoriously inappropriate actions. In the African context, and thus for NEPAD,[1] the reconstruction of the concept of family farming is a particular challenge given that competing and often pejorative concepts are used to describe it: small-scale agriculture, subsistence farming, and peasant agriculture among others. By allowing us to clearly identify the contours – as well as the

---

[1] *The New Partnership for Africa's Development.*

variability – of family farming models, this book provides us the means to unmask the ideological underpinnings or the limitations of each of these concepts. Clearly identifying the specificity of these issues will be an important first step to be able to answer questions on the ability of family farming to cope with global challenges, as compared to other types of agriculture. This work will need to be put to the test in real-world situations, especially in Africa, where the notion of family itself takes multiple forms.

This book's recognition of family farming as one of the main drivers of rural transformations currently underway and its conviction that this form of farming is worth promoting in many regions of the world is encouraging, especially to us in Africa. It reinforces our vision of agriculture which accords more responsibility to those who produce themselves, of an Africa proud of its agriculture and its farmers. This book presents an overview that reinforces our commitment towards constructing a plural Africa which assumes and hopes for a better future.

This collaborative work is decidedly an important milestone in CIRAD's long-standing commitment to address fundamental issues for the sustainable development of a changing world.

Dr. Ibrahim Assane Mayaki
Executive Secretary
NEPAD Planning and Coordination Agency

Acknowledgments

This book is a real collective achievement. In addition to the authors of chapters and boxes acknowledged at the end of the book, many people were involved in it. All their contributions, whatever their shape and size, are part of the richness and the diversity of this work.

We express our sincere thanks to:

Jacques Avelino, Vincent Baron, Hubert de Bon, Muriel Bonin, Julien Capelle, Alexandre Caron, Véronique Chevalier, Marc Corbeels, Jean-Philippe Deguine, Stéphanie Desvaux, Sophie Devienne, Jean-Marie Douzet, Noël Durand, Sandrine Dury, Guillaume Duteurtre, Céline Dutilly, Bernard Faye, Getachew Gari, Michel de Garine-Wichatitsky, Régis Goebel, Flavie Goutard, Vladimir Grosbois, Hubert Guérin, Jean-Luc Hofs, Ferran Jori, Rémi Kahane, Rabah Lahmar, Renaud Lancelot, Luc de Lapeyre de Bellaire, Fabrice Le Bellec, Matthieu Lesnoff, Geneviève Libeau, Jacques Loyat, Lucia Manso-Silvan, Thibaud Martin, Pierre Montagne, Didier Montet, Paule Moustier, Krishna Naudin, Mathilde Paul, Fabrice Pinard, Michel Rivier, Éric Scopel, Samira Sarter, Renata Servan de Almeida, François Thiaucourt, Bernard Triomphe, Jean-Michel Vassal, Jean-François Vayssières.

We also want to thank Patrick Caron, Pierre Fabre, Étienne Hainzelin, Bernard Hubert, Denis Pesche and Emmanuel Torquebiau who commented and proposed improvements on an early version of the text. Last but not least, we are most grateful to Régine Chatagnier; this book would not be what it is without her precious assistance.

Contents

1 General Introduction 1
    Jean-Michel Sourisseau

Part I Defining and Understanding Family Farming

2 Family Farming: At the Core of the World's Agricultural History . . . 13
    Bruno Losch

3 Defining, Characterizing and Measuring Family
    Farming Models 37
    Pierre-Marie Bosc, Jacques Marzin, Jean-François Bélières,
    Jean-Michel Sourisseau, Philippe Bonnal, Bruno Losch,
    Philippe Pédelahore, and Laurent Parrot

4 Families, Labor and Farms 57
    Véronique Ancey and Sandrine Fréguin-Gresh

5 Family Farming and Other Forms of Agriculture 71
    Jacques Marzin, Benoît Daviron, and Sylvain Rafflegeau

Part II Helping to Feed the World and Territories to Live

6 Contributing to Social and Ecological Systems 95
    Laurène Feintrenie and François Affholder

7 Contributing to Territorial Dynamics 111
    Stéphanie Barral, Marc Piraux, Jean-Michel Sourisseau,
    and Élodie Valette

8	**Contributing to Production and to International Markets** .... Sylvain Rafflegeau, Bruno Losch, Benoît Daviron, Philippe Bastide, Pierre Charmetant, Thierry Lescot, Alexia Prades, and Jérôme Sainte-Beuve	129
9	**Contributing to Innovation, Policies and Local Democracy Through Collective Action** .... Pierre-Marie Bosc, Marc Piraux, and Michel Dulcire	145

Part III Meeting the Challenges of the Future

10	**Challenges of Poverty, Employment and Food Security** .... Philippe Bonnal, Bruno Losch, Jacques Marzin, and Laurent Parrot	163
11	**Energy Challenges: Threats or Opportunities?** .... Marie-Hélène Dabat, Denis Gautier, Laurent Gazull, and François Pinta	181
12	**Health Challenges: Increasing Global Impacts** .... Sophie Molia, Pascal Bonnet, and Alain Ratnadass	199
13	**Challenges of Managing and Using Natural Resources** .... Danièle Clavel, Laurène Feintrenie, Jean-Yves Jamin, Emmanuel Torquebiau, and Didier Bazile	217

Part IV Research and the Challenges Facing Family Farming

14	**Co-constructing Innovation: Action Research in Partnership** ..... Eric Vall and Eduardo Chia	237
15	**Innovations in Extension and Advisory Services for Family Farms** .... Guy Faure, Michel Havard, Aurélie Toillier, Patrice Djamen Nana, and Ismail Moumouni	255
16	**Support for the Prevention of Health Risks** .... Sophie Molia, Pascal Bonnet, and Alain Ratnadass	267
17	**Agricultural Biodiversity and Rural Systems of Seed Production** ... Danièle Clavel, Didier Bazile, Benoît Bertrand, Olivier Sounigo, Kirsten vom Brocke, and Gilles Trouche	285
18	**Lessons and Perspectives of Ecological Intensification** .... François Affholder, Laurent Parrot, and Patrick Jagoret	301

General Conclusion 313

Boxes 321

Liste des auteurs 327

References 331

Chapter 1
General Introduction

Jean-Michel Sourisseau

One does not become a Cassandra just by acknowledging the magnitude of the risks and challenges confronting the world. Statistical data, various studies and reports originating from across the world, as well as empirical evidence, overwhelmingly show the impact and negative consequences for our societies and their environment of current development models.[1] Poverty and inequality; under-employment, unemployment and job insecurity; food insecurity; energy transitions; existing and emerging health risks; scarcity or depletion of natural resources (including of land); loss of biodiversity; climate change; and more – the list of these risks and challenges keeps on growing, even as the links between them are revealed, greatly complicating the search for measures to confront them. But at the same time, reports of the Millennium Development Goals offer some hope, showing that concerted action, driven by a strong political will, can be effective and influence catastrophic trends (UN 2013).

Food security, and hence agriculture as a means to achieve it, along with all the possible strategies that can be used, figure prominently in most national debates on development and is on the agendas of global governance entities. Because it is the primary sector contributing to food security and employment (both globally and in the majority of emerging countries and those of the South), because it defines landscapes and shapes territories over the entire planet and because it feeds mankind, agriculture cannot be excluded from any serious future perspectives. After being neglected for over twenty years – at least in terms of investments –, its importance is now once again being recognized (World Bank 2007). Even

---

[1] We can mention here six of the most significant and the most publicized expert reports covering agriculture, poverty, the green economy, climate and environment: UNCTAD (2013), IAASTD (2009), IFAD (2011), UNEP (2010), IPCC (2007) and MEA (2005a).

J.-M. Sourisseau (✉)
ES, CIRAD, Montpellier, France
e-mail: jean-michel.sourisseau@cirad.fr

© Éditions Quæ, 2015
J.-M. Sourisseau (ed.), *Family Farming and the Worlds to Come*,
DOI 10.1007/978-94-017-9358-2_1

though it does contribute to increasing the risks to humanity through its negative externalities, it also provides at the same time solutions derived from farmer knowledge and know-how, technological advances, and also – as it has always done – through enlightened public choices and proactive policies.

By choosing to declare 2014 the International Year of Family Farming, the UN and political and civil society actors who campaigned for this declaration invite us to address these global challenges, and to examine them through the prism of agriculture and its various models and forms of production. While the goal of the International Year and its expected extensions may well be to arrive at an ideal form of production and to promote it, it is also a matter, implicitly, of comparing it with other forms which may be less virtuous and whose hegemony may be detrimental to the sector and, by extension, to the whole of humanity.

This message becomes even stronger since it is delivered in a context marked by the well-publicized highlighting of the limitations and excesses of various current agricultural models, as revealed by their competitions and confrontations.

- Thus the large-scale land grabs – for agriculture as well as for environmental preservation and mineral exploration – reflect both the strategic nature of the land and its control as well as the appropriation of the use of natural resources that are found on it, the first and foremost being water. But these land grabs also reveal asymmetries between countries and between economic and social actors, especially in the agriculture sector (Rulli et al. 2013).
- Thus agriculture is explicitly implicated in many parts of the world for its impact on the integrity of the environment, its dangerous management of natural resources and its inability to control public health risks, many of which it has itself helped initiate or worsen. Criticism is particularly strong for agriculture which benefited from post-war technological revolutions: the accelerated modernization in developed countries and its "developmental" variations in the form of the Green Revolution in many developing countries and those described today as "emerging."
- Thus the staple-food price crises of 2007 and 2008 in the South, resulting social upheavals and weak State and institutional responses highlighted the limits of regulations by the market and showed them as failures of national and international agricultural policies (HLPE 2011). These crises showed that wealth disparities continue to widen to the detriment of the poorest countries and their rural zones.
- Thus the capacity of agriculture to generate sufficient employment and income to allow rural zones to exit from poverty is called into question. Some even prefer instead an urban development model based on a reduction of agricultural jobs resulting from its modernization (Collier and Dercon 2013).
- And thus, finally, agriculture's identity crisis is regularly brought to the fore. In the developed and emerging countries, it takes the form of a growing distance from the rest of society, which increasingly perceives this vocation negatively.

The psychological dimension of this disconnect[2] and the sense of an activity undertaken by default are clear signs of the depth of the crisis. To exit the profession would mean aligning oneself with the rhythms of other economic sectors and with urban consumption levels. In the South, statistical evidence shows that the refusal to become a farmer and instead according preference to urban life usually results from forms of patriarchal family farming which do not give sufficient importance to young people and women.

All these signs alert us to the need for promoting, renewing or rediscovering production models that are more suitable to face the challenges confronting agriculture. These efforts are essential for redefining the place and role of agriculture in society. The signs also invite us to reflect on the sustainable limits of agricultural externalities and to discard approaches strictly oriented towards the production of commodities. These approaches remain currently dominant and have demonstrated their ability to expand rather than to respond to difficult and uncomfortable questions about current agricultural models.

The International Year of Family Farming can be a catalyst for encouraging such reflection, necessary before concrete steps can be taken to change the course of agriculture. However, we have to ensure that the proposals and implications of changes in thinking prompted by the International Year – as too the resulting adapted public policies – manage to overcome the caricatural divide sometimes reflected in political messages and scientific controversies. A naive defense of local agriculture as an undertaking which encompasses all economic, social and environmental virtues appears as risky and counter-productive as recurring proclamations of the end of farming and farmers, especially family farmers and peasants, supposedly too unproductive and thus incapable of surviving the competition of the sector's industrialization and its inclusion in globalization. Scenarios which reject the use of any mineral or synthetic fertilizers and fossil energy appear as unrealistic as a world with only 5 % of the world population working in agriculture and the rest living in urban concentrations with standardized levels and patterns of consumption.

Box 1.1. Some conventions used in this book.

Even though this geographic distinction has lost its meaning, throughout this book we use the generic terms "the countries of the South" or just "the South" to refer to the poorest countries. These are countries with large agricultural bases and which are often dependent on international aid for implementing public policies.

We call "the countries of the North" or "the North" the older industrialized countries, with the highest levels of wealth. These are countries in which agriculture's weightage in employment and GDP has fallen sharply due to the diversification of their economies.

And, finally, we call "transition countries" or "emerging countries" those that still have a significant agricultural base and whose development has accelerated over the last two decades. These countries are now approaching overall levels of wealth comparable to the countries of the North.

The text often refers to consultations of international databases. The references to these sites are not systematic, but all were accessed (and accessible) in 2013.

---

[2] According to the French Institute for Public Health Surveillance, 485 farmers committed suicide in France between 2007 and 2009.

1.1 This Book's Objectives

This book aims at a problem-oriented review of research on family farming systems. It examines the methods and results of research for development on family farming systems, and their likely place and roles in the face of global challenges. It is therefore not a plea – development actors from outside the world of research have that aspect in hand – but rather an effort to problematize the diversity and specificities of these forms of production through the prism of issues of agriculture and rural development. Through this problematization, the book more generally explores social choices and the path of development at national and international levels, as well as the roles that agriculture plays and should play.

Family farming systems have long been the subject of research for development – especially at CIRAD – but they reflect differentiated approaches. Some research focuses directly on family farming systems, which are then the work's main subject, others on the production of knowledge or processes intended for these forms of family farming, which are then one of the beneficiaries of the work but are not the subject, while still others explicitly enter into partnerships with family farming systems, which are then true actors of protocols. Providing an overview of the research on the topic requires distinguishing between research on, for and with family farming systems.

The family farming category is also researched by many disciplines and a wide variety of studies at all observation and analysis levels, ranging from the genome to global governance. This book incorporates, as widely as possible, these different disciplines which contribute, alone or in interdisciplinarity, to put family farming systems into perspective with respect to major development challenges.

Based on these goals, the book sets out four specific objectives, which determine its structure:

- by taking care to differentiate the normative registers between common sense, academicism and policy, it helps to define family farming systems and to analyze the issues underlying this effort of definition;
- by adopting an often critical posture, it assesses the overall and specific contributions of family farming systems – positive and negative, spelling out their limitations and weaknesses – to the economy, environmental management, territorial structuring and social equilibria;
- by being wary of systematically misery-centric representations and by emphasizing the importance of public policies which are (or are not) addressed to them, it explores global challenges that confront forms of family farming, and responses the latter have (or do not have) to overcome or mitigate these challenges;
- through representative illustrations of the diversity of work, it finally shows how targeted research conducted by CIRAD and its partners can help improve family farming systems' responses to these global challenges.

1 General Introduction

The book ultimately aims to show that these farming systems cannot be reduced to a coherent set of forms of production and that their diversity requires that their performance not be idealized, much less stigmatized as archaic or as synonymous with poverty. It nevertheless stresses that if the right conditions are met (both in terms of public goods as well as of appropriate measures of public policy for support and capacity building), family farming systems can be credible instruments of action. Often through linkages – but also in competition – with other forms of production, they can meet the global challenges of the rural world and help design more sustainable production models.

1.2 A Collective Writing Process

Some 50 researchers – agronomists, economists, sociologists, geneticists, animal scientists, anthropologists, political scientists – mainly from CIRAD, were mobilized for writing this book's 17 chapters.

A first task was to inventory research conducted for and with family farmers and then to organize it. The majority of work identified relates to the first research approach – "on" family farming systems – and mainly involves the human and social sciences. But relevant research by geneticists, animal scientists and agronomists was also revealed. Since their subjects were the genome, pandemics, the plant or cultivation and production systems, some initial reflection followed by work of formalization was necessary to link these subjects with family farming. This has imbued the book with some originality in how its technical content is introduced. Using these research postures has allowed the various disciplines to be linked and inter-combined as much as possible in the chapters.

On the basis of the research thus identified, a working group arrived at the book's overall rationale and structure, and drafted the synopses of its parts and chapters. Each part was coordinated by a team which defined that part's messages and arguments and structured them into chapters. Then each chapter, with its contours now predefined, was written by a group of authors. These authors also included the knowledge and expertise of other researchers, mainly in the form of text boxes. The book's scientific editor and the coordinators of its parts ensured the coherence and progression of each part, then of the whole book. Finally, the first manuscript was reviewed and commented on by a scientific committee, and then iterated through the same process.

1.3 Navigating the Book

The book is divided into 4 parts and 17 chapters. Because of the way it was written, the chapters can be read independently: each reveals and explains the linked results of research and studies, each has its own message, arrived at through research on a

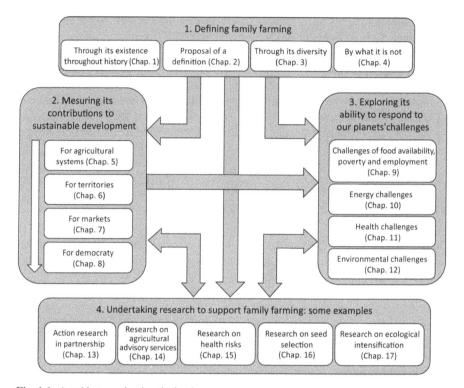

Fig. 1.1 A guide to navigating the book

thematic or disciplinary coherence. Grouping them into parts is the second level of structural coherence, of the development of arguments linking various disciplines.

- Chapters 2, 3, 4 and 5 explore the definition of the various forms of family farming (Part I).
- Chapters 6, 7, 8, and 9 assess the contributions – whether positive or negative – of these forms of family farming on sustainable development (Part II).
- Chapters 10, 11, and 13 discuss the global challenges ahead and the ability – or inability – of families to confront them (Part III).
- Chapters 14, 15, 16, 17, and 18 show how research can support the familial forms of agricultural production in response to these challenges (Part IV).

Figure 1.1 shows the rationale behind these groupings and offers suggestions for navigating the book.

The first part, which defines and positions the book's topic, frames and clarifies the subject and places it in context. The vocabulary and conceptualization used in this first part makes it easier to read the other parts.

The other three parts interact more amongst themselves. Thus examples of research work on supporting agriculture presented in the book correspond to lever points found in Parts II and III. But in return they call into question the methods for measuring and increasing the contributions of the many types of family farming to development, and provide information on or strengthen their ability to respond to challenges.

Part I
Defining and Understanding Family Farming

Coordinated by Pierre-Marie Bosc, Laurent Parrot, Christian Corniaux

If some commentators are to be believed, the future food needs of the world will not be met by peasants and family farmers, but instead by a growing number of large companies working in conjunction with upstream and downstream industries. Family farming systems, especially those that have kept their distance from agricultural revolutions and which today account for the bulk of the world's farmers, would be doomed to relegation – in the words of Hervieu and Purseigle (2011). Needless to say this is not an opinion that the authors of this book share!

This first part lays out our conceptual framework and issues, mainly by identifying the key topics that will be referred to later in the book and by placing them in the context of current debates on the future of different forms of agricultural production. This requires defining what we mean by family farming but also to list, briefly, the other forms of organization of agricultural production that can be found on our planet.

Chapter 2 provides a long-term historical perspective of agriculture. Looking beyond the names used in each era and each society's own family structures (Todd 2011), we realize that over a long period it is the various family forms that have been able to fulfill humanity's food and non-food needs. Indeed, family forms of production are found on all continents. They occupy all environments, from the most favorable to the most hostile. They have proven their ability to domesticate plants and animals and to transform natural resources to make them more suitable for production than they originally were. These family forms of production were able to meet increasing demands, initially evolving technically in quite an autonomous manner with respect to the rest of society. Then, with the changes brought about by the industrial revolution, they were able to use more elements produced by upstream industries and fulfill the diversity of standards required by downstream operators. Market integration and selective dissemination of conventional modernization occurred more or less gradually depending on the context. However, at the global level, there has been a very uneven modernization of family farming systems with the majority of them still being based on manual cultivation.

It is on this very inegalitarian basis that these family farming systems of the world have to compete, especially since the creation of the World Trade Organization (WTO) some 20 years ago. This purely trade-oriented opportunity and opening provokes reactions of rejection from rural and farmer organizations and it raises new challenges for research and societies on the reproducibility and sustainability of conventional technical models.

Therefore, given this strongly contrasted global perspective, it is important to try to make clear what is meant by family farming or family farm. This is what Chap. 3 intends to do. "Family farming" or "family farm" are terms used by farmers in West Africa using manual labor or animal traction as well as by farmers in the U.S. and Latin America who have adopted mechanization. It is therefore important to provide a definition that goes beyond differences in socio-political contexts and allows a better understanding of what this form of production is. Hence, Chap. 3 examines the origin of the vagueness surrounding the names and definitions of family forms and models of agricultural production and its consequences in terms of the discourse and representations of agriculture. For this, it is imperative to specify what viewpoint one adopts while defining and labelling: that of research, of the farmers themselves, of politicians and policy-makers or of society? The chapter explains the "ways of naming" a multifaceted reality and offers keys to grasping a diversity which has to bear some responsibility for the naming difficulties. It finally offers a strict definition of family farming based on family labor and the inseparability of the farm from the family. Such a definition leaves open the possibility of an improved identification and quantitative assessment of family farming systems and their contributions through agricultural statistics.

The diversity of families and the way they operate is the key to introducing Chap. 4. The authors accept the challenge of showing the usefulness of thinking in terms of family farms but by going beyond the requirements of a strict definition and by discussing its principles. This chapter highlights the complexity of the processes at work resulting from family pluriactivity, mobility and multifaceted strategies in which farming can play a central and strategic position or occupy a more modest role but still remain important in terms of food security for family groups. The reference to the social functioning of family farmers is central to the issue as it allows economic functions to be positioned in their rightful place. Family farmers are first and foremost social actors and only then are they agricultural producers. But their role certainly cannot be reduced to the one of "raw material suppliers." What finally seems to be the best key to understanding the tremendous resilience of this form of production is the interweaving of social and patrimonial rationales on the one hand, with productive and economic rationales on the other. To understand the dynamics of family farming, this chapter launches a methodological challenge which consists of assuming at the same time the robustness of a global generic definition, the need for contextualization (geographical and historical) and the need for expanding analyses to the non-market and non-sectoral dimensions.

Chapter 5 returns to the characterization of other forms of organization of agricultural production and stresses through this contrast the specificity of family forms such as those described in Chap. 3. This chapter complements thus the analysis of forms of organization of production by the family business and enterprise models, emphasizing their distinctive features and highlighting the diversity of the types of relationships they have with family forms. The debates continue to be animated on the strengths and limitations of these differences, both for supplying populations with food as well as for the provision of international markets. The different positions adopted are strongly dependent on social and public policy choices.

Chapter 2
Family Farming: At the Core of the World's Agricultural History

Bruno Losch

The diversity of agriculture in the world reflects the immense variety of societies and natural environments on the planet. Indeed, agricultural systems range from various types of shifting slash-and-burn practices – sometimes very similar to those of the first sedentary human groups – to quasi-automated agricultures in some regions of the world. These systems present huge gaps in terms of modes of exploitation of natural resources, levels of capital use, productivity and market integration. They reflect various stages of transformation of agriculture depending on their technical level, their integration into globalized markets and the structural changes of national economies around the world. They also echo the transition from agrarian societies – organized around the relationships between rural communities and with their natural environment –, to predominantly urban ones characterized by a high degree of division of labor, where agricultural production is increasingly implemented through processes of artificialization of cultivated areas and the industrialization of the food chain. And yet, in absolute terms, there have never been as many farmers globally as there are today.

A historical perspective is necessary to understand the multiplicity of agricultural situations existing today and the very specific and central role of family farming systems. Family agriculture is embedded in agrarian history, a history that has played a key role in the overall evolution of economies and societies. In recent centuries, it has been intrinsically linked with the major agricultural and industrial changes that have taken place, at very different speeds in different parts of the world.

B. Losch (✉)
ES, CIRAD, Montpellier, France
e-mail: bruno.losch@cirad.fr

This chapter[1] discusses the major stages of technological advances that have marked the world's agricultural history. It then shows their inclusion in broader processes of structural change that have characterized the world's different economies and societies. Finally, it addresses the emergence of agricultural policies and the way they have dealt with peasantry and then family farming. This review is intended to help better understand the origin of the gaps of productivity between world's agricultures, gaps which lead to growing asymmetries contributing to increasing challenges related to poverty, employment and use of natural resources. These challenges will be addressed later in the book.

When limited to a few pages, such a goal is necessarily extremely reductive. Therefore, this chapter is primarily intended to provide a useful overview of the different analytical perspectives that will be developed in later chapters of the book. It also seeks to encourage questions and challenge our beliefs on technical and organizational configurations which are considered achievements and models to be replicated, but which are, above all, the result of economic and social power relations built over time, and whose local and global sustainability remains open to question.

In this regard, the emphasis we place in this chapter on the process of modernization of European agriculture, inseparable from the industrial revolution and its gradual global spread, is not due to any tropism or analytical bias of the book's authors. It is a matter rather of a specific choice to suggest keys to help interpret current challenges facing family farming systems around the world. This does not imply that the history of other agricultures elsewhere in the world – long viewed through the prism of a Eurocentric historiography – are any less important (Goody 2006; Bertrand 2011).

2.1 A Brief Review of Agriculture's Long History

Access to nutrients necessary to meet the physiological needs for survival and reproduction is a fundamental imperative that the human species cannot avoid. The manner in which this access is organized has helped structure the functioning of the first human groups, initially through direct extractions from ecosystems, later by the domestication of plants and animals. The origins of agriculture are part of this process that has contributed to the gradual settling down of nomadic hunter-gatherers. The domestication of species and cultivation of the land have, in fact, involved localized management of productive assets and harvests.

The organization of the family is the core of social dynamics, and today's diversity of family types contributes to and shapes the many forms of family

---

[1] This chapter has benefited from the collective input of and specific feedback from V. Ancey, P. Bonnal, P.-M. Bosc, J.-F. Bélières, B. Daviron, J. Marzin, D. Pesche and J.-M. Sourisseau.

farming (Chap. 4). Indeed, the very basis of family agriculture expresses the embeddedness of agricultural activities with family dynamics (Chap. 3).

Over time and in different regions, family types have taken very different contours. In Eurasia alone, Todd (2011) identifies 15 types which he groups in three main classes: nuclear family, stem family, and community family. In ancient Rome, the *familia* included the entire household: parents, children, servants, slaves and "clients." This family configuration went beyond the direct line of descent, which also raises the question of its scope of reference: that of the founding ancestor of the clan or lineage, or in the most restrictive version, of the direct ascendants and descendants. Thus, as shown in particular by the work of Godelier (2004) or Meillassoux (1975), the family is shaped by diverse practices and references. It manages activities and assets whose outcomes and transmission are at the heart of complex rules and alliances. Depending on the context, its actual functioning often results from the overlapping of different units whose contours and organization (including decision making) are dependent on specific objectives: residence, consumption, production, or even accumulation. Single or multiple family affiliations induce rights and obligations related to moral or economic solidarity.

Thus, there are numerous "family variations." They range from the parental-couple type which has developed in urban societies by updating a sort of genuine nuclear family[2] – itself now challenged by single parenting and blended families – to extended families, with, for example, more than 50 members in Sahelian Africa. Family forms, just like those of farming, have historically been at the heart of civilizations whose "grammar," as Braudel (1993) reminds us, expresses the embeddedness of spaces, societies, collective mentalities and economies – a process where population density has long shaped regularities, provided the tempo to changes and triggered ruptures.

2.1.1 Major Steps in the Evolution of Productivity

The history of agriculture belongs to the great process of technical change of human societies.[3] These advances have deeply transformed their ecological impact, their economic performance and their social and political identities. They have consisted of combinations of innovations, triggered by multiple drivers of change, which have led to many technological and organizational changes (Chauveau and Yung 1995).

---

[2] Todd (2011) develops the idea of the nuclear family as the original model for all humanity, which subsequently led to more complex forms, including the appearance – in certain specific circumstances and in an unsystematic way – of patrilineality, a form involving the coexistence of several nuclear families from different generations. According to Todd, patrilinearity places severe constraints on individuals and is less stable than the original nuclear forms.

[3] This section provides an overview of numerous studies on the evolution of agricultural productivity. It relies in particular on the work of Mazoyer and Roudart (1997) and of Bairoch (1989).

Agriculture was invented in the Neolithic era. It appeared 8,000–10,000 years ago, depending on the region, and spread from a few population centers: Central America, the Andes, Mesopotamia, China, and New Guinea. Depending on the natural environments and population conditions, it has developed along three main forms. Where demographic pressure was high, slash-and-burn cultivation systems in temperate and tropical forest areas have led to a complete deforestation resulting in new anthropized environments (including, however, ebb and flow movements of forest cover). Pastoral systems spread over savannas or steppes (high-altitude regions, Central Asia, Middle East, the Sahel). Irrigated systems were developed in the drier regions (oasis and large valleys: Nile, Euphrates, Indus). At the historical scale, these original forms have changed extremely slowly, like the civilizations from which they spring, which "take an infinite time to emerge, to develop their habitat, to bounce back" (Braudel 1993).

If long time periods and progressive shifts are the rule, the major agricultural regions have, however, experienced extremely wide-ranging rates of change. Major farming systems with favorable natural conditions and sufficient labor have been able to improve performance in terms of crop yields and labor productivity. But more marginal areas, often subject to higher physical stresses, were also able to engage in their own processes of change. At all latitudes and altitudes, whenever faced with excess water, drought or steep slopes, the creativity of farmers has also helped invent "extreme agricultures" which are surprisingly varied and unique (Mollard and Walter 2008) and which have been able to adapt over time.

There exist several periodizations of agricultural transformations. Historians and specialists of agrarian systems have identified several "revolutions" that have marked milestones of technical progress, organization and agricultural performance. Some authors such as Gordon Childe (1949) consider the Neolithic the first revolution, whereas others like Duby (1962) highlight the revolution of the Middle Ages. Mazoyer and Roudart (1997) focus on the agricultural revolution of the early modern period, which they consider started consolidating in the 1700s – the century in which a veritable jump in productivity occurred (Bairoch 1989).

This accelerated process of agricultural change in the eighteenth and nineteenth centuries cannot be understood in isolation from what constitutes the real revolution in the history of human societies: the transition from a system founded on solar energy (based on biomass, wind and water) – which represents the cornerstone of agrarian societies – to a system based on fossil fuels, which has led to the emergence of industrial and urban societies (Wrigley 1988). While the energy regime of agrarian societies was constrained by biomass production (land availability, vegetation's seasonality and fertility), the one of industrial societies has access to abundant resources, without annual limits and available at very low costs[4]: all that is necessary is to extract underground resources (Krausmann 2011).

---

[4] Concerns about the depletion of fossil-fuel resources will only make a hesitant and late appearance in the last quarter of the twentieth century.

2.1.1.1 Before the Energy Revolution

While agricultural performance improved only slightly over the long term in all the great centers of agricultural development, technological advances were significant and kept pace with the slow population growth. The domestication of animals, selection of species, use of equipment and persevering land management resulted in significant accumulations of capital (infrastructure and know-how), as exemplified by the rice terraces of Asia or Madagascar.

In the case of European agriculture – which will later experience the most spectacular developments –, animal draught cultivation with light plowing practiced since antiquity, based on the use of the swing plow with fallow and biennial rotation, gradually gave way to heavy plowing in the Middle Ages (between the end of the tenth and thirteenth centuries). Its use permitted rapid tillage, helped fight weeds and significantly increased the cultivated area per worker. It was accompanied by the dissemination of other tools, such as the harrow, the widespread use of the wain for transport of hay, litter and manure, the development of stalling and a better integration of animal husbandry in farm activities. These changes in technologies and practices led to a transition to 3-year crop rotations and improved yields.

But these advances, which lacked the sophistication of Asian rice systems, remained spatially uneven and always precarious. There were periods of instability and decline (wars or pandemics), sometimes resulting from agricultural crises caused by the overexploitation of the environment, as was the case for example in France in the fourteenth and fifteenth centuries.[5] Globally, despite heavy investments in labor and improvements in technology, productivity gains did not exceed a rate of 0.01 % per annum from the Neolithic to the seventeenth century (Bairoch 1989).

Changes that started taking place in the eighteenth century in the temperate regions of Europe were characterized by a rapid increase in yields and especially in productivity. They were the result of a hybridization of multiple processes of change rooted in earlier periods, which were self-reinforcing and relied both on market dynamics – with the gradual development of cities which changed the fundamentals of agricultural demand[6] – as well as on the evolution of ideas, which challenged the social and political order and slowly modified the economic balances of power.

---

[5] In this regard, the population/natural resources ratio has often been presented as a major determinant of technological change. Boserup (1965) has thus highlighted demographic pressure as a driver of innovation, challenging the position of Malthus who postulated, on the contrary, a constraining determinism founded on the relationship between population level, resources and technical systems. The history of agricultural change reveals mechanisms which are much more complex.

[6] The ratio between the agricultural and non-agricultural populations continues to fall. On the whole, each producer is responsible for feeding an ever-increasing number of mouths.

Changes of a legal nature, such as the progressive abolition of livestock grazing on common land,[7] the reduction of various taxes related to manorial rights[8] or the removal of other barriers to full use of the land (collective rotations, joint ownership),[9] unlocked technical progress and gave a strong boost to accumulation and investment processes. This was a matter, in particular, of the development of continuous crop rotation facilitated by the replacement of fallow by forage crops (legumes mainly), thus contributing to the development of animal husbandry. Performances were boosted by the use of improved seeds and animals and development of farm equipment.

Born in the Netherlands and the United Kingdom, this vast movement spread all over Western Europe in the period leading up to the early nineteenth century as well as in the English colonies of America (Taylor 2001), soon to became the independent United States. It then developed more slowly in Central and Eastern Europe (mid-nineteenth century).

The productivity gains achieved during this period (about a century and a half) – relatively short on a historical scale – were phenomenal. They were as large as those made in the previous eight or nine millennia.[10] Growth of productivity outstripped that of the population; it facilitated the trend towards urbanization and allowed allocation of labor to other economic activities – two of the most significant changes in the world's long history.

2.1.1.2 The Energy Revolution and Its Consequences

The energy revolution was not a sudden occurrence. It smoldered in the background during the eighteenth century and then started by facilitating various changes before emphatically causing new ones. With energy efficiency per surface unit of fossil resources in the order of 10,000 times greater than that of biomass (Smil 1991), it brought about an upheaval for human society. The industrial revolution took place and agriculture in the industrializing countries took full advantage. The advent of the steam engine revolutionized human labor and the transport of merchandise.

A second agricultural revolution resulted. It took over from the first and was characterized by mechanization and the use of fertilizers (mineral fertilizers and

---

[7] Grazing on common land (or on the "commons") was the right of access of herds from the entire community on fallow land and also on post-harvest cropland.

[8] For example, in France before the 1789 Revolution, and depending on region, manorial land rights were as high as 10–25 % of agricultural produce, to which the tithe due to the clergy (7–10 %) had to be added. Furthermore, there were various royal taxes of 10–20 % (Moulin 1992) such as the "taille."

[9] Reference is often made to the enclosure movement in England (gradual appropriation of open fields and communal lands by fencing) to illustrate the beginnings of this general trend. This movement stretched from the sixteenth to the nineteenth century.

[10] Productivity increased almost a 100-fold, growing 0.9 % per year with the adoption of new techniques (Bairoch 1989).

new organic fertilizers). It began in the middle of the nineteenth century and progressed slowly until the Second World War, spreading by varying degrees in different regions of the first agricultural revolution, i.e., mainly in Europe and the United States as well as in European settler colonies (Canada, Australia, New Zealand, the southern part of Latin America).

The development of agricultural machinery, which started in the 1850s and made rapid progress in the early twentieth century, was characterized by the mechanization of animal traction. New tools (reversible plows, seeders, hoe-cultivators, etc.) and harvesting equipment (harvesters, reaper-binders, threshers) removed, one by one, the main bottlenecks of the most time-consuming operations in the agricultural cycle. Mechanization progressed rapidly in the "new" countries, where large farms made possible by the expropriation of land belonging to indigenous peoples and the relative scarcity of labor favored its spread. Its development was markedly slower on the Old Continent, where the conditions were reversed, i.e., small farms and abundant labor.

Dramatic advances in land and sea transport caused by the rapid development of steamships, railways and the cold chain had profoundly transformative effects on European economies and their overseas offshoots. Agricultural products benefited from potentially unlimited outlets since they could be sold on local or rapidly integrating national markets – stimulated by urban growth – as well as on faraway international ones. Remote areas were opened up and the "new" countries very quickly turned into significant actors in international agricultural trade by becoming major suppliers of raw materials. At the same time, there was an expansion of tropical export crops in Latin America and the new European colonies, located mainly in Africa and Asia. This process led to an explosion of trade with the tropics – regions whose contribution to agricultural markets had been limited to sugar since the seventeenth century, with the sugarcane specialization of the Caribbean and Indian Ocean islands.

New transportation systems also made it easier to bring labor to cities and to new countries by facilitating both rural depopulation and European migrations. Starting in the late nineteenth century, they also led to improved soil fertility and yields by bringing mineral and organic fertilizers (nitrates, phosphates, potash and guano) to the farms.

This second revolution intensified with further modernization of techniques. The development of the process for the industrial synthesis of ammonia at the beginning of the twentieth century ushered in the era of chemical fertilizers. Similarly, the development of the automobile led to the appearance of the first tractors and their gradual popularization after the First World War. The rapid increase in motorization, chemicalization (fertilizers, pesticides) and selection of species (varietal improvements and, more recently, genetic modifications) after 1945 is often considered a third agricultural revolution. It is, in fact, primarily an intensification of processes previously unleashed. It gradually spread from its origins in Europe and the former settler colonies and is progressively taking hold, most often partially (geographically and technologically), in the rest of the world through national agricultural policies which promote modernization programs supported by international institutions.[11]

A study of the progress made by motorization is illustrative of the magnitude of technological leaps that agriculture has taken. There has been a rapid growth of traction power, which went from 10 to 30 HP after Second World War to between 150 and 300 HP today. This growth in power was accompanied by the modernization of tools and the development of self-propelled equipment (such as the combine harvester) and flexible multitasking tools which allowed a large increase in cultivated areas. Meanwhile, modernization of farm buildings equipped with new tools (such as milking machines) has helped rationalize farming activities, especially for animal husbandry. This meteoric technological progress has put agriculture at the forefront of sectoral productivity gains.[12]

Consequently, between manual cultivation without any fertilizers and the most sophisticated levels of motorization and chemicalization, cultivable area per worker increased from 1 ha to between 150 to 200 ha and labor productivity from 1 to 1,500 t of grain-equivalents per worker. Productivity growth, which struggled to reach 1 % per year during the first agricultural revolution, a threshold it barely crossed in the second, reached 5 % during the third (Bairoch 1989). These successive gains were achieved in increasingly shorter time periods (respectively about 150, 100 and 50 years).

2.1.2 Specialization, Differentiation and Widening Global Disparities

These dramatic changes in technology have taken place in progressive shifts, whenever conditions have been favorable. While natural and demographic conditions and opportunities for accumulation of capital and investment have played a key role, the adoption and development of new technology were also largely driven by the role of States. Governments were sometimes clever enough to create favorable economic and institutional environments, not only in terms of organization of markets, relative prices, information, training and advice, but also with respect to incentivizing credit and insurance schemes – an essential step in encouraging investments (see Sect. 2.3.2 later in this chapter).

---

[11] These include international research centers specialized in major field crops, established after the Second World War in the context of the Cold War, and coordinated by the Consultative Group on International Agricultural Research (CGIAR), established later in 1971. These centers, whose original purpose was to fight world hunger, were the main vectors of the "Green Revolution," based on the generalization of the use of inputs and improved seeds.

[12] In France, between 1950 and 2010, hourly agricultural labor productivity increased 32 times, whereas the corresponding increases for industry and services were 14 and five times respectively (INSEE data).

This radical change in farming techniques has had a significant impact on family farming – and global agriculture in general – at two levels: radical changes in the nature and characteristics of agriculture of the most technologically advanced regions, and a widening disparity between world regions.

2.1.2.1 Integration and Specialization of Modernized Agriculture

Wherever it has taken place, agriculture's rapid modernization has resulted in its integration with the rest of the economy and the generalization of the division of labor. The traditional downstream connection to the market for products finds itself reinforced by the development of the agro-industrial sector and is complemented, upstream, by markets supplying equipment, inputs and services.

This process, often driven by public policy and the reorganization of food systems, tends to push family farms towards specialization. Farmers become less pluriactive and the supply of goods and services in rural economies become more professional, focused and relocated to rural towns and small cities. Indeed, it then becomes possible to acquire intermediate and other farm-consumption products without having to resort to self-supply (manure, equipment, fattening of young animals, fodder and feed). This change reflects the progressive giving up of growing food for home consumption, a pattern common to agricultures engaged in a rapid market integration. This withdrawal from self-consumption is also linked to increases in purchasing power and changes in lifestyles of farmers and their families.

Through this specialization, family farms lose autonomy and become part of new value chains. On the upstream side, new methods of production (materials, inputs) are conceived and implemented, as well as related activities of training, extension and financing. Downstream, there is an explosion of the agrifood sector (with activities of initial, secondary, and even tertiary processing) and of agrochemistry (pharmaceutical industry and now biofuels). This is accompanied by the development of the modern retail sector, with the gradual worldwide spread of the "supermarket revolution" (Reardon and Timmer 2007).

In this movement for agriculture's industrialization, multi-tasking peasants become farmers. This radical change is accompanied by a new mix of agricultural production factors resulting in more capital and less labor.[13] The nature of farm work undergoes profound change, evolving towards a sort of Taylorization.

Over the last 60 years, in industrialized countries, this gradual shift has resulted in first the marginalization and then the phasing out of farms that lack the investment capacity to adopt technical improvements and be sufficiently profitable to ensure labor income comparable to that in other sectors. It has increased agriculture's financing needs, resulting in a gradual decline of family contribution and an increasing role of other stakeholders (association with other farmers or other

---

[13] Which is itself becoming more expensive in high-income countries due to labor regulations.

economic agents). In most cases, the outcome has been a shift towards a managerial type of agriculture (Chap. 5). Finally, there has been an acceleration of the exit of workers from the agricultural sector. A movement of progressive concentration of production structures and increased surface areas per farm are the consequences.

Agricultural production itself has also been subject to specialization, since farmers can focus on the most profitable activities allowed by local conditions (climate, markets, economic environment, relative prices). It has resulted in a gradual regional specialization at the expense of multipurpose crop-livestock systems which had long existed. Consequently, in regions with the least favorable natural conditions and often poorly developed infrastructure, rural households have become impoverished and rural depopulation has become the rule. This has led to new challenges of territorial development. At the same time, the global demand for agricultural products has been met through the intensive use of new transport networks – including very long distance ones[14] –, the global integration of markets that their liberalization has allowed and the increasing role of processing and distribution macro-actors. This globalization is accompanied by the widespread dissemination of food quality norms and standards.

2.1.2.2 A Profoundly Asymmetrical Global Agriculture

However, this final stage of agriculture's motorization and chemicalization, which has led to a radical change in methods of agricultural production, concerns only a small part of the world's agricultural population today. The majority of farmers still exclusively use manual equipment and "modern agriculture is far from having conquered the world" (Mazoyer and Roudart 1997).

The distribution of tractors by major regions is a useful – even though reductive – indicator of the magnitude of the differences between the very diverse agricultures of the world. According to the FAO, there are fewer than 30 million tractors in the world for 1.3 billion agricultural workers (Fig. 2.1).[15] Sixty percent of the world's tractor fleet is used in Europe and in the European offshoots (United States, Canada, Australia, New Zealand), nearly 10 % in Japan and South Korea, and the rest of the world sharing the remaining 30 % (23 % in Asia, 6 % in Latin America, and less than 1 % in sub-Saharan Africa).

Reducing the diversity of situations in the world to a few major types of agriculture classified according to their technical characteristics, Mazoyer (2001) points out that two-thirds of the world's agricultural workers are still using manual techniques. This effectively limits cultivation to a maximum of 1 ha per worker per

---

[14] This long-distance trade of agricultural products, which allows, for example, the consumption of strawberries from Chile during the European winter, has been made possible by the low cost of energy. As the cost of fossil fuels rises, such trade is likely to suffer.

[15] FAOSTAT data are derived from national agricultural censuses. As far as tractors are concerned, the last year with complete information for all countries is 2003.

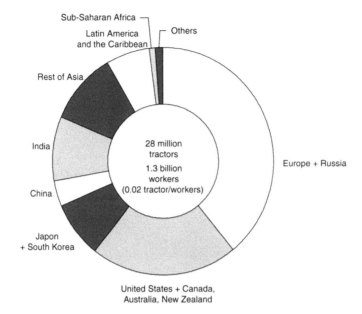

Fig. 2.1 Geographical distribution of the world's tractor fleet in 2003 (Source: FAOSTAT 2013)

Table 2.1 Stylized productivity gaps across technical systems

Type of agriculture	Hectares/ worker	Production in tonnes[a]/ hectare	Production in tonnes[a]/worker
Motorized traction and Green Revolution	100	< 10	1,000
Animal traction and Green Revolution	5	< 10	50
Manual and Green Revolution	1	< 10	10
Manual	1	< 1	1

[a]In grain-equivalent tonnes. This table is a stylized representation of differences in global productivity. Yield per hectare and surface area per worker have no statistical value; they are only used to indicate orders of magnitude and refer to the highest values of each technical system
Source: Author (based on Mazoyer 2001)

year for yields of, at best, one grain-equivalent tonne[16] per hectare per year. Half of these farmers have adopted the technical package of the Green Revolution (improved seeds and chemical inputs), allowing up to a five-fold jump – even as high as ten-fold in some cases – in the level of per-hectare yields and per-worker productivity (Table 2.1 and Fig. 2.2). Only one-third of workers have the benefit of animal traction, an asset that can multiply by about five the cultivated surfaces and

---

[16] Grain is the food category that is most consumed in the world but it only forms part of human diet. Short of conducting an analysis for all productions, in kilocalories for example, the use of grain-equivalents still provides extremely useful orders of magnitude.

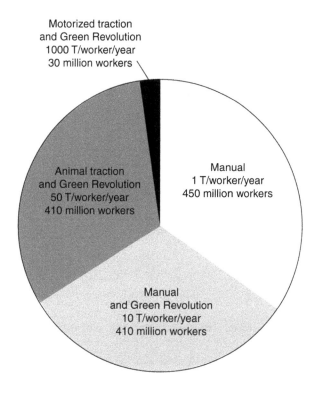

Fig. 2.2 Distribution and productivity of various types of agriculture worldwide. (Source: Author, based on Mazoyer 2001)

productivity per worker (at the same level of intensification). For these manual farming models or those using animal traction, the use of irrigation can often allow two crops a year – sometimes three – and significantly reduce climatic risks. But because of the sharp increase in working time, the arable land per worker diminishes greatly and the productivity per worker taken over the entire year improves but does not change dramatically.

Only a very small proportion of the world's farmers, some 2–3 %, use motorized equipment. The cultivated area per worker varies depending on the power level of the traction (and, of course, the topography), but by assuming 100 ha cultivated per worker without changing the yields, the difference with manual agriculture without technical package is already 1,000 to 1. On the Great Plains of North America, the surface areas cultivated per worker can reach 200 ha but with much more extensive practices. In parts of Europe and Japan, yields can exceed 10 t/ha.[17] In both these situations, the productivity gap with the most rudimentary agriculture can reach 1,500 (or more) to 1.

---

[17] Maize (between 9 and 10 t/ha in the United States and Western Europe) and rice (between 7 and 10 t/ha in China, the United States and Egypt) have, for cereals, the highest average yields. Wheat yields are lower: 8 t/ha in Western Europe, 3 t/ha in North America and less than 2 t/ha in Australia or Argentina (FAO data).

This stark diversity of world agricultures pits, at first glance, the old industrialized countries – high-income countries in Europe, North America and Oceania – against the rest of the world. But agricultural realities are much more nuanced since highly technical agricultural systems also exist in many Asian, African and Latin American countries. In most cases, though, the number of farmers concerned is not significant, especially when compared to the total farm population. They are usually large companies, mainly agro-industries, which are enclaves within existing farming systems, or a small proportion of farms that were able to access the capital required for modernization. But this relatively atypical character of these highly technical agricultural systems does not mean that they have no impact on their socio-economic environment. On the contrary, they are often associated with land expropriation or capturing the market share of certain products, especially when they do their own processing and have their own marketing channels. However, they can also create synergies and facilitate the access of other producers to market networks and certain techniques (Chap. 5). Furthermore, through their use of modern techniques based on chemicalization and by reducing access to certain resources, these new farming systems can have significant impact on the physical environment and on the production and living conditions of family farmers. Nevertheless, some regions have seen this "modern" agriculture develop considerably. Notable examples are Brazil and the Southern Cone of Latin America, the northern and western regions of Mexico and also parts of southern Africa, where the development of the entrepreneurial sector has resulted in a dual agriculture with composite effects of boosting some parts of the farm sector while marginalizing others.

These differences in technical levels and the consequent increase in productivity gaps lead to a profoundly asymmetrical global agriculture. The diversity of factor endowments, of government support and of performances provides unmatched capacities to adapt to evolving natural and economic environments. However, the increased integration in the value chain and the amount of capital invested also brings with it weaknesses, whereas the most "rustic" agricultures have a much greater resilience when compared to agriculture that depends on hypermotorization and chemicalization.

2.2 Agricultural Changes Embedded in the Many Economic and Social Transitions

The acceleration of the processes of change and the growing divides between countries and regions represent a situation which is unprecedented in world history. Not only is agriculture impacted but so is, more generally, the entire economic system. At issue is the management of the new imbalances that have been and are being created locally, nationally and internationally. Putting in perspective the diversity of mechanisms of structural change across countries will help us better assess the extent of the challenges that different regions of the world – and their farming systems are facing.

2.2.1 The "Statistical Evidence" of the Exit of Workers from Agriculture

A study of the processes of economic change in different regions of the world and the continuation of trends observed during the last two centuries would theoretically allow the hypothesis of a world "without agriculture" (Timmer 2009) or one "without farmers" (Dorin et al. 2013). In fact, the trajectories followed by today's richest and most technologically advanced countries after the energy revolution of the nineteenth century reveal the transition from an agriculture which occupied a predominant place in their economic aggregates to one that is now marginal. The proportion of the labor force working in agriculture in European countries in 1800 – at the time of the first agricultural revolution – ranged from 65 % to 80 % (Bairoch 1989). Today, the share of agricultural workers in the total labor force stands at less than 5 %. The agriculture sector's share in national GDPs shows even sharper decline in the majority of high-income OECD countries (Organization for Economic Cooperation and Development): it is below 3 %.[18]

Such a trend was mirrored in other regions of the world, although at a generally much faster pace – a few decades instead of two centuries – due to technological and organizational leaps stemming from the adoption of innovations from the most economically developed countries. Thus, in many Latin American countries, the contribution of agriculture to GDP is less than 10 % (5 % in Brazil and less than 5 % in Chile and Mexico). The change is slower in Asian countries, where for most countries this figure ranges between 10 % and 20 %.[19] But in Africa, agriculture is still prominent in national economies: in 17 out of 53 countries, agriculture's contribution to GDP exceeds 30 %; in 10 countries it is between 20 % and 30 %; in Egypt, Morocco and Senegal it is around 15 %; and it is 10 % in Tunisia. African countries with economies dominated by mining or oil exports are a special case, with agriculture's contribution to GDP being below 10 % or even 5 %.

The declining share of agriculture in national wealths is only one dimension of structural change since the decline in agricultural workers is much slower than changes in GDP. Indeed, even though the OECD countries can be viewed as having structurally "exited" agriculture – a meaningless perception since agriculture still retains its economic,[20] social and environmental importance – agriculture remains the world's largest employer (Chap. 3). According to FAO data, it still accounts for – on average and with significant national differences – 15 % of the workforce in

---

[18] Less than 2 % for Western Europe, the United States and Japan, and even less than 1 % for some countries such as Germany, the United Kingdom and Belgium. The data presented in this section are taken from *World Development Indicators* (World Bank).

[19] About 10 % in China, Malaysia and Thailand; 15 % in India and Indonesia; and 20 % in Vietnam and Pakistan.

[20] Even though agriculture's economic importance has declined drastically in numerical terms, the activities upstream and downstream of production (agrifood industries and services) have developed rapidly since the 1960s and the agrifood sector currently accounts for around 15 % of the European Union's GDP.

Latin America, about 50 % in Asia and 60 % in sub-Saharan Africa. In Africa, particularly in the Sahel, more than 75 % of the workforce is employed in agriculture in some countries. South Asia too is a major agricultural region in terms of agricultural employment. The case of China is less precisely known due to statistical shortcomings, but there too agricultural workers could still represent between 50 % and 65 % of the total labor force.

These differences in agriculture's contribution to national GDPs and employment can be explained by productivity gaps between agriculture and other sectors, but also by the fact that, in many rural societies, livelihoods are also, partly at least, based on agricultural activities which are not taken into account by strictly economic criteria. As shown by the analysis of the modernization process, agricultural work is not very productive when it is largely manual or has a low level of mechanization. It then becomes quickly decoupled from other types of activities and results in lower agricultural incomes. The phenomenon is exacerbated by changes in relative prices between agricultural and non-agricultural goods. Consequently, the value addition of other sectors rises much faster than for agriculture which, nevertheless, continues to employ a significant proportion of the working population (McMillan and Rodrik 2011). Given the importance of agriculture in rural areas, these processes explain the income gap between towns and the countryside and the extent of rural poverty (Chap. 10). This process is illustrated in Fig. 2.3 where the change of each country can be considered its "signature" illustrating the diversity of trajectories of structural change.

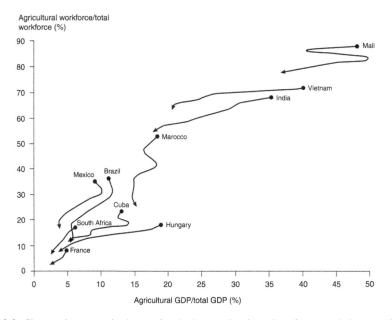

Fig. 2.3 Changes in economic shares of agriculture and trajectories of structural change (1980–2010) (Source: World Bank, *World Development Indicators 2013*, for GDP; FAOSTAT, 2012, for the labor force. Note: up to 2005 for France and Mali, starting from 1985 for Vietnam)

The selected examples show the different dimensions of these changes (Bélières et al. 2013). To begin with, the continuously decreasing share of agriculture in GDP and the labor force is reflected in the general move from the upper right quadrant to the lower left quadrant. In addition, the speed of change is expressed by the length of the trend line: the slower changes may reflect a structural inertia or the existence of older transitions (such is the case of France, for example, where intersectoral restructuring took place before 1980, i.e., outside the graph's time period). Finally, the comparison helps differentiate between countries according to their trajectories: countries engaged in a process of widespread economic diversification where the share of agriculture in GDP and employment is decreasing (Mexico and Brazil); countries on the path of diversification where the share of agriculture in GDP is declining but without a proportional transfer of labor to other sectors (Mali, Vietnam, India); and countries where agriculture retains an important macroeconomic role, but with a rapid decrease in agricultural workers, which illustrates the rapid gains in agricultural productivity (Morocco).

2.2.2 The Importance of Historical Sequences

This shift from agriculture-based economies to those that are more diversified is at the origin of an evolutionary vision of change, which postulates stages that can "naturally" be followed by all parts of the world. This vision goes hand in hand with the hypothesis of a certain standardization of lifestyles associated with urbanization and driven by globalization. These stages, apparently confirmed by the changes of the last two centuries, must however be assessed in a historical perspective which alone allows one to take the full measure of current global challenges.

The evolutionary approach, which was formalized after the Second World War (Rist 1996), is the source of mainstream thinking on development. It is based on the idea of a step-by-step catching up (Rostow 1960) with the most advanced countries in terms of technical, economic and social progress, progress that is generally measured through standards of living and often more prosaically reduced to per-capita GDP.

A stylized summary of the structural evolution of European economies (and of countries of European settlement) and their main determinants shows that the gradual transition from an agriculture-based economy to one based on industry and then on services, and hence from rural to urban areas, was made possible by the energy shift to fossil fuels. This shift is indeed the cause of profound technological changes and impressive productivity gains, which led to wealth accumulation and then the transfer of labor and capital from one sector to another. This process was accompanied by an increase in income and demand and of its diversification. It has benefited from the demographic transition at the origin of an improved ratio between the working[21] and non-working populations and was facilitated by mass education. In this process of change, agriculture played an initial role and was the first driver for accumulation (see the example of China in Chap. 10). Productivity gains have been accompanied by a massive exit of workers from agriculture, their migration to the cities, to other regions or to other countries.

This view of the stages of development is obviously mechanistic but it still finds prominent place in international discussions, development aid and public policy. It is reinforced by the similarities observed in the trajectories of some Latin American and Asian countries. Discussions and claims related to the "emergence" of some countries – i.e., their transition from underdevelopment to development – are a perfect illustration of the mechanistic visions underpinning the current debates (Gabas and Losch 2008).

This postulate that past transitions should be reproduced, however, tends to ignore history and the very specific paths followed by different regions of the world in their trajectories of transformation. Each region has undergone changes specific to not only its own endogenous combinations of natural, economic, social, political and institutional factors, but also to its relations with the rest of the world. And these relationships between internal and external processes, between national and international patterns, and the specific times at which they occurred are critical to understanding the dynamics of change and power relations (Losch 2012a). They are part of the historical construction of markets and the gradual spread of capitalism (Braudel 1979; Wallerstein 1989) and they underscore the point that identical replication of past sequences is not possible.

Thus, the European transitions that took place in the late eighteenth century benefited greatly from the hegemonic status of Western Europe, largely based on the "capture of America" at the turn of the sixteenth century (Grataloup 2007; Pomeranz 2000). Resources obtained from the Western Hemisphere funded European growth and its subsequent conquest of the rest of the world. Imperialism and colonization helped European accumulation based on the "unequal exchange," while providing adjustment opportunities critical to the structural transformation of European economies through mass migration to the "new worlds."[22]

The transitions observed in Latin American and later in Asia are not of the same nature as European ones. While they do have some similar characteristics – their economic diversification and the changes in their labor forces, for example –, they do not duplicate European transitions, mainly because they took place at another "moment" of world history, a moment characterized by the implementation of proactive policies of modernization.[23] Indeed, from the inter-war period (when the transitions in Latin America began) to the liberalization phase initiated in the

---

[21] The demographic transition is the result of improvements in health and lifestyles which translate into the reduction in mortality and birth rates.

[22] Between 1850 and 1930, about 60 million Europeans emigrated, helping European nations overcome the problems of underemployment and poverty (Hatton and Williamson 2005).

[23] What Giraud (1996) calls "self-reliant national development."

1980s, the international regime was characterized by the preponderance of autonomous national policies aimed at modernization through strong State intervention and import substitution. These transitions – in which a large part of Asia participated after the Second World War – benefited from the technical and organizational progress made earlier, but also from high levels of national protectionism. Often significant too were large amounts of capital transfers, especially to Latin America and Asia, in the particular context of the Cold War years from 1950 to 1980 between the United States and the USSR.[24]

For those countries that are still agriculture-based – mainly in sub-Saharan Africa and a few in Asia – and which have not yet begun their effective transitions to more diversified economies, the challenge is to succeed in their structural transformations in the new international regime of a liberalized global economy where competition is the rule. These countries have to manage new constraints related to struggles over resources, but without benefiting from the same economic policy options that other countries before them did – a consequence of new international regulations.[25]

2.3 Family Farming Emerges on the Political Stage

In this long history, the emergence of family farming as a political subject and object – as actor and objective of policy – happened late. Indeed, until very recently, due to the overwhelming share of the agricultural population in every society, family farming was never perceived as having a specific status; it merely expressed the ordinary position. In every corner of the world, as soon as the first forms of government appeared, decisions of the prince (mainly focused on collecting taxes) were directed primarily at the great mass of his "country dwellers" – namely the "peasants"[26] – who tried to make a living from both the natural environment and resources of the land they cultivated with their families (often without owning them fully).

In Europe, it is the slow emergence of nation-states from the mid-seventeenth century – consolidated over the next two centuries by the upheaval of the three orders of the *Ancien Régime*, i.e., the clergy, the nobility and the Third Estate (commoners) – which saw the gradual appearance of the first national public policies: on the unification of legislative and taxation systems and the development

---

[24] Funding for international agricultural research should especially be viewed in this specific context.

[25] Chang (2002) emphasizes the difference in status between countries according to their hegemonic or subordinate position. In particular, he recalls how the richest countries now wish to prevent others from applying the policies they had themselves implemented (especially those of protections and subsidies) and which they sometimes continue even today.

[26] From Old French *paisent* (country dweller) and *pais* (country), based on Latin *pagus* (country district).

of education and conscription. These policies contributed to the territorial consolidation of nation-states (Gellner 1989) and provided the basis for the implementation of targeted geographical and sectoral policies. This was the framework in which agricultural policies were developed, focusing then and increasingly on production and on producers.

2.3.1 From the Peasant Question to Family Farming: A Slow Transition

The peasant question was the guiding principle behind the development of family farming policy. Indeed, even though agricultural production was the result of the work of farmers and their families, for a long time family farming was rarely referred to as a category (Chap. 3) but indirectly or occasionally. On the contrary, the peasant has always held an important social position. Irrespective of his status – slave or colonist, serf, or even laborer or sharecropper –, the peasant has long been the backbone of economic activity. He has fulfilled the economic function, one of the three functions that are specific to the organization of most Indo-European societies[27] – the other two being the religious and military ones. The evolution of the peasant's status and its consideration by public policies is discussed here with a special reference to the case of France, whose characteristics, despite their specificity, find echo in other parts of the world.

The trilogy of *sacerdotes, bellatores* and *laboratores* of the Roman world, the monks, knights and peasants of the Middle Ages or the three orders of clergy, nobility and the Third Estate of the *Ancien Régime* in France established a permanent and historically dominated category. Peasants were commoners and farmed to feed the two noble orders, which in turn provided spiritual and military services but also required compensation in the form of taxes and free labor, especially because of their control over land. The Third Estate also encompassed other categories of workers –artisans, merchants, usurers, lawyers and administrators – and its representation remained urban in nature. This meant that peasants – who accounted for the bulk of the population – were doubly marginalized.

This dominated status did not preclude deep inequalities relating primarily to land ownership. Some were peasant-owners, others were tenants and sharecroppers (paying a rent).[28] These differences were reflected in livelihoods which ranged from situations of survival for the poorest peasants to those of relative opulence for

---

[27] This tri-functionalism, as a core common to Indo-European societies, was put forward by Dumézil (1968) from a comparative approach to history and mythology, and then taken up by Duby (1978) in his work on feudalism.

[28] In the France of the *Ancien Régime*, peasants owned between 30 % and 50 % of the land depending on the region, a much larger proportion than peasants in England did, for example (Moulin 1992).

large landowners at the head of farms of several tens of hectares and employing a large workforce. This differentiation led to a gradual social stratification which saw the emergence, alongside peasants, of rural notables with the attributes of "capitalist" bosses and exploited farm workers. Thus, and in particular due to the dispersion of the rural population, peasant protests took place only occasionally, usually in the form of limited revolts and rebellions when the tax burden became too high. However, these disadvantages did not prevent the European peasantry from participating in political revolutions and societal transformations at the turn of the nineteenth century which were, for the most part, initiated by the urban classes and the bourgeoisie (itself comprising many landowners). Peasants derived benefits from these transformations[29] but the gap between rural and urban incomes grew rapidly with the advent of industrial employment and wages, a significant step in the history of structural change.

In Europe, the history of the peasantry, which for very long formed the demographic majority, then merges fully with that of the process of agricultural modernization. As Moulin (1992) notes, peasants strove to improve their status by perfecting their techniques – often following in the footsteps of rural notables. The most successful earned the title of cultivator, then of agriculturist, echoing the development of agronomics. Meanwhile, governments started to pay attention and to offer support in order to facilitate technical changes and improve economic conditions. After all, peasants did form the electoral base of the new representative democracies. Peasant demands directed towards the State focused on the regulation of the new national markets and on prices of agricultural products. This is the case in France during the inter-war crisis where the protests arising from the collapse in grain prices led to the creation of the Wheat Marketing Board (*Office du blé*) in 1936.

The movement towards professionalization that accompanied the pursuit of agricultural modernization gradually formed the basis for a broad process of change broadly supported and encouraged by public policies. After the Second World War, the "farm" took center stage. The farmer gradually specialized from a technician to a manager-entrepreneur, a process that increasingly disconnected farming from the peasant's way of life, rooted to his rural setting. This process led Mendras (1967) to proclaim the "end of peasants" and Shanin (1974) to advance the concept of "agriculturization" (in the sense of agricultural industrialization).[30] In that perspective, the loss of agriculture's special status during the negotiations leading to the liberalization of international trade – ending with the creation of the World Trade Organization (WTO) in 1994 – corresponds to the culmination of a process of normalization that had been going on for the past two centuries. Agriculture was

---

[29] In the French case, it was on the milestone night of 4 August 1789 that, within a few hours, the nobility and clergy lost their privileges. The Civil Code of 1804 subsequently enshrined the "national ideal of a land-owning peasantry" (Laurent and Rémy 2000).

[30] Paradoxically, this process took place in the 1950–1970 period while an intellectual debate, in which Mendras and Shanin participated, "discovers" or rediscovers the peasantry, with amongst others Redfield and several Marxist economists (Chap. 3).

henceforth to be treated like any other economic activity for the purposes of international trade.

In this long history, the "family farming" category has only lately found place in the public debate. Its recognition is similarly recent at the international level, although in practice public policies often took into account – and even focused on – farmer families and family farms.[31]

The peasant question, however, is still present in the political debate and in policy concerns for three main reasons. The first is that the peasantry – and by extension, the family farm category – has resisted the processes of modernization and standardization, thus denying a complete victory to the agro-industrial business model. It has stood firm and has even partially assimilated these processes by demonstrating effectiveness and flexibility in the use of all resources (natural, technical, social and family-based ones) in a way that the employer wage-based model has been unable to (Chaps. 5 and 8). The second is that the peasantry, due to its demographic significance, has been an integral part of many liberation and independence struggles against colonial rule in Asia and Africa in the 1960s and 1970s (Friedmann 2013). Even though the "political expropriation of the rural masses" (Copans 1987) has often been the rule because of the balance of power and dominant influence of urban classes, this expropriation tends today to fuel movements of identity. It sometimes leads to the meeting of diverse indigenous movements (such as those in Latin America) which are challenging the dominant political and economic order. The third reason is that the peasant question has become part of the debate, at least in the richest countries with modernized agriculture, questioning the productivist model and its downward slide. Health problems, the ecological crisis, food quality, the dependence on the agro-industrial and modern retail sectors, the headlong rush for mechanization and the related bank indebtedness and Taylorization of agricultural activity are all leading to a search for a new place for agriculture in its rural territories and its local settings and to a reinvention of new social and environmental linkages.[32]

Those rallying to the banner of family farming – endorsed by the United Nations which has dedicated 2014 as the Year of Family Farming – therefore form a composite group. Several social movements in different parts of the world, such as Roppa in West Africa,[33] declare themselves to be primarily advocates for

---

[31] In the case of France, Laurent and Rémy (2000) show the emergence of the concept of the family holding in statistics and in law, starting in the inter-war period, and the first signs of the family farm model during the Vichy regime (1940–1944). Its consolidation had to wait until after the Second World War and, in particular, the advent of the 1960 and 1962 orientation laws for agriculture which established the keystone of the "two Man-Work Unit (MWU)" farm (i.e., with two full-time workers, typically a working farm couple) – as the basis for an alliance between "modernist family farmers" and the State.

[32] The creation of Via Campesina in 1993, which brings together farmers and farm workers in 70 countries from all regions of the world, is in line with these multiple perspectives.

[33] French acronym for *Réseau des organisations paysannes et de producteurs de l'Afrique de l'Ouest* (Network of peasant and producer organizations of West Africa) (Roppa 2013).

peasants or for family holdings rather than family farmers' movements – though Brazil is a major exception. Family farming, despite the large audience of Via Campesina which promotes a peasant alternative to agricultural industrialization, could gain prominence over peasant agriculture, at least in international governance bodies. Because wording counts, this prominence in the way of naming reminds us the importance of family farming in global agriculture. But it is also a matter of perhaps, and especially, the result of a hybridization and a compromise between the desire of farmers to professionalize and the search for an alternative model to the excesses of market-driven productivism. Such hybridization is reflected in a widespread recognition of the importance of family forms of production (agricultural or otherwise) in the world. This family model, a priori threatened by and incompatible with industrial and commercial concentration – but nevertheless resistant –, seems able to provide alternatives to the deteriorating employment conditions and the distance which is growing between modes of increasingly artificialized and financialized industrial production and the consumer-citizen. Family farming is also promising in defending the interests of agriculture in the South, largely threatened by the growth of agribusiness.

2.3.2 Invention and Differentiation of Support Policies

Agricultural issues have always occupied a prominent place in government agendas. The strategic nature of food makes agriculture a true "affair of State." Agriculture has, after all, contributed to the creation and rise of the State in its various forms. Agricultural policies were, indeed, along with fiscal policies, among the first interventions of modern States (Coulomb et al. 1990).

Several major objectives have historically structured State action, with obvious political aims: feed the people, accumulate for growth and development, and increase farmer incomes. The first objective concerns the primary function of agriculture: feeding of farmers and supplying food to the non-agricultural population, whose share in the total population has been growing with urbanization and economic diversification. It is a necessity for social peace and even for the State's very survival. The second objective is promoting the transfer of capital and labor from primary activities, foremost among which is agriculture, to other sectors of the economy through direct and indirect taxes and labor mobility. This objective goes hand in hand with the third objective of increasing farmer incomes. This was the compensation of the direct costs of modernization – i.e., the exclusion of some farmers resulting from productivity gains. But increasing farmer incomes also contributes to the reduction of rural poverty in contexts where the countryside has long been – and still is – the home to the majority of the population. Sometimes this third objective has required compromise between representatives of farmers and the State in establishing agricultural policies and defining their framework.

These three objectives have been grouped together in a broader context of economic and social progress. They have led to the implementation of wide-ranging supply-side and modernization policies, without targeting any particular farmer category due to the wide initial homogeneity of technical levels (Bélières et al. 2013). The main objectives have focused on increasing the available supply – most notably through higher yields – and, at the same time, on improving labor income by means of productivity gains. Increased volumes and improved incomes were also a necessary step to offset the downward trend in prices resulting from growth in supply and transfers of value to other sectors.

The modalities and rates of implementation across countries were based on two main options: on the one hand, support for processes of change through market integration and competition and, on the other, a break with the existing economic order by changing the distribution and ownership of the means of production.[34] These transformational policies have had a more or less durable impact but, in historical terms, they were "moments" attempting to change the balance of power and trying to manage economic and social transitions. The range of instruments brought to bear was quite similar across countries. They aimed for a better functioning of markets (by more efficient movement of goods and management of supply), improved production structures and an increase in performance through technical progress. These instruments can be divided into two broad categories. The first pertains to public goods provision, namely the basic infrastructure, the rule of law (including land rights), education, training, information and research. The second concerns market support and protections to address the important issue of risk, which constitutes a major obstacle to investment, and the issue of the financial resources needed for modernization (Chap. 10).

Nevertheless, even though supply-side and modernization policies are the foundation of agricultural policies and their field of historical development, these policies have also diversified into two non-exclusive directions: their integration into a more comprehensive approach to rural and territorial development, and the emergence of policies targeted at specific categories of agricultural producers.

In the first case, the overall economic and social transformation has given rise to other requirements related to territorial balances and to the management of the dynamics between rural and urban areas. Policies have therefore focused on planning and on the diversification of rural activities, including looking for and encouraging intersectoral linkages. The negative impacts of the growth model on the environment and natural resources have also led to corrective interventions and a search for other "ways of producing." This new outlook has taken the form of policies promoting the multifunctionality of agriculture,[35] going beyond mere agricultural production and taking into account the production of environmental

---

[34] This is the case of agrarian reforms aimed at redistribution of land with the objective of social justice and economic efficiency, undertaken in a more or less authoritarian manner, and, of course, of collectivization with the abolition of private ownership of the means of production.

[35] Multifunctionality and its development figured prominently in the political debates in the 1990s and 2000s, especially in European countries. This approach, however, was largely derailed by its instrumentalization in the context of the debate on agricultural liberalization. As a result, the search for alternative development models has suffered (Barthélémy et al. 2003).

services and the integration of activities in a broader territorial perspective – a process that has been strengthened in many countries by the movement towards decentralization. Also part of this vision are the "new rurality" approaches, especially in Latin American countries (Bonnal et al. 2004).

In the second case, the progressive countrywide differentiation of agricultural structures and the growing performance gaps between different types of agriculture – a consequence of the unequal distribution of modernization and its technical packages – have led to targeted policies, specific to each farm type and dependent on regional settings. This movement has taken the form of dual policies, implicit or formal – as in the case of Brazil (Chap. 10) –, with the implementation of specific extension systems and support for boosting incomes and modernizing production structures.

This evolution, these inflections and diversification of agricultural policies should be analyzed in light of the economic and social patterns of each country, since sectoral and territorial policies are primarily the outcome of structural realities that are gradually evolving in tune with global changes. It is these realities that determine the priorities of public interventions.

Ultimately, the main agricultural and rural policy differences between the major regions of the world depend of course on the means available for their implementation, i.e., the ability of states to undertake actions. This has resulted in significant gaps between the richest countries, the "emerging" ones and the others. Differences also depend on the global economic and institutional environment and the international climate, which determine the types of policies acceptable between States. The current liberalized regime established by the WTO is unfavorable to market protections and extremely restrictive in terms of support: various types of support are either allowed, acceptable or banned – a matter that the WTO goes into in detail – according to market distortions they are expected to create. This observation obviously leads to the question of the difference in treatment between countries that have historically been able to use the full range of public interventions and others who came later but find that the former group has "kicked away the ladder" (Chang 2002) which they themselves used to facilitate modernization and manage structural change.

Chapter 3
Defining, Characterizing and Measuring Family Farming Models

Pierre-Marie Bosc, Jacques Marzin, Jean-François Bélières,
Jean-Michel Sourisseau, Philippe Bonnal, Bruno Losch,
Philippe Pédelahore, and Laurent Parrot

3.1 A Definition for Each Domain

Chapter 2 shows that the wide diversity of agricultural forms stems from the political and social structures rooted in historical trajectories, where representations have been forged by power relations and the dissemination of technical progress.[1] This diversity and its reasons invite us to make an effort, necessarily reductive, to define, characterize and measure family farming models, and to clarify what makes them a political and analytical category. To name the production units[2] of the agricultural sector, several categories are mobilized by actors, all of which pertain to different professional spheres but do so in interaction with each other.

There are four broad domains in interaction within which categories are generated and used to describe agricultural production actors. These representations are

---

[1] This chapter incorporates some of the elements presented in an expert report (Bélières et al. 2013) due for publication in 2014 as part of the *À savoir* collection of the French Development Agency.

[2] We make a deliberate choice to focus the analysis on the dimension of the "organization of agricultural production." Other approaches, discussed in Chap. 4, choose instead to focus on the multiplicity of functions that coexist with agricultural production: consumption, residence, accumulation, etc.

P.-M. Bosc (✉) • J. Marzin • Jean-Michel Sourisseau • P. Bonnal • B. Losch • P. Pédelahore
ES, CIRAD, Montpellier, France
e-mail: pierre-marie.bosc@cirad.fr; Jacques.marzin@cirad.fr; jean-michel.sourisseau@cirad.fr; philippe.bonnal@cirad.fr; bruno.losch@cirad.fr; philippe.pedelahore@cirad.fr

J.-F. Bélières
ES, CIRAD, Antananarivo, Madagascar
e-mail: jean-francois.belieres@cirad.fr

L. Parrot
Persyst, CIRAD, Montpellier, France
e-mail: laurent.parrot@cirad.fr

not fixed, they are constantly changing under the reciprocal influence of exchanges between these strongly interconnected domains. We present them here in a segmented manner in an attempt to find the reasons for ambiguity in the designation of forms of organization of agricultural production.

- The *cognitive domain* is the source of analytical categories which allow actors of agricultural production to be designated. It pertains mainly – but not exclusively – to academia and research. Its purpose is to improve the understanding of agricultural realities and the changes taking place. This requires going back and forth between production of concepts, collection of empirical data and development of models to represent reality. Categories thus generated relate to disciplines that tend to segment the reality according to the theories and frameworks mobilized.
- The *policy, administration and public action domain* (public and private actors, collective civil society actors) generates normative categories by defining beneficiaries of (and therefore also those excluded from) public policy measures. The definition of normative categories thus pertains to choices which depend on the objectives sought by the policies. The categories defined have to identify the target audience of the policies, and, subsequently, assess the effects of these policies.
- *Societies*. Citizens and their organizations have more or less direct and proximity links with agriculture, based on their personal histories but also influenced by the place agriculture occupies in society. These links are, of course, highly variable and any generalization of them would be perilous. However, agriculture is specific in that its productions are mainly intended for human consumption and they are therefore "constituent" of individuals. In addition, the large extent of land devoted to agriculture still strongly shapes territories. Here, the peasant as a class occupies a special place, between a "naturalist" image carrying with it historical and cultural representations,[3] and a socially shaped and constructed image. This concept used by historians, economists and sociologists is also part of everyday language and of social imagination, especially in a country like France, but also in India, China and Latin America (*campesinos*). While the peasant character has been historically dominant, it is now part of the diverse denominations used with reference to the farm (sometimes capitalist), cooperatives and other associative forms, commercial agriculture or, more rarely and recently, to the family farm.
- The *professional domain*. The actors of production define by themselves how they should be called, which, in some ways, also relates to the normative and political dimensions. By choosing its name, the group constitutes itself as a social entity in order to interact with others, especially with public authorities, but also more generally to communicate with society. For example, the term

---

[3] As example, we refer to the naturalist novels of the nineteenth century, among them Honoré de Balzac's *The Peasantry* or Émile Zola's *The Earth*. Brazilian writer Jorge Amado too described class struggles in the cocoa plantation region of his native Bahia.

"peasant" is claimed at the international level by Via Campesina, a movement that is in favor of agriculture based on family groups as opposed to entrepreneurial farming. The question of professional identity is also addressed by the Network of Peasant and Producer Organizations in West Africa (French acronym: *Roppa*), which brings together "peasant organizations" representing family farmers and "organizations of agricultural producers."

These different domains are not distinct and separate, neither through the many individual affiliations, nor through the ideas of and values represented by different actors and social groups, which are constantly changing and influencing each other.

3.1.1 Limitations of Current Denominations

Neither in common language nor in academic works have we found a foolproof way of denoting family farming, especially as translations from one language to another help introduce inaccuracies. But it seems possible, as first approximation, to distinguish four parameters commonly used to define family farming.

- The *size of the farm*, in hectares, is often used (Eastwood et al. 2010). It can also be expressed in heads of cattle or by turnover or sales volume (United States). This approach of distinguishing farms by size pertains to number of expressions, generally relative (large, medium and small farms), which can, of course, be combined with other characteristics (for example, small family farm). This classification by size in hectares has its usefulness because the land is a strategic factor for the development of agricultural production. This data is often the most systematically collected through agricultural censuses and therefore most accessible for analysis. However, the surface area is only one element of the capital mobilized by rural households. It is a criterion which depends on the type of production system and physical investments (irrigation facilities, plantation, buildings for livestock, etc.). And, finally, the area farmed is largely dependent on national contexts and is meaningful only at that level: a small farm in Argentina has nothing in common with a small farm in Kenya or in Asia. Differentiating by size often excludes pastoral farms – where access to common property resources is vital – and, more generally, shifting and itinerant farming systems. It also excludes production based on activities of harvesting "nature's products." Designating by size tends to overvalue the large farms and undervalue small ones, thus implicitly contributing to polarizing the debate between "large farms," perceived as "efficient," and "small farms," which are thought to be less so. This vision pertains to a representation of progress which is dominated by economic functioning. In practical terms, it translates into the desire for increasing the size of farms, especially through the development of motorization and intensification through artificialization of the type advocated by the Green Revolution, in an economic environment in which price competitiveness is king. It is also not surprising that the term "small agriculture" or smallholder agriculture has entered the lexicon of international institutions, and that it has, over the last two decades, become widely used in the cognitive domain. Yet, the functions that family farms fulfill are far more diverse.

- The understanding of *farm rationalities or strategies* is often based on the destination of the productions or only on the economic dimension of the unit, but is also often combined with identities. The tendency is thus to contrast subsistence (or semi-subsistence) farming with commercial or capitalist agriculture. However, leaving aside the case of farms which have chosen to specialize or even hyper-specialize, farms are, in general, found in an "in-between" state between food production for the family and for marketing. Furthermore, the relationship with the market can also take the form of the sale of labor, the purchase of food, products and services, or via the market-oriented development of non-agricultural activities. This is what is meant by the terms "pluriactivity" or "non-agricultural diversification." Similarly, reference is often made to the manner of production, especially in terms of how intensive a particular production system is. We can thus refer to intensive agriculture, often associated with large farms, or to extensive farming, which may correspond to an economic advantage (livestock rearing in New Zealand) or to a social practice (extensive *latifundia*). But these are not invariable correspondences: farms that are small in surface area often prove to be very labor and inputs intensive. Finally, the literature also refers to capitalist logic or rationalities, as opposed to other types of rationality – peasant, subsistence, etc. –, which shed light on the behavior of some entities in given contexts. It is also difficult to arrive at a satisfactory and accepted definition of capitalist agriculture[4] because, besides the traditional definition of a capitalist venture which separates owners of capital from workers, an effective capacity of capital mobility will be required (Petit 1975), which is what the dynamics of extreme financialization at work today may be attempting to bring about (Ducastel and Anseeuw 2013). While the rationalities approaches are very commonly used in the cognitive domain, their criteria of differentiation are not robust enough to define stable categories.

- The *identities* of those who work on farms are represented by various terms: peasant, farmer, producer, etc. These identities are constructed using concepts originating from militant social movements (unions, associations, cooperative movements, etc.), as well as from disciplines such as sociology, history and the rural economy. In many parts of the world, especially in Asia, the peasant has a stable identity, without any pejorative connotations, at least until recently, because there has been no questioning of this identity in the absence of diversification of forms of production (see the case of India and China). In some contexts, such as in Latin America, peasant agriculture (*agricultura campesina*)

---

[4] According to Bergeret and Dufumier (2002), "in capitalist farms, the owners of the means of production do not work directly themselves; they only contribute the capital. These farms are often run by employed managers whose task it is to adopt production systems that maximize return on capital."

makes sense for many in academic and political circles in response to the historic agrarian dualism, as opposed to commercial agriculture or to *latifundias*. In the European context, the farmer has taken over from the peasant (Mendras 2000), even though the "peasant" denomination is now once again being claimed collectively by some farmers who do not recognize themselves in the terminologies of conventional modernization. Identity is sometimes associated with farm size: small peasants are often contrasted with large-scale intensive commercial agriculture, also called "capitalist." Categories based on identities reveal a sometimes demand- or protest-oriented dimension, associated with value judgments and grounded on ideological or political positions.

- The *judicial status* and *legal forms* pertain to the normative domain of public policy (what legal, judicial, fiscal statuses for "farmers"?). They are therefore highly dependent on institutional contexts and social and political recognition of farmers within societies. In the case of France, statuses of type "shareholder" may be more attractive to family farmers for reasons of taxation, social security, inheritance and transfer, especially since they can be acquired without losing the family character of the farm. The legal forms thus generate some confusion because a family farm can be registered as a company… and some companies are mostly family-run or -owned.[5]

The manner of denominating farms refers to how professional and social identities are constructed or deconstructed (Rémy 2008). It also depends on the orientation of agricultural policy that will decide the organizational model of production to support, such as the transition from the farm to the enterprise in France with the laws of 2005–2006, or the emergence of family farming as a target category of dedicated policies in Brazil in the late 1990s.

3.2 Proposal for Definitions

We place ourselves in the cognitive domain to propose a definition of family farming which makes sense and which is "robust" across different institutional and political contexts that shape and are shaped by the actors of the agricultural sectors across the world. Our aim is also statistical in the sense that the proposed definition is intended to allow family farms to be counted, to assess their contributions and identify them so that dedicated public policies can be implemented. Our definition must also allow the comparison or aggregation of situations. It is not a choice which is exclusionary for agriculture which does not correspond to this category. It is a matter rather of proposing a way of naming that is able to justify specific and possibly differentiated public policies.

---

[5] Thus GAEC (Common Grouping of Farms) in France, or GFA (Grouping of Agricultural Lands), which are most frequently associations formed between parents and children or between siblings.

3.2.1 The Agricultural Holding

To define forms of agriculture, we choose to start at the concept of the farm, as it is at this level that decisions are made regarding the organization of agricultural production. We mobilize the definition established and validated by the FAO, which forms the basis of its recommendations on agricultural censuses, and which have worked well over several decadal census cycles.

> An agricultural holding is an economic unit of agricultural production under single management comprising all livestock kept and all land used wholly or partly for agricultural production purposes, without regard to title, legal form, or size. Single management may be exercised by an individual or household, jointly by two or more individuals or households, by a clan or tribe, or by a legal person such as a corporation, cooperative or government agency. The holding's land may consist of one or more plots, located in one or more separate areas or in one or more territorial or administrative divisions, providing the plots share the same production means utilized by the farm, such as labor, farm buildings, machinery or draught animals. (FAO 2007)

This definition captures the observable diversity of functionings, but as regards family farms, four observations can be made:

- Non-agricultural activities are part of the strategies developed by farmers and they have to be taken into account in order to assess the real functioning of production units. Possibilities to develop agricultural activities or, conversely, their limitations will directly depend on non-agricultural investment choices. It is therefore essential to inventory these activities in the same manner as agricultural activities. This pluriactivity has historically been prominent – even dominant – in the developed countries (Mayaud 1999), and generally plays a stabilizing role for small holdings. This includes efforts to increase the added value of agricultural products. It also provides opportunities for change to family farms (Gasson 1986) based on its members' professional and personal aspirations. Pluriactivity and professional mobility are consistent with the perspective developed by Tchayanov (1990) that it is strategies for production and for the use of family labor that guide choices, not economic rationalities directed towards the pursuit of profit or marginal productivities of commercial agricultural activities alone (Shanin 1986).
- The non-monetary dimensions of agricultural and rural activities are also key. Self-sufficiency of food, gifts and return gifts hold great importance to many family farms because they help to reduce food costs and operate as "social safety nets." Such activities can be a source of protein at low-cost, offset the risk of exposure to volatile markets or help pass through difficult periods resulting from economic shocks or social obligations Lamarche (1991).
- This definition of the agricultural holding includes the multi-localization of activities for a same farm, which can manage plots and animals in different locations. Spatial contours of the farm are therefore flexible and, in some situations, even if the definition does not mention it, the farm is "multi-located"

3 Defining, Characterizing and Measuring Family Farming Models 43

beyond just agricultural activities. Its members' mobility is integrated into the strategies implemented (temporary or permanent migration).
- Finally, the diversity of possible legal forms or statuses of the farm is taken into account: households, firms or other social forms of production.

Nevertheless, this definition of an agricultural holding is too generic and therefore insufficient to fully and specifically characterize family farming and farms. We will need to identify specific and robust criteria in order to do so.

3.2.2 Putting Family Farming in Perspective with Other Types of Farming

Political and activist discourses make a rough but schematic distinction between family farming models and enterprise farming models. This distinction is usually based on the place occupied, on the one hand, by the family organization and, on the other, by the modalities of control of productive capital.

To us, this contrasting vision does not seem to correspond to the realities in place. While there are two major ways that agriculture is thought to be organized – entrepreneurial farming and family farming –, by placing ourselves in the cognitive domain necessary to understand the process, we come to the conclusion that there exists a gradient of situations. We also think it is important to introduce the question of the wage relationship, especially of permanent hired labor, to better describe this gradient of situations. Indeed, the use of permanent hired labor profoundly alters the characteristics and the rationales of a farm's functioning (status of workers and relationships between farm workers, regular monetary expenses and cash flow requirements).

This choice leads us to propose three organizational forms of agricultural production which translate into three very different types of farms, themselves highly diverse, but which can be further distinguished on the basis of finer criteria and operational characteristics (Table 3.1).

Besides the nature of the labor relationships, five non-exclusive criteria of differentiation take into account the different dimensions of activity: origin and ownership of the capital, the modalities of decision making (administration and management), destination of production (i.e., the share for home consumption), legal status and land rights. These combined criteria also provide information on the level of economic independence of the technical system.

The table shows that criteria other than the labor relationships are not robust enough. Too dependent on local and national contexts, they do not allow the definition of stable forms of agriculture. Nevertheless, the three types of agricultural holdings thus correspond to a gradient of situations that goes:

Table 3.1 The different types of family farms

	Entrepreneurial forms – family forms		
	Types of enterprises	Types of family business farms	Types of family farms
Labor	Exclusively salaried employees	Mixed, presence of permanently salaried employees	Family dominance, no permanently salaried employees
Capital	Shareholders	From family or family association	From family
Management	Technical	Family/technical	Familial
Home consumption	Not relevant	Residual	Ranging from partial to full
Legal status	Limited liability or other company form	Farmer status, associative forms	Informal or farmer status
Land-rights status	Property or formal rental – property, or formal or informal rental		

- from the family's exclusive role in the mobilization of production factors and their management to its complete absence in entrepreneurial forms;
- from an informal legal status (corresponding to an exclusive domain of the family or community) to the recognition of different formal legal forms, with an intermediate form of the recognition of the specific status of the farmer by certain public policies;
- from home consumption of the entire production to exclusive reliance on the market, i.e., from a non-monetary purpose to a fully monetary one.

In order to characterize agriculture objectively and robustly, we find that holdings can best be differentiated based on the origin of production factors, especially labor, and less on other, more ambiguous criteria listed above. On this basis, we distinguish three forms of agriculture:

- forms of family farming, which rely on the labor of family members – usually exclusively, sometimes partially, with occasional temporary recourse to some proportion of non-family labor;
- forms of entrepreneurial agriculture corresponding to an exclusive use of hired labor with no link between capital (means of production) and the labor mobilized;
- forms of family business agriculture – an intermediate form – corresponding to a situation with many variants, but whose business status stems from the permanent recourse to permanent hired labor, which assumes here a structural nature.

In many cases, this recourse to permanent hired labor can be avoided by a significant substitution of family labor by capital (mainly mechanization) with similar effects on the farm's financial operations (expenses and cash flow requirements). Because of their importance in the history and development of agriculture in some national situations, these family business forms deserve greater attention, both from research entities and development policymakers. Taking them into account will, in particular, enrich the debate on production models suitable for the future and those worth promoting.

Each of these three forms of agriculture covers, of course, a wide variety of real situations. However, our proposition based on the family-entrepreneurial gradient and focused around the issue of labor has the advantage of a robustness that transcends production systems and the size of farms. Our characterization – represented by a grid applicable to all situations – allows us to view the dynamics of agricultural transformation and the effects of policies on these transformations in a new light. It also allows us to overcome the normative definitions adopted in different countries and contexts, without preventing the definition of typologies more specific to local situations within each of these forms. Our approach is similar to that adopted by Otsuka (2008) when he defines the peasants. It is also similar to that developed by Hayami when he defines the plantation (Hayami 2010) by building on the work of Jones (1968), and also when he contrasts the same plantation with smallholders (Hayami 2002).

But in our approach, we have opted for a strict demarcation, pertaining to a clearly measurable structural variable in the context of standardized agricultural censuses: the use of permanent family labor.

3.2.3 A Stand-Alone Definition of Family Farming

> Family farming designates one of the forms of the organization of agricultural production which encompasses farms characterized by organic links between the family and the production unit and by the recourse to family labor with the exclusion of permanent hired workers. These links are formed by the inclusion of productive capital in the family patrimony and by the combination of domestic and economic rationales, both monetary and non-monetary, in the process of allocation of family labor and its remuneration, as well as in choice of product distribution between final consumption, inputs, investments and accumulation. (Bélières et al. 2013).

By including this definition in the debate, we position ourselves in the perspective put forward by the rural economist Tchayanov (1990) in the early twentieth century. Organic links are formed by the inclusion of farm capital in the family patrimony and by the combination of domestic and economic rationales, both monetary and non-monetary, on the one hand, in choices of product distribution between final consumption, inputs, investments and accumulation and, on the other, in the process of allocation of family labor and its remuneration.

The close link between the family and the farm marks the relationship between the social sphere (domestic and patrimonial) and the economic sphere (Cirad-Tera 1998). This relationship between the family patrimony and the farm capital partly explains the resilience of family forms because it permits adjustments to be made to limit the effects of shocks (economic, climatic, etc.). Given these links, choices in product allocation are made as follows. Once the inputs and any loan interest are paid off, priority is given to family consumption, then to the accumulation of social capital, and only finally to productive accumulation, with both these forms of accumulation often being intrinsically linked because of the farm's family nature (Chap. 4). And, conversely, when economic, social and climatic hazards so require, family patrimony can be mobilized to overcome the farm's difficulties.

The second criterion is the use of family labor. In the literature, there are several qualitative terms to describe the ratio between family labor and hired labor: primarily, essentially, almost exclusively, in a dominant manner, etc. These expressions leave too much room for interpretation between what is the familial form and what is not. At best, they establish thresholds that are very context-dependent (Hill 1993) and can certainly allow any particular definition to be adapted to different national contexts. However, in our view, the definitions that do so result obscure two elements. First, it becomes important to distinguish occasional or temporary hired labor (but which can acquire a regular character over time) from permanent hired labor. Only the latter is structural in the sense that it permanently modifies the farm's productive structure, such as the development of a specific component of the production system (for example, the raising of small cattle) or an expansion of the cultivated area, in a way that would not be possible without this permanent labor. Second, the presence of one or more permanent hired agricultural workers supposes the creation of wage relationships within the production unit. This relationship changes significantly the farm's productive rationale insofar as it becomes necessary to ensure that part of the production generated globally at the farm level is permanently earmarked for revenue generation to allow the worker(s) to be paid. The logic of this fixed remuneration is clearly different from the logic of the remuneration of family labor which can be adjusted downward or upward, depending on the level of production reached.

We note that this definition of family farming is in line with the one previously used by researchers from CIRAD (CIRAD-Tera 1998). It referred to "the central and privileged link between agricultural activity and the family organization, especially with regard to patrimony, the means of production, mobilization of labor and decision-making."

3.3 An Inclusive Definition that Singles Out the Family Farm

Defining family farming "strictly" in this way allows the inclusion of similar denominations – usually mobilized in the literature on rural or peasant studies[6] – in a metacategory.

---

[6] We refer here to studies that refer to "peasant agriculture" or "peasant" categories. These studies pertain to different schools of economics and sociology, even political sociology, and have, in particular, appeared in the Journal of Peasant Studies and in the Journal of Agrarian Change.

For Friedmann (2013), "the term 'peasant' has entered the political field" and it assumes very different meanings in different contexts and domains in which it is used. It is difficult to use it to define a category rigorously on the basis of structural data alone. The proposed definition of family farming potentially allows the inclusion of "peasant" or "smallholder"[7] forms of agriculture based on a simple criterion – family labor versus "non-family" labor.

Shanin (1986) defines peasants "as small farmers who use simple equipment and the labor of their families to produce primarily for their own consumption, directly or indirectly, and to meet their obligations towards holders of political and economic power." Ellis (1993) also refers to family labor: "Households [...] use mainly family labor." He takes into account the diversification of activities, and the difficulties in inserting them in upstream and downstream markets: "The peasants derive their livelihood mainly from agriculture. They primarily use family labor in agricultural production activity, and are characterized by partial engagement with input and product markets, which are often imperfect and incomplete." Finally, Otsuka (2008) combines farm size and family nature of the work: "The peasants are considered primarily oriented towards production for subsistence, full-time or part-time in the case of small farms, and towards production of food and cash crops while developing non-agricultural activities. Peasants can thus be defined as small family farmers who either own their farms or rent them from others." The criterion of home consumption – as well as the notion of subsistence – is called upon frequently to characterize the peasant, even if the share of resources devoted to the production of food for the family is subject to fluctuations induced by market instabilities (Ellis 1993). Mendras (1976) likes to define a peasant based on his being part of wider community: the peasant society. But this criterion too is very difficult to assess or qualify through structural surveys or censuses.

In all these cases, over and above the other – highly variable – criteria used, the common denominator is family labor as the main source, if not the exclusive source, of farm labor. We find this same presence of family labor in most definitions in studies on smallholders (HLPE 2013).

The proposed typology also distinguishes family farming from other forms of agriculture. This proposal refers to a long-standing debate that preoccupies societies and academia on how to organize the production of food and non-food products from agricultural activity. In these debates, the family-labor agricultural model is challenged by advocates of enterprise models using hired labor (Collier and Dercon 2013) in the name of efficiency. The classic agrarian debate initiated by Marx, then taken up by Kautsky and Lenin in the late nineteenth and early twentieth centuries, is thus revisited. Chap. 5 seeks to clarify the characteristics of other forms of agricultural production and their relationships with family farms.

Finally, our approach is partially consistent with that used by Hayami (2002). For him, the farm size is not a discriminating factor, at least not for family farms in developed countries, but does remain one for developing economies. This is a

---

[7] The reference to size in this denomination inserts this concept into a highly contextual relativism.

distinction that we do not share here. The importance of "small structures" in family models means revisiting the thinking and the discourse on "agricultural modernization." Can we even consider a modernization which will lead to a drastic reduction in employment in contexts of high population growth and the relative lack of opportunities in other economic sectors? "The family farm [is] defined here as the production unit controlled primarily by the farm head, resorting primarily to the labor of his family members [...]. The farm can be large in terms of arable area since a farm of several hundred hectares in high-income countries can be cultivated by one or two family workers relying heavily on mechanization."

The aim is also to overcome certain assumptions pertaining to rationalities which have been "assigned" to farmers. For example, family farms are not synonymous with poverty, although some situations are marked by economic or food insecurity. Indeed, family farming may, in favorable agricultural policy conditions, engage in dynamics of economic accumulation. Family farming systems are not focused on home consumption; they are in the market but they can produce for subsistence or for non-monetary exchanges (which tends to increase their resilience as a safety net). Family farms are not necessarily synonymous with small size or "small-scale agriculture" as assessed by the amount of land resources put to use. Rationalities of family farms cannot be analyzed with the economic tools of theory of the firm.

3.4 An Approach of the Diversity of Family Farming Models

In order to fully understand family farming models, we have to go beyond the invariants and principles that distinguish this category from others and understand their diversity (Bélières et al. 2013; Sourisseau et al. 2012).

Our approach to the diversity of family farming does not, unlike the previous approach, consist of establishing a "closed" typology of these farms. It is a matter rather of identifying and then discussing the criteria of differentiation which we deem essential and which make sense in light of the challenges agriculture in its diversity has to face. These issues pertain and may be specific to national contexts. To this end, we adopt the approach proposed by World Agriculture Watch (WAW) (FAO 2012). Table 3.2 lists key criteria and their possible modalities which provide a broad idea of the main forms of family farming. They can, of course, be broken down to each local situation, depending on important issues specific to family farms.

We have identified six first-level criteria to explain – more through the functioning of families than through only farm structures – most of the diversity of family farming models. In what follows we will not go into details of the possible modalities of the various criteria listed in the table. For this, see Bélières et al. (2013).

Table 3.2 Main criteria of differentiation of family farming models and possible modalities

Criteria	Modalities
Security in access to resources	Unsecure access
	Secure acess (legal or not)
Investment capacity	Reduced
	Increased
Home consumption	Yes
	No
Type of insertion into downstream markets	Little insertion/insertion only in proximity markets
	Insertion into local supply chains with local standards
	Insertion into niche international markets
	Insertion into international commodity markets
Pluriactivity/system of activity	Agriculture only
	Agricultural and non-agricultural activities
Level of agricultural diversification or specialization	Specialized agriculture
	Diversified agriculture, including on-farm processing activities
Additional composite criteria	
Substitution of family labor with capital	Family labor only, no substitution
	Moderate substitution by non-family labor
	High degree of substitution by non-family labor
Strategies and objectives of the activity and usage of the result	Simple reproduction (priority for family consumption)
	Family and social accumulation
	Productive and social accumulation

Source: Bélières et al. (2013)

Our reasoning relies on the security of access to land resources, including common property resources (not to be confused with free access), as a basis for agricultural activities or as a source of extractions from "nature" production (fishing, hunting, gathering, etc.). There are two main reasons for this choice. First, private ownership is less important than the guarantee of the right to exploit the resources through farming or other primary activities and transmission rights.[8] Second, access to common property resources contributes significantly to the food security of some rural households, both during normal and critical periods and, in particular, allows them to obtain animal proteins at low cost (hunting, fishing).

While the level of capital is significant for differentiating family farms, the issue of access to appropriate sources of credit for investment in all its forms is a key

---

[8] See Courleux (2011) for France or the work of the Land Tenure and Development Committee for the countries of the South (Colin et al. 2009).

determining factor for the future of family farming. The challenge, indeed, is of connecting financing systems to the needs of family farmers by considering not only the diversity of possible investments in agriculture but also those outside agriculture for family farms which are pluriactive in their vast majority (HLPE 2013).

A farm with self-supply of food for home consumption is not synonymous with "backward" farming at a primitive stage of development, as the dominant discourse advocating market integration and the importance for family farmers to make their production part of major supply chains would have us believe (Shanin 1988). In fact, the overwhelming majority of family farmers are connected to the market economy across multiple markets in which they participate through labor, inputs, access to land and, of course, the agricultural products they sell and the food they purchase. The conditions of their participation in different markets are the issue and require regulation (HLPE 2013). But farming and non-farming families also produce food for their own subsistence or for non-commercial trading systems (see, for example, Cittadini 2010; ENRD 2010).

Taking non-agricultural activities or pluriactivity into account is not inconsistent with the development or maintenance of agricultural activities. Rather, this "coexistence" describes the vast majority of situations in the South and a significant proportion of the rural populations in countries that have developed their agriculture through a very intensive model.

For farms, as far as agriculture is concerned, it is important to distinguish between those which have specialized (with a view to improving economic performance but also assuming an increased risk) and those which have chosen to diversify production, accompanied with specific methods of production to improve returns (processing, direct selling).

We also believe it necessary to introduce a specific criterion for differentiation on the basis of the real practices that link – or are complementary between – family labor and physical capital. This is the core of the issue of family farming and of the preference accorded to family labor relationships when taking major strategic decisions. The decision to rely exclusively on family labor – which therefore is usually limited in size – may come into conflict with the expansion required to accumulate capital or to support a larger family. In our opinion, how this conflict is resolved indicates specific strategic orientations.

Finally, the nature of the organic links between family and production is a distinguishing criterion which expresses the fungibility of patrimony and family capital. It is a source of flexibility and allows the farm to withstand shocks better than if labor remuneration is in a monetary form. The remuneration of family labor expresses the use made of the farm's income, after deducting the fixed costs and expenses. This is a composite criterion, which allows family farms to be located on a gradient going from, at the one end, a peasant ideal-type to, at the other end, agriculture that is capitalist in terms of the end-use of production. It is similar to the criterion on the ability to invest, by qualifying the concretization of this capacity. It can also be considered by taking into account the strategies of transmitting agricultural patrimony.

3.5 Contributions Which Are Difficult to Measure and Quantify

Not only is family farming hard to define, measuring its size and effects remain a cognitive and methodological challenge. The data available today are unsuitable to allow us to "count" family farms, estimate their share of the agricultural labor force, their share of the area farmed and their share in total production and thus, ultimately, to assess their political significance for each country. The attempts to do so (below) and the limitations encountered underline the need for a renewal of national and international statistics to better weigh family farming systems.

It is at the country level that information about farming is obtained and recorded through agricultural censuses. The FAO has created a common reference which it recommends States use for their agricultural censuses. The responsibility for their implementation, however, as well as the ownership of the data remains with the States. Censuses are held regularly since 1950 on common basis with a time frame of 10 years, a pattern followed by a growing number of countries: from 81 in 1950 to 122 in the 2000s. As far as the last round of censuses (1996–2005) is concerned, 122 countries have completed their census and 114 of them have submitted their reports to the FAO (FAO 2010). Furthermore, while there are a limited number of countries which have held several such censuses each, allowing changes in production structures to be analyzed on a chronological basis, the data available, though not exhaustive, still covers 83.5 % of the world's population. These include 81 countries which, on the basis of data compiled by FAO for 1996–2005, are comparable by surface size class (HLPE 2013). We can consider these data as representative from the viewpoint of orders of magnitude, especially as they include the most populous Asian countries (Fig. 3.1).

As far as numbers of family farms are concerned, "structures" of a size smaller than 2 ha encompass almost 85 % of farms worldwide. This proportion rises to almost 95 % if we consider holdings of less than 5 ha.

In the case of the 27 countries of the European Union, the available data show that 70 % of the number of farms are smaller than 5 ha and this proportion rises to 80 % when considering holdings of less than 10 ha (HLPE 2013). In the United States, which measures the farm sizes by volume of sales, 87.3 % of the total number of farms are small family farms and among large and very large farms, only 4.1 % are entrepreneurial (i.e., non-family) farms.[9] Even in countries with "developed" agriculture, the issue of family farming systems looms large, both for their proportion of large farms (8.6 % in the U.S.) as for their dominant presence in small farms (Fig. 3.2).

As for the proportion of the population and the labor force involved in agricultural production by type of farm, the quality and availability of census data varies widely and we must content ourselves with agricultural labor data originating from

---

[9] http://www.epa.gov/agriculture/ag101/demographics.html, retrieved 28 January 2014.

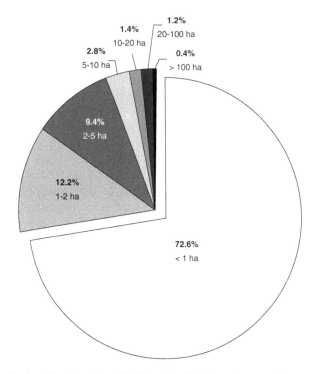

Fig. 3.1 Distribution of total number of holdings classified by surface area (81 countries) (Source: FAOSTAT data, graph by authors)

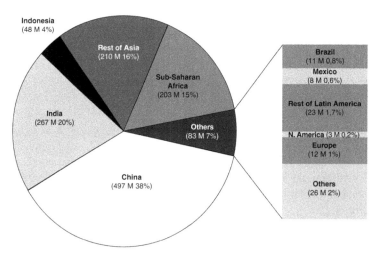

Fig. 3.2 Distribution of the agricultural labor force by continent and country (Source: FAOSTAT, graph by authors)

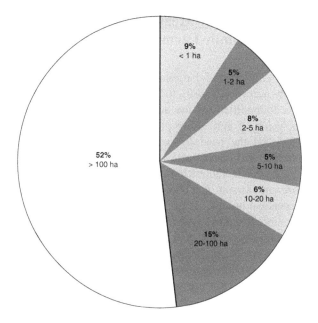

Fig. 3.3 Distribution of agricultural surface areas by size class (81 countries) (Source: FAOSTAT, FAO 2010, graph by authors)

population census data (rather than agricultural censuses). Nevertheless, given the large number of farms we can safely assume to be family farms, at least for the vast majority which are smaller than 10 ha (about 98 %), the bulk of family farms are located in Asia. India and China, along with other major Asian agricultural countries – Indonesia, Vietnam, Thailand and the Philippines –, account for 78 % of the global agricultural workforce.

The other continent that is significant in this geography of family farms is Africa, sub-Saharan Africa to be precise, which has about 15 % of the global agricultural workforce. Africa has the distinction of having not yet completed a demographic transition, unlike some Asian situations which will find themselves affected earlier than Africa by the issue of an aging agricultural population.

As for the area used by family farms, the agricultural census data by size class is not comparable across countries. They do not allow us to draw a clear and definite conclusion, given the incompleteness of the data (Fig. 3.3).

On the basis of our estimation in Fig. 3.3, we see, however, that farms of more than 100 ha occupy a little more than half of the cultivated land, while they represent only 0.4 % of the number of farms worldwide. It is certain that in many countries with an agrarian structure that includes large farms, many of these are family farms or family business farms. Finally, little is known of the surface areas farmed and used by enterprises or firms. Similarly, common lands, primarily used by family farming – especially for livestock rearing purposes – find little if any mention in the statistics. The available data, however, show an unequal distribution that must be analyzed in each country separately in order to take historical and institutional conditions into account.

3.6 For Policy Measures Adapted to the Characteristics of Family Farming Models

Discourses on and the policies implemented in the agricultural sector have historically accorded priority to developing the sector's production function, the supply of commodities for the agri-food industry (Gervais et al. 1965) and, since the inclusion of agriculture in GATT negotiations and the creation of the WTO in 1994, on the need for greater integration of farmers into markets, especially international ones (Vorley et al. 2007; Biénabe et al. 2011; OECD 2011).

These orientations have resulted in an ever-increasing specialization of agriculture and pursuit of productivity. Farms are becoming larger, with revenue targets dependent on other sectors (Dorin et al. 2013), but they require fewer workers and are more and more dependent on the prices of agricultural commodities, which even the most organized and largest of them can no longer influence. These policy choices are accompanied by a considerable concentration of agri-food industries (Rastoin 2008) which have captured a significant share of added value of the global food system (Rastoin and Ghersi 2010; Mc Cullough et al. 2008). These farms, touted as "hyper-efficient" from the point of view of strictly economic efficiency (as measured by labor productivity) are the direct outcome of public policies of direct or indirect support they continue to receive. The conventional model of technical intensification on which they are based not only induces risks at the individual level but also generates negative externalities at different community levels such as the economic (employment and income) and the environmental (greenhouse gas emissions). The long-term management of ecosystems and resources is also adversely impacted. These negative externalities find little or no place in economic calculations, as this would lead to a drastic rethinking of the production model.

Thus, the proposed definition of family farming should help establish an alternative rationale, relying in particular on the notion of autonomy present in the definition of the peasant by Mendras (1976, 2000) and taken up by, among others, Van der Ploeg (2008, 2013) in a context of agriculture's continuing market integration. The notion of autonomy is, however, not synonymous with autarky or regression. It is a matter of a reasoned approach to production systems and their market insertion in order to make the practice of agriculture more resilient in technical terms – self-regenerating fertility, taking account of biodiversity – and more remunerative of family labor. Adopting this definition also means revisiting the issue of subsistence production, which does not reflect an archaic vision of agricultural practices, as some would have us believe, but a way to ensure food security and improve nutrition in a decentralized manner, through the use and localized strengthening of social networks. While subsistence agriculture no longer finds mention at the global level – except in the discourse of some entities who compare it pejoratively to commercial agriculture –, production for subsistence is a true social reality which extends beyond the agricultural sector in the North (Deléage and Sabin 2012) as well as in the South (Cittadini 2010) in the context of improving the food security of agricultural, rural and even urban households. In this, subsistence production is in line with the idea of the economy of proximity, where autonomy is considered not only at the farm level, but also through its role in the surrounding territory.

This reflection on the definition of family farming also allows us to revisit the issue of the substitution of family labor by physical capital. This issue can thus be approached through the perspective of the generation and distribution of added value and agricultural and rural employment in a context of diversification of production systems.

This contribution around a proposed definition can finally be an invitation to revisit the historical trajectories and to provide some basis for specifying the development options available for the future of the world's family farming systems.

Chapter 4
Families, Labor and Farms

Véronique Ancey and Sandrine Fréguin-Gresh

The rural worlds in the South have changed dramatically in recent decades. In addition to transformations brought about by globalization (Losch et al. 2012), they have also been affected, from near or far, by the major trends which have taken place in most contemporary societies: demographic transition (in Latin America and Asia), family reconstructions, individualization of spouses' professional activities, questioning of patriarchal authority and the emancipation of the "dependents" (the youth, women). Even if rural societies are not all affected to the same degree, the social relationships existing in the South have evolved, resulting in profound and lasting changes in the conditions of production and reproduction of family farming.

Yet, most research on family farming[1] in the South has not taken into consideration social relationships in its analyses. Whereas, since the 1960s (Bourdieu and Sayad 1964), rural sociology has sought to analyze the broader evolution of societies through the "family-labor-farm" interlinks (Bessière et al. 2008), making domestic structures part of general political and economic developments (Friedmann 1978), most of the ways of looking at family farming (Sourisseau et al. 2012) have attempted to "better understand the functioning of farms [in order to] improve production" (Brossier et al. 2007). These studies have focused on the links between farms and the productive potentials of the environment at various scales (plant, animal, plot, herd, farm, region, value chain, etc.) using

---

[1] In this chapter, we use the terms "agriculture" and "farming" to refer to all activities undertaken on a farm: cultivation of crops, livestock rearing, gathering and on-farm processing of farm products.

V. Ancey (✉)
ES, CIRAD, Montpellier, France
e-mail: veronique.ancey@cirad.fr

S. Fréguin-Gresh
ES, CIRAD, Managua, Nicaragua
e-mail: sandrine.freguin@cirad.fr

various prims of analysis (functioning, performance, etc.) and with several objectives (finding technical solutions, making recommendations for rural development, recognizing farmer knowledge and know-how, involving farmers in the conservation of natural resources, etc.). The agricultural family is then understood as a "social workforce," led by the head of the farm and implementing "Production factors" (Mazoyer and Roudart 1997; Cochet 2011), even if some research studies further analyze the functionning of consumption units (Gastellu 1980), the modes of decision-making (Ancey 1975), and re-place family farming into sustainable rural livelihoods (Chambers and Conway 1991), which can include spatial mobility. Research on family farming has thus often remained disconnected from the rationales of the groups and individuals behind it, restricted to its productive activities anchored in one territory, and focused on the exclusive nature of agriculture (Sourisseau et al. 2012).

This chapter complements and refines – but also questions – the effort of definition undertaken in the previous chapter concerning the links between the family, the farm and mobilization of family labor. To this end, it invites us to revisit the complexity and diversity of the "family-labor-farm" triple inter-links from about 15 viewpoints based on work undertaken by CIRAD teams and their partners. Case studies in different contexts in the South structure the chapter and its reasoning. Using them, the first section will show how labor in family farming is embedded in social and productive and habits to meet the requirements necessary for living (from agricultural production), reproducing, maintaining (ascendants and descendants) and transmitting (heritage). This chapter's second section focuses on the links between the agricultural family's labor and the farm, analyzes the social responses to the difficulties experienced and the search for off-farm opportunities, especially through pluriactivity and spatial mobility. Finally, in the third section, the links between the farm and the family are discussed in terms of strategies of creation, accumulation and transmission of heritage. We conclude this chapter with some thoughts and perspectives for future research.

4.1 An Embedding of Social and Productive Rationales

Forms of labor in family farming depend primarily on the characteristics of societies in which they evolve. These themselves are in constant evolution: gender relations, domestic structures and broader social configurations of agrarian societies. Thus, the major trends of evolution in contemporary societies (declining birth and mortality rates, urbanization, divorces and resulting family reconstructions, fragmentation of domestic units, economic independence of women, etc.) challenge the stereotypical Western model of the concept of family: a couple and their offspring (the head of the family as the head of the farm, "helped" by his wife and children). This definition of the nuclear family, defined by its links of kinship and alliance, has been called into question for a long time in the South where, as elsewhere, the "diversity of family forms is such that the search for a universal type of family is as futile as idealistic" (Seccombe 2005). Indeed, it is first in the South that the historical concentration of anthropological research has revealed that "contrary to what conventional anthropology [teaches...], biological links [...] count for much less than the nutritional capacity of those who look after children [...and] when social cohesion is formed, it is on other grounds" (Meillassoux 2005). This is also why transposing statistical units such as the "household" indiscriminately in all contexts has been criticized for a long time now (Charmes et al. 1985).

However, the diversity of family farming forms is now part, as it has always been (Polanyi 1983), of patterns of production, distribution and exchange which are more complex than merely commercial ones. Perceiving the recurrence and diversity of these patterns in countless family farming forms is essential in order to contribute to development policies. Box 4.1 provides an example of the links between family structure and productive organization.

> **Box 4.1. Pastoral camps in the Sahel: a way of life beyond the farm.**
>
> Christian Corniaux
>
> The Fulani or the Moors generally structure their lives around the pastoral camp. Each pastoral camp corresponds to a lineage segment and brings together many adults and their offspring under the tutelage of a father or other elder. A camp's population can vary from a dozen to several dozen people. One should not, however, equate a camp to a production unit or *a fortiori* to a farm (Corniaux, 2006). It corresponds, rather, to a lifestyle, a residential unit, and to the manifestations of solidarity and mutual assistance. The production unit is at a lower level – that of the *gallé* – which is organized around small-ruminant flocks or a part of the camp's cattle herd. The pastoral camp then represents the family's framework of activities and of cooperation in labor. It is at this level that the majority of decisions are taken. The most radical decision is that of a break with the "mother" camp when a son wants to be strike out on his own, especially when his father dies. It is in this form of social organization, with its dynamics and tensions, that pastoral families in the Sahel manage – collectively (cattle) or individually (small ruminants) – one or more herds essential to food security in an environment where uncertain and dispersed resources impose seasonal spatial mobility.

While social organizations are complex and as diverse as agrarian societies, it is possible to identify a common feature of family farming. Indeed, as Barthez (1984) pointed out in the French context in the 1970s and in accordance with the definition proposed in this book, the family farming workforce consists exclusively of "a group of individuals, neither hired nor selected from the labor market, but chosen according to a rationale of family development." It is the social ties binding the family, and which need careful characterization and analysis, that distinguish family farming from other forms of agriculture. It is a matter therefore of characterizing the family structure (family composition and roles of its members), its social functioning (links and circulations), its economic functioning (in particular, on-farm and non-farm activities as we shall see in the next section) and spatial functioning (location of activities where some members diversify their sources of income).

However, defining the contours of family farming requires some sociological or ethnological skills, as shown in Box 4.2, which presents a social organization illustrative of Western African Sahelian societies. This form of family farm involves a large number of individuals in social and economic relationships that go beyond the functioning of the farm. In this case, group and individual rationales guarantee social reproduction through labor in family farming.

Box 4.2. Agricultural families in Office du Niger: an organization at the heart of rice production and social reproduction.

Jean-François Bélières and Jean-Michel Sourisseau

In the Office du Niger in Mali, agricultural families are organized according to a patriarchal model centered around agricultural labor (Sourisseau *et al.*, 2012). The unit of residence (the homestead) is the same as the unit of consumption and includes a main production unit where all the family members from several generations work under the authority of the patriarch, who is the primary decision-maker (allocation of activities, stocking of the granary which other members can draw on for food, management of production and sales). The contribution of everyone in the farm's activities leads to significant production, which extends the range of strategic options for the head of the farm. In addition to ensuring subsistence, the head manages a negotiated balance between rights and obligations, especially to meet the social needs of the dependents and evaluates and approves their individual projects (vegetable gardens, small business). However, the accumulation takes place at the level of the extended family. This social organization is subject to tensions and conflicts. If the farm generates accumulation, the equilibrium of the system of rights and obligations shifts towards greater demands from sub-units (dependent households and individuals). The collective shares of the crop are reduced, as is the head's room for maneuver. Reproduction of the system also requires a land expansion for every separation of a residential unit which exceeds a critical size. If access to new land is not forthcoming, the fragmentation of the family results in a reduction of surface area per worker and the inability of the head of the new unit to meet his duties and obligations towards his dependents.

In addition, defining social contours of families on the basis of kinship and alliances can be problematic. How to include those who are "part of the family," i.e., servants, laborers, shepherds, apprentices, retainers, etc. who are clothed, fed and housed by agricultural families who employ them but without compensating them monetarily? These individuals, whose statuses are often of confiage or modern slavery, play a central role in the links between labor and family by replacing at low cost the labor contribution of a family member who has migrated for a short or long period or gone on transhumance. Should these individuals who participate fully in on-farm activities and in the social reproduction of the family – but not within a strictly non salary-based relationship – be excluded from the social contours of the agricultural family?

Box 4.3 provides an illustration of the Sahelian pastoral economy where the shepherds are fully part of the social rationale of the family farm. Should they be considered employees? If yes, should we exclude pastoral camps from the ambit of family farming? Is this mixed mode of remuneration not evidence instead of the monetization of social relationships in the pastoral economy, compatible with the family functioning of these farms?

> **Box 4.3. The shepherds of the Ferlo: evidence of the monetization of the pastoral economy.**
>
> Véronique Ancey
>
> Currently, more than a quarter of pastoral farms in the Ferlo in Senegal employ one or more shepherds, often on a temporary basis for transhumance, sometimes permanently (Wane et al., 2010). The reasons for the recourse to external labor vary: the lack of sons of age for transhumance, the necessity of the care of small ruminants, the growth of the herd, or the migration of a family member. The relationships that the pastoral family has with these shepherds go well beyond a merely salary-based one. Their function is as old as the pastoral society itself and their statuses are differentiated, from a young handyman (*surga*) to the qualified shepherd who is entrusted with a transhumant herd for several hundred kilometers (*gaynaako*). In past times, their remuneration was mainly in kind: sandals, a stick, a bull-calf per year, board and lodging. In recent decades, however, a meager monetary amount is additionally offered. This partial monetization of the labor relationship does not upset the social links within the family camp, but it has become necessary to ensure the livelihoods of the pastoral camps.

Family labor in agriculture is not therefore identified on the basis of a non-salary relationship (Boucher 1990). In many agrarian societies, working on the farm is not a free choice but an obligation within the family social relationship. As Barthez (1984) described for the French context of the 1970s, the "concepts of working hours, qualifications, and social mobility found normally in labor markets have little meaning here. [...] On farm labor results from an unavoidable family affiliation, pertaining to a natural necessity and not a social one. Labor merges with the family obligations, whence its timeless character, whence the inability to define a market value of this elusive labor, a coextensive with life itself, a work that acquires the meaning of a destiny."

This effort, normally not taken into account, characterizes the family farmer. After all, no hired laborer would make stone walls or dig drains during his leisure hours, but a family farmer, regardless of age, would. This effort – we can also describe it as work that is not "valued" through a capitalist relationship of production – not only contributes to the production but also to investment and accumulation so that an improved heritage can be transmitted or sold. This recalls the definition of the organic link between family, heritage and farm capital of Chap. 3. But organic does not mean "natural!" On the contrary, the labor done by women remains invisible. They do not have property rights and are removed from the processes of transmission of heritage. These exclusionary features are integral to most family farming systems and stem from the long history of patriarchy as an autonomous social, economic and political construction (Delphy 2013).

To illustrate this observation, Box 4.4 shows how, in some societies of the Maghreb, women's labor continues to be overlooked despite their essential role on the farm.

Box 4.4. The invisibility of women's work on livestock farms of southern Tunisia.

Nathalie Cialdella

Women participate fully in family livestock breeding activities in southern Tunisia. Nevertheless, their contribution is rarely recognized. The progress report on the program for managing natural resources in Médenine is emblematic in this regard: "Women's activities consist only of activities traditionally managed by women: poultry farming, beekeeping, rabbit breeding and craft activities." And yet, the role of women in livestock breeding is essential. On farms with sedentary livestock herds, women's work has increased and almost all tasks necessary for livestock breeding have been assigned to them: guarding the herd on the pasture, feed distribution, and gathering of pastoral resources and crop residues for transferring them to troughs near the habitations. Their invisibility is explained by several factors. On the one hand, women rarely claim property rights as established in the Code of personal status concerning heritage. A desire to avoid the fragmentation of land is the reason generally advanced for this reluctance. On the other hand, once married, a woman and all that belongs to her come under her husband's control.* Finally, there exists strong social pressure to keep women out of the economic sphere: "The concept of salaries for women threatens the man's guardianship role, and is a potentially subversive factor in decisions concerning the family" (Cialdella, 2005).

* This also explains endogamous marriages practiced by 21% of Tunisians in 1991 (Darghouth Medimegh, 1992).

As a result, greater importance should be accorded to the issue of the convergence between feminist emancipation movements and professional aspirations in the agricultural world. Indeed, the functioning of a family and relationships of its members with the farm evolve hand in hand with fundamental changes in the balances of power and intra-family domination (Guétat-Bernard 2007). These transformations also eventually call into question the conditions of social reproduction of agricultural families and farms.

In other contexts, economic and social realities of agricultural families run up against international regulations and lead to other recompositions of the social organization of families. These may even include forms of exclusion of certain members of the agricultural families, which may mean a break in the transmission of knowledge and know-how. Thus, while measures encouraging market integration in a value-chain approach propose – or impose – new social rules in the organization of farm labor, the consequences of implementing these rules can be quite unexpected. This is illustrated in Box 4.5 on coffee certifications in Costa Rica.

Box 4.5. Unexpected inclusions and exclusions: effects of coffee certifications in Costa Rica.

Nicole Sibelet

Since the 1990s, the social and solidarity economy movement has affirmed that labor is not a commodity. This proposition implies calling into question the value of agricultural labor which may be an element of social inclusion – if it does not include exploitation (children, local or migrant seasonal workers) or the invisibilization of labor (women, son, caregiver). Costa Rica has incorporated environmental and social dimensions in agriculture (Sibelet, 2013) through the development of coffee certifications (FLO Cert, UTZ Certified, Rain Forest Alliance, CAFE Practices, Fairtrade, etc.) which, *inter alia*, prohibit child labor. Nevertheless, in adhering to the principles of these certifications, Costa Rican coffee producers are confronted with a paradox. Thus, some adolescents have never worked on their parents' coffee plantations, and therefore did not acquire the knowledge and experience necessary for agricultural work. These young people find themselves unemployed and sometimes, according to the adults, "fall prey to drugs," a recent phenomenon affecting Costa Rican rural areas. The exploitation of children through labor is, no doubt, unacceptable and should be fought against wherever it is found. However, in the case of Costa Rica, learning and preparing for the future through adolescent labor – with respect accorded to him as a person – in the family farm would be a positive factor for social inclusion.

4.2 The Links Between Work and Farm Affected by Pluriactivity and Spatial Mobility

Family farming's objective must be to provide the agricultural family with the means to live and to do so with dignity. In a context where public policies target with varying degrees the goods and services needed to ensure food security, an acceptable standard of living and basic services, the family farming forms as "exclusive occupations of the family" are no longer recognized in the North, nor in the South. Now, as earlier – even if this reality has long been obscured by decades of a promotion of a commercial-productivist model encouraging agricultural specialization and professionalization –, agrarian societies have remained diversified, and it is rare nowadays for agriculture and the exploitation of natural resources to be the only ways of making a living for rural families in the South. Similarly, almost all rural societies are connected to products, services, and labor markets, as they need to obtain from the market what cannot be produced on farm nor exchanged. Thus, pluriactivity is a structural component of the rural worlds in the South and has developed in many societies over a long period. This is illustrated in Box 4.6, which shows how Cameroonian cocoa farmers have developed activities outside the family farm which are central to the functioning of their family plantations.

Box 4.6. The diversification of activities: an old strategy brought up to date by the Cameroonian cocoa farmers.

Philippe Pédelahore

Before the colonization of Cameroon, activities of rural families consisted of agriculture, fishing, hunting and gathering. With the introduction in 1924 of the *Code de l'Indigénat* (Indigenousness Code), families were forced to work in other sectors (porterage and construction of infrastructure) and seek employment in the cocoa plantations of local chieftains and colonizers. The abolition in 1948 of the *Code de l'Indigénat*, the imposition of taxation and the monetization of trade stimulated the rural diversification of these families. In fact, the growth of cocoa plantations saw the development of salaried agricultural employment. At the same time, in addition to farming on their own or others' farms, many adult men simultaneously took on other jobs during certain periods of their lives: local artisans, security guards in cities, cooks, construction workers, soldiers, etc. Independence (in 1960), education, vocational training and urbanization led to an increase in the non-agricultural pluriactivity of these families. While this trend led a small number of them to break completely with their family farm of origin, for most Cameroonian rural families, agricultural and non-agricultural activities remain permanently intertwined (Pédelahore, 2012).

While rural diversification in the South often refer to survival strategies (Losch et al. 2012), these individual and collective rationales also reflect other deeper transformations. The individualization of spouses' professional activities, the challenging of patriarchal authority, the empowerment of the "dependents" (women, the youth), etc. challenge forms of family farming based only on an farm-labor relationship which defined each member's role on criteria of gender, age, social status or birth order. This raises the short-term issue: How to keep on running a family farm when it is less and less based on the social organization of the family? And over the long term? Is it still possible to ensure the sustainability of the family farm when children that will inherit the farm are engaged in diverse activities out of the farm?

Spatial and social mobility of the young sometimes helps directly and permanently stabilize the operation of the family farm. Rural diversification does not mean that the work-farm relationships are anchored to the family on a single spatial and economic territory, that of the farm. Indeed, family social cohesion do not require all family members to be engaged in agricultural mono-activity, nor to be only localized around a single residence: the farm. It is then that social links at the heart of the functioning of family farming express the proverb that "blood is thicker than water," as illustrated in Box 4.7.

> **Box 4.7. Agricultural families in Nicaragua: social cohesion in agricultural work and anchoring to the *terroir* maintained through multi-localization.**
>
> Sandrine Fréguin-Gresh
>
> In Nicaragua, in the dry Pacific coast, agricultural families are structured around the patriarch, who exercises authority over the rest of the group, makes decisions and assigns roles to each member, and around his wife, on whom converge the social links uniting the family. Usually, families consist of three generations. There are many social ties which unite the family into a "system" (borrowings, loans or discounts, property rights, herds, labor, decisions, responsibilities, experiences, solidarity, affection, etc.). Within families, roles are assigned by age, gender, marital status and birth order. Only some members are involved in on-farm activities: married men and older sons; women and other youth help during peak work periods if there is not enough cash to allow the hiring of temporary workers. For these families, the residence is not at the core of the social rationale and of the farm's functioning. Indeed, these families are characterized by the spatial dispersion of their members: members live individually or collectively at different places at various distances from the farm and practice various activities, for varying durations. The cohesion around the family farm is ensured by links and circulation of material and inmaterial goods and visits, which allow investments in agriculture that would not be possible through income generated only locally. In this manner, the farm's sustainability and the family's social reproduction are ensured.

However, these responses in the form of rural diversification and mobility do bring with them some risk of breakdown of family farming. Indeed, social changes (lessons learned, changed values, new ways of life, emancipation, etc.) which accompany these reconstructions and transform work-farm social relationships over time can also weaken farms. Box 4.8 shows how the evolution of migration from rural areas of Mozambique, historically linked to the South African labor market, is today calling into question the durability of links to agriculture of mobile populations and is undermining the reproduction of family farms.

> **Box 4.8. Mobility in southern Africa: heading towards a break with family farming?**
>
> Sara Mercandalli
>
> In South Africa, the profound restructuring of the agricultural and mining sectors, changes in working conditions and new policies implemented since the end of apartheid have led to major changes in the forms of mobility in rural Mozambique. In fact, at present, more than the increase in the amount of mobility, it is the diversity of its forms and the expansion of its range that mark major changes for young migrants, as compared to previous generations. This has to be viewed in the context of changes in the movement of people and goods (existence of networks built over several generations, the diversity of livelihoods, improved remuneration and communications infrastructure, etc.). These changes have serious implications for Mozambican family farms. Indeed, while the migration of some members – most notably young adults who would have taken over as heads of farms – tends to become permanent, values and lifestyles are changing, agricultural knowledge, know-how and practices are being lost, and professional aspirations are being oriented towards other sectors. The very continuation of family farming is being called into question since the farm is no longer the main source of income. And when they return to their region of origin, many migrants no longer participate in the activities of the family farm (Mercandalli, 2013).

At the intra-familial level, when migration cannot be justified by any food needs, it becomes even harder to accept. Box 4.9 illustrates this ambivalence.

> **Box 4.9. Mobility of young herdsmen in the Ferlo in Senegal: between strategies of security, emancipation and changing lifestyles.**
>
> Claire Manoli and Véronique Ancey
>
> The links maintained between migrants from pastoral camps and their families through the movement of people, goods and services (transfers, shared control) show a reality more complex than the simplistic statement: "For bush people, everything revolves around livestock rearing; for city people, it is making money that matters." The flow of youth between the city and the bush shows that the herdsmen also want to "make money." This spatial mobility results from individual and collective strategies for securing livelihoods of pastoral camps by the diversification of activities, and from individual aspirations for changes in lifestyle. The discourse of migrant youth (20–35 years) and their perceptions differ significantly from the patriarchal discourse which justifies this mobility by the requirements of transhumance or by the tradition of letting young men go and "experience adventure" before endorsing their roles in the family economy. These young migrants endorse both family socio-economic strategies and attempts at individual emancipation. The change in the social meaning of money may lead to deep changes in the familial organization (Manoli and Ancey 2014).

Finally, familial forms of agriculture cannot escape the processes of territorial transformation, particularly in emerging countries with a proactive policy environment. In Brazil, under the policy of "territorial integration," the federal government has funded major entrepreneurial economic projects in southeastern Amazonia and encouraged the migration of the "landless" or farmers with small surface areas by helping in the displacement of families. The government has exhibited little concern for the indigenous peoples during this "colonization" of the Amazon. They are very few but they are already there: Amerindians, *quilombolas* (descendants of runaway slaves) and *riberinhos* ("inhabitants of river banks"). Family farming that has resulted therefore today takes many diverse forms in which portions of spaces (*ramals*) with families from the same state or a municipality coexist with others of more mixed origins.

4.3 Strategies for Creating, Accumulating and Transmitting Heritage: The Pillar of "Farm-Family" Relationships

Family farming must also be analyzed through the links between farm and family in terms of the creation and the accumulation of capital sufficient to ensure the social reproduction of the farming family, and of the continuation of on-farm activities by the transmission of the agricultural holding to offspring. Active adults on the family farm are also parents, and resources and capital are subject to a patrimonial rationale. This proposition is illustrated in Box 4.10. It shows how, in Indonesia, the accumulation on the family farm in plantations deals with heritage, as well as with production rationale.

Box 4.10. The long-term building up of patrimony by small Indonesian farmers.

Éric Penot

Since the introduction of rubber in Indonesia in the early twentieth century, the small rubber growers have acquired extensive experience in perennial crops, in long-term management of rubber agroforestry systems and in the establishment of a patrimony (Penot, 2006). The 1970-2000 period saw the development of rubber clonal plantations which led to the tripling of production. In the 1990s, oil palm plantations appeared on the scene. The adoption of a system based on 2 hectares of oil palm (for a period of 20 years) and the gradual transformation of jungle rubber* into clonal agroforestry plantations (2-4 four hectares per family for a period of 35 years) have allowed Indonesian agricultural families to, on the one hand, build a legacy of significant value and, on the other, ensure a long-term future in terms of a decent retirement and transmission of highly productive heritage, with crops of different lifespans to stagger investments and production. Price changes in one product tend to be offset by the other (rubber and palm oil) thanks to a steadily increasing demand since a century for rubber, which is sold at a very profitable price since 2003, and for palm oil since 1995, whose price has remained relatively stable since 2000. The creation of a family patrimony is also accompanied by an increase in farm incomes.

* Jungle rubber is a complex rubber-based agroforest, with fruit trees, timber trees and other usable products (rattan, medicinal plants, etc.). The planting material of jungle rubber traditionally consists of non-clonal rubber. Its average yield is 500 kg/ha/year of dry rubber against 1000-1500 kg/ha/year with clonal plant material.

Mendras (1967) asserted that "one is born a peasant and remains a peasant, one does not become a peasant: as a peasant, one has no profession," and that it is recognized in most agrarian societies that one enters into farming through patrimonial inheritance or marriage. But becoming a family farmer through other means is possible though difficult, and requires a number of conditions to be met. This is what Box 4.11 shows. It describes the conversion of miners in South Africa to agriculture and explores the conditions necessary for them to embark on family farming.

Box 4.11. The conversion of black miners into farmers in South Africa.

Sandrine Fréguin-Gresh

Among the disruptions that mark the post-apartheid period in South Africa, two concern the links between farm and family in agriculture and involve the black population. On the one hand, blacks are now allowed to live in territories that were previously prohibited to them. On the other, they are entitled to re-enter the commercial agriculture sector from which they were excluded as far back as the early twentieth century. While subsistence agriculture always existed in the homelands to provide a livelihood to unskilled workers, opportunities for blacks to develop commercial agriculture remained limited. The end of apartheid led to the idea that the development of black commercial agriculture may provide an employment solution to those unemployed and laid off from other sectors, especially mining. Agriculture was then perceived as a sector in which conditions of political stability and socio-economic security could be created, conditions necessary for the success of the post-apartheid government. However, the conversion of black miners to commercial agriculture was not without difficulties. Only a minority aspired to a professional conversion to agriculture, which requires financial capital, access to services (credit) as well as integration into a network allowing access to land. Even though the technical and economic performances and return on capital invested in agriculture are relatively high, the investments required exceed the initial capital available to the majority of blacks wanting to convert (Anseeuw, 2004).

Furthermore, the creation of a farming heritage *ex nihilo*, when it is driven from outside the social group, as in land reform processes that lead to families without any agricultural experience being installed on a farm, is not without difficulty. In this case, society may reject new farm-family relationships from being created, as in Nicaragua during the land reform of the 1980s (Box 4.12).

Box 4.12. The partial failure of Nicaraguan land reform: the imposition "from above" of forms of land tenure at odds with rural social relationships.

Pierre Merlet

During the 1980s in Nicaragua, the government implemented land reform which significantly altered the country's agrarian structure within a decade. This land reform affected 28% of the agricultural surface area and benefited more than 70,000 families, generally consisting of former farm workers who had worked on holdings of landowners – who were often close to the old regime – and poor farmers with very little or no land of their own (Merlet and Merlet, 2010). The land reform was based on the creation of state farms and production cooperatives. In cooperatives, a threefold rationale was at play: distribution by the State of collective property rights, the obligation to work collectively on large farms, and the insertion of production units in a centralized and planned agricultural production system. This logic found itself at odds with practices that had developed historically in the countryside according to which the security of land tenure reflected the social relationships in a defined space and time. The imposition "from above" (by the State) of a mode of land tenure was not consistent with local practices and did not meet basic needs. It was a factor that impelled many beneficiaries to quickly resell the rights they had received through agrarian reforms. This marked the beginning of a new process of land reconcentration.

But these processes of inserting or reinserting families into agriculture have consequences that go beyond the links between family and farms. They can indeed generate social phenomena. Using the creation of a land heritage in particular, farmers can gain access to a previously inaccessible political sphere. Or, they can constitute a captive electorate for the political classes within a framework of clientelist relationships, as shown in Box 4.13, which illustrates the case of Mexican land reform.

Box 4.13. The Mexican land reform: a family farming project.

Emmanuelle Bouquet and Éric Léonard

Land concentration that prevailed in Mexico in the early twentieth century helps explain the implementation of land reform starting in 1915. The reform had three objectives: the return to villagers of lands expropriated from them in the nineteenth century; the breaking up of large holdings, considered a source of social and economic exploitation; and land allocation to landless peasants. Land reform thus helped the creation of new family farms under a new land tenure system: the *ejido*, which is a collective entity legally entitled with its own land endowment. Over 70 years, nearly 103 million hectares (50% of the country's land area) have been distributed to 3.5 million beneficiaries, helping to consolidate and protect a new form of family farming, characterized by moderate holdings (a few hectares), exclusive use of family labor, and inalienable and unforeclosable property rights, giving them *de facto* status of family holdings (Bouquet and Colin, 2009).

Starting in 1970, questions began to be increasingly raised about this arrangement since land resources were becoming exhausted and would soon be unable to meet the demands of new generations. These tensions intensified in 1982 with the financial crisis and the policy of structural adjustment and liberalization, which deprived the *ejido* of most of the public support which it enjoyed. From 1992 onwards, a reform of the *ejido* system put an end to land distribution and redefined property rights by allowing market transactions. Nevertheless, the *ejido* has continued to be part of a family farming project attached to a specific regime of property rights different from private ownership, in which the State retains a regulatory prerogative. It also receives special attention in the contexts of rural policy and development projects (Léonard, 2011).

4.4 Some Perspectives for Further Research

Revisiting the outcome of research from the inter-links between labor, family and farms allows us to show the interweaving of social and productive rationales at the heart of family farming. Families are not only a production factor. Their labor goes beyond solely the agricultural and a single location. Their rationales, at the individual and group level, are heterogeneous, following multiple objectives and combining market and domestic production relationships both inside and outside. Not to take into account these structural elements, which go beyond the merely technical (agronomics, animal production sciences) or economic, leads to an incomplete understanding of family farming. To match this substantial challenge for research, this chapter suggests an interdisciplinary approach to a review of existing field studies and invites us to go further by undertaking new research into the social rationales underpinning family farming forms.

Understanding the diversity, the differing perceptions and compromises at work in a social system is always useful to an understanding of the dynamics and strength of the whole, and such an approach is not incompatible with the choice of definitions made in the previous chapter. This is why it is especially necessary, even if only heuristically, to characterize the dynamics involving youth and women, who comprise a strategic workforce often pushed into the background or into invisibility by relationships of authority and domination (Guétat-Bernard 2007) that exist today in family farms, and more widely in the rural world – and perhaps also in the scientific community. To the same end, the case studies presented in this chapter argue for integrating pluriactive and spatial mobility dimensions in family farming studies. As we have seen, in many cases, these structuring elements ensure the durability of family farming models in the South. However, approaches to analyze these dimensions may fail (Sourisseau et al. 2012). Studying these phenomena of family farming means neither encouraging the exit of families from agriculture or a rural exodus, nor to resign oneself to a diversification which is often only a survival response to rural poverty (Losch et al. 2012). Any improved policies which do not address these dimensions inherent to family farming today will be unlikely to manage to keep farming families on their territories in acceptable living conditions.

Chapter 5
Family Farming and Other Forms of Agriculture

Jacques Marzin, Benoît Daviron, and Sylvain Rafflegeau

As a counterpoint to Chaps. 3 and 4, which focus on "stand-alone" definitions of family farming and its diversity, this chapter takes a brief look at non-family forms of agriculture. It thus helps define family farming by what it is not and also explores the conditions of existence and evolution of family farming models through their relationships with these other forms. These topics are addressed in three ways: through a historical overview of the emergence of competition between family and non-family forms of agricultural production – thus complementing the analysis of Chap. 2, through a characterization of the diversity of non-familial forms of production, and, finally, through the complementary or competitive interaction between family and non-family forms of agriculture.

5.1 A Coexistence that Extends into the Past

5.1.1 The Role of the Market in the Emergence of Non-familial Forms of Production

While the various models of family organization of agricultural production encompass its vast majority, agrarian histories remind us that they complement or compete with other forms of production. The following provides a brief historical perspective of the non-family forms of organization, whose main characteristic, besides the

---

J. Marzin (✉) • B. Daviron
ES, CIRAD, Montpellier, France
e-mail: Jacques.marzin@cirad.fr; benoit.daviron@cirad.fr

S. Rafflegeau
Persyst, CIRAD, Montpellier, France
e-mail: sylvain.rafflegeau@cirad.fr

fact they rely on employed labor, is that the entire production is destined for the market.

Three historical phenomena have contributed to this at various times in the past: the gradual urbanization of societies, different periods of globalization of agricultural trade and the advent of marketing standards (certifications, labels, etc.) related to the development of the industrialization of agrifood systems. This results in a wide variety of market integrations and, consequently, of configurations of family and non-family forms of production.

Complementary to Chap. 2, which provides a more general historical overview, our focus is intentionally Euro-centric, is "limited" to the modern period and ignores pure subsistence farming – if it even exists anymore. Despite these self-imposed restrictions, our exercise remains ambitious and open to criticism considering the wide range of literature on the subject. We will thus limit ourselves to the main points.

In order to describe the situation in the late eighteenth and early nineteenth century, it may be useful to make a distinction between local markets and distant markets, given that the distance constraint comes very quickly into consideration: anywhere beyond a day's walk in that period was considered distant. Weber (1991) suggests that the density of cities, in other words, the distance to the market, explains much of the observed differences in agricultural structures between eastern and western Germany in the seventeenth century: "As the density of cities decreased on the map, those of farms[1] increased." This contrast between peasant agriculture participating in local markets and large farms relying on limited labor availability can be extended well beyond Germany. The first configuration is, of course, one of the finest illustrations of this observation in France with its dense network of towns, whose function, as explained Braudel (1986), "can be summarized by the local market, the common covered marketplace of villages." The second configuration is embodied – alongside the *junker* (landowner in Prussia and eastern Germany) and the second serfdom in Central Europe – in the figure of the planter who, based on a tropical island, produced "spices" using African slave labor to supply distant European markets.

However, the early nineteenth century saw the emergence of another form of organization of production, distinct from the previous two types: English agrarian capitalism in the form of large farms managed by farmers imbued with a capitalist spirit, who leased out lands and employed salaried workforces. Two English expressions can be associated with the emergence of this form of agrarian capitalism: "enclosure" and "high farming." The first designated the long-term movement in which communal lands were gradually privatized, and which led to the open-field landscape being converted into farmland. The second expression refers to changes in production practices and, in particular, in the management of soil fertility. These changes consisted of a very tight integration of crop cultivation with livestock rearing and the systematic introduction of legumes into crop rotations. Liberal

---

[1] Large-sized agricultural structures, which he contrasts with peasant agriculture.

thought predicted that these two forms of agrarian capitalism would gradually replace family farming.

Similarly, in Marxist thinking of the same nineteenth century, the future of family farming was called into question. In the capitalist system, the accumulation of land is assumed to gradually exclude small producers, with the lack of cooperative structures preventing them from generating economies of scale needed to be competitive with commercial agriculture (Kautsky 1970). In the prospective visions of a collectivist system, the future called for the industrialization of agricultural processes and thus the creation of large collectivist entities (Lenin 1899). Finally, it should be noted that the concept of private ownership of the means of production came into ideological conflict with the development of revolutionary processes. This would culminate in the exile of many affluent peasants to Siberia under Stalin.

During the second half of the nineteenth and the early twentieth century, the large agricultural unit seemed to be an inevitable necessity, both for the socialists as well as for the liberals, whether for long-distance trade or in the context of the industrialization of European economies (England and later the USSR). Looking outside Europe in the twentieth century, we thus found forms of agricultural production built on these visions of the large farm. Four of them, stemming from different ideological conceptions, are worth recalling:

- Large plantations in South and Southeast Asia (India, Sri Lanka, Indonesia, Malaysia, Vietnam), which emerged at the turn of the twentieth century in the colonial context. They benefitted from access to disciplined labor pools, especially in the case of India and China;
- State farms, imposed in the framework of "real socialism" on different continents throughout the twentieth century. They were set up at the cost of sometimes extreme violence,[2] and regardless of the agrarian histories of the countries concerned;
- Large latifundiary areas in Latin America, and their South African equivalents, products of a colonial history of land expropriation, again often very violent, by an oligarchy of European origin;
- Partially collective forms of agriculture, combining market integration and land use rights in many different ways. We can refer to the Mexican *ejidos* or farming communities in India.

But history shows that, contrary to the predominant ideas of nineteenth century elites, family farming has had an unexpected ability to integrate into markets and to actively respond to the surge in demand that resulted from rapid industrialization and urbanization. This new configuration first appeared in Europe (Denmark, France, in particular; Servolin 1989) but also took hold in America. It was facilitated by improvements in long-distance transport by trains and steamboats. Indeed in the US, the agricultural frontier, powered by the multitude of European migrants,

---

[2] For example, dekulakization in Russia or people's communes in China. Also the short-lived experience of *Ujamaa* in Tanzania.

created in its wake a family-labor based agriculture tightly integrated into the market and producing for distant consumers in New England or Europe (Friedmann 1978; Kulikoff 1993). The profoundly extractive nature of this agriculture required the colonization of new "virgin lands" but gave it a formidable competitive advantage over European agriculture systems (Duncan 1996).

This model of pioneer family farming spread starting in the late nineteenth century to multiple locations around the world, first to the British dominions with European migrants (Canada, Australia, New Zealand), then also to the tropics (southern Brazil). These family farming models sometimes replaced large plantations, such as those for coffee cultivation in Colombia. Even more significantly, this shift towards a pioneer agriculture based on family labor tended to spread even in the colonial context. Ghanaian cocoa farmers, who within a few years managed to push out the large Portuguese planters of Sao Tomé from the international market, were vanguards of this movement.

Finally, family farming producing for distant markets also arrived in old European agriculture, with spectacular successes in Denmark, where livestock production grew significantly in order to supply the English market.

We can therefore consider that, challenged by the dominant visions of the late nineteenth and early twentieth century, commercial family farming imposed itself almost as the norm in international agricultural markets and more generally in world agriculture.

5.1.2 The Persistence of Family Farming in the Face of Capitalist Agriculture

The persistence of family farming in the face of capitalist agriculture is explained differently depending on the analytical perspective and agrarian histories studied. Five major aspects are considered.

5.1.2.1 Resilience in the Face of Poor Economic Conditions

This interpretation is advanced by Tchayanov (1990) and those authors who have taken his work further. It is based on a simple observation: family farming combines a household and a business within the same entity. Therefore, the family regulates its propensity to work through a function of utility that is stable over the long term. It has the liberty of working less when the work is more remunerative during periods of high prices, and is willing to work at a lower level of remuneration than in other jobs when prices are low. This feature gives it greater flexibility in the allocation of resources[3] and especially an improved resilience as compared to capitalist agriculture during a period of low prices.

---

[3] Including the labor force which may be allocated on or off the farm.

5.1.2.2 The Cost of Monitoring Labor

This interpretation is summarized very clearly by Hayami (1996). He argues that any economic activity using employees is faced with the problems of monitoring the activities of the workforce to ensure that its behavior is in line with the enterprise's goals. In the case of agriculture, this monitoring problem is two-fold. On the one hand, the spatial dispersion of activities increases the cost of monitoring as compared to industry, particularly in very large agricultural farms. On the other hand, the low predictability of the effects of agricultural practices (lack of mastery over biological phenomena, climatic risks, and heterogeneity of agroecological conditions) makes it difficult to evaluate the workers' efforts by simply measuring inputs and outputs. Therefore, monitoring at close hand becomes necessary.

Family agriculture is able to use other means to monitor and punish family members (including through coercion) and also to mobilize other psychological motivations. According to Hayami, this lowers the cost of labor monitoring for family farming as compared to capitalist agriculture. In addition, when working together as a family at the same job, adults monitor the work done by younger family members and gradually train them to perform various tasks.

5.1.2.3 Opportunities and Constraints Pertaining to Technology

Technical constraints, regarded as an exogenous variable, are often used as an argument to explain the prevalence or importance of large units for certain crops such as sugarcane, palm oil, rubber and tea. This technical constraint can take the form of a specialized task, a maximum period between harvesting and processing of the product, a minimum surface area to achieve economies of scale or the requirement for a high enough production to justify very powerful and expensive primary-processing equipment.

Some authors such as Rowe (1965) highlight the requirement for a specific and precise task during the cultivation or the harvest to explain the advantage large farms enjoy. However, it should be noted that the same argument, this time on need for skilled labor is sometimes advanced to explain the predominance of family farming in crops such as tobacco. And thus, the same Rowe uses this argument to explain the dominance of large units in tea cultivation in the tropics, and that of family farms in European agriculture.

Other authors emphasize the existence of economies of scale in cultivation or primary processing. The importance of economies of scale in agriculture has been subject to vigorous debate, in the past as well as the present (Boussard 1986). The dominant position now seems to err on the side that believes that they are not all that important, especially as family farming nowadays manages to achieve economies of scale through professional organizations and the organization of the downstream sectors.

However, it is a fact that the most efficient primary-processing units (palm oil mills, rubber factories, etc.) require a minimum volume of raw material to be processed in order to recoup their costly investments. According to Binswanger and Rosenweig (1986), economies of scale, access to powerful processing units and tight scheduling requirements between harvesting and processing due to the perishable nature of the product explain the superior efficiency of large plantations for crops such as sugarcane or oil palm.

5.1.2.4 Access to Markets for Products and for Inputs

As we have seen above, market access (dependent on the presence of nearby cities) was advanced by Weber to justify the existence of peasant agriculture. This argument was echoed by Daviron (2002) to explain the growing importance of family farms in long-distance trade starting in the late nineteenth century. The concurrent and related development of product standards and futures markets have helped bring markets closer to the farmers. Market transactions can take place even in the depths of the countryside, even for products destined for distant markets. Indeed, standards and futures markets have transformed the profession of the merchant from a broker to a trader by providing him the opportunity to buy at "farm gate" prices. The trader can therefore hold the product, with an acceptable price risk, for the duration necessary for its delivery.

5.1.2.5 Relations Between the Core and the Periphery

In his historical review of the global capitalism system, Wallerstein (1974) defends the thesis that the world is structured between a core – which changes over the long term: Venice, Amsterdam, London, New York – and a periphery. The core is where the concentration of wealth and power is found and it imposes its rhythms and standards on the periphery. Wallerstein also supports the idea that as the distance to the core increases, so does the use of coercion in the organization of work. Thus, slavery in the Caribbean islands and the second serfdom in Central Europe can be explained by the peripheral position of these territories. Similarly, Arrighi (1994) highlights the differences between the successive models of economic organization espoused by the United Kingdom and the United States in their roles as hegemonic powers. According to this reading, the model of the large farm with hired or coerced labor is associated with English hegemony, whereas the commercial family farm pertains to American hegemony. The presence of large plantations in the twentieth century can then be seen as a form of path dependency, i.e., as a relic of the British hegemony manifested in the territories where the UK still enjoys an unchallenged dominance – in its former colonies (Malaysia and Kenya, for example).

5.2 Characteristics of Non-family Forms of Agriculture

Entrepreneurial farming models differ from familial forms in their use of exclusively hired labor, forms of ownership of their farm capital and in their management. In order to describe them more precisely, we distinguish two main types, capitalist firms and managerial enterprises, differentiated by their access to land, technical means and human resources (Table 5.1). As was presented at the beginning of the book, the family business farm – a family farm that employs permanent salaried workers – is located at the interface between family and entrepreneurial forms of agriculture.

In the following sections, we analyze the characteristics of the forms of capitalist, managerial and family business production, through five main criteria: ownership and management of capital, access to land, use and management of technology, management of human resources and integration into the socio-economic environment.

5.2.1 Capitalist Firms

Agricultural capitalist firms, called agro-industries in certain sectors or firms by other authors (Hervieu and Purseigle 2011), are often subsidiaries of multinationals with significant resources which farm thousands of hectares at one location and set up industrial facilities for primary processing or for export packaging. They

Table 5.1 Typology of the different forms of agriculture, from the capitalist firm to the family farm

	Entrepreneurial agriculture – family agriculture			
	Capitalist firm	Managerial enterprise	Family business farm	Family farm
Labor	Exclusively salaried employees		Mixed, some presence of permanently salaried employees	Family dominance, no salaried employees
Capital	Mobile and held by shareholders	Not mobile and held by shareholders	Held by the family or a family association	Held by the family, rarely by a family association
Management	Technical		Family/Technical	Family
Home consumption	Not relevant		Residual	Ranging from partial to full
Legal status	Limited liability concern	Limited liability concern or other company forms	Farm status, associative forms, sometimes company forms	Informal or farm status, sometimes company status in Europe
Land status	Property or formal rental		Property, or formal or informal rental	

sometimes supply a secondary process factory which is also a subsidiary of the same multinational.

5.2.1.1 Capital and Its Management

The capital of firms, especially in recent times, is owned by shareholders often disconnected from the sphere of production itself. The financial profitability of this capital is the priority. It guides the shareholders, who may decide, depending on market opportunities, on an annual or shorter change of product mix or even to sometimes direct resources outside agriculture. Thus, even though some firms are in agriculture for the long term, the transferability of investment choices allows for greater capital mobility, a source of vulnerability for the employees concerned.

5.2.1.2 Access to Land

Land is generally accessed through long-term leases or purchases, depending on local legislation. It is characterized by farms of several thousand hectares, often contiguous. Such large sizes can have two implications: at the time of the grant of a lease or the sale of land, expropriation of private lands often results in either the loss of rights to use common lands by local people, the eviction of a part of the population, or, in equatorial zones, deforestation of high conservation value forests with the attendant carbon emissions and biodiversity losses. In addition, several years or even decades after the award of the concession, the unfulfilled commitments translate into social demands (provision of schools, health centers, etc.) or demands for the return of the land from descendants of the original owners or from local chiefs who had initially negotiated the access to land for the capitalist firm through intermediaries.

5.2.1.3 Technical Management and Performance

Capitalist firms specialized in cattle rearing in Australia, the United States and Argentina, for example, depend on extensive technical practices and mobilize their capital to acquire production units of several thousand hectares and several thousand heads of cattle. As far as crop production is concerned, other specialized firms use expensive techniques that require significant volumes or surface areas. These can include aerial application of fertilizer or pesticide, on banana, for example, or all that concerns primary processing (rubber factories, oil mills, sugar mills, fruit juice factories, pineapple canneries, etc.) and the profitable use of waste and by-products (Box 5.1) (Figs. 5.1 and 5.2).

Fig. 5.1 Small-scale processing work site

Fig. 5.2 Industrial oil mill

> **Box 5.1. Processing of palm oil.**
>
> Sylvain Rafflegeau
>
> Only capitalist firms are able to invest the millions of dollars necessary for setting up an industrial oil mill to improve extractive performance and with a high level of compliance with environmental requirements. Whereas, the Caltech model of small-scale hand mill, which costs a few hundred dollars in Cameroon, is within the reach of family business farms and even of family farms. From a tonne of selected palm bunches, a small-scale mill produces about 150 kg of artisanal red oil whereas an industrial oil mill produces about 250 kg of palm oil and palm kernel. Investors who hold shares in these capitalist firms regularly monitor the technical and economic performance in order to possibly adjust their investment strategy. Therefore, private agro-industries backed by large multinationals typically obtain good agronomic and technological yields which are stable over time. Lower performance is characteristic of problematic situations, such as a land dispute with indigenous populations leading to theft of production, fires or sabotage in the fields, or a lack of investment to maintain production equipment in the absence of a private buyer of an old state-owned development company.

5.2.1.4 Human Resources

Multinational capitalist firms usually employ human resources necessary for the proper technical and financial management of their production units by hiring executives and managers with experience in the sector concerned. In their hiring, they accord priority to the country of operation. In sectors requiring little technical expertise, capitalist firms are less stringent in the recruitment of executives and managers.

Highly contrasting situations are observed in the recruitment of workers by capitalist firms, depending on the availability of labor and the technical requirements. Other factors that affect their hiring practices are the level of corruption in local government, the ability of the State to enforce the law and, finally, the influence of local civil society and that in the country of origin of the multinational firm. Some firms invest in their human resources: compliance with labor laws, training, social investments or creation of public goods (roads, etc.). Others prefer to limit their commitments by using temporary labor or by outsourcing part of the work to local entrepreneurs.

> **Box 5.2. Paternalist agricultural capitalism in Indonesia.**
>
> Stéphanie Barral
>
> Over one third of the world's palm oil is produced in Indonesia. More than nine million hectares of oil palm trees are grown there, primarily on the main outer islands of the archipelago, such as Sumatra, Kalimantan and Irian Jaya. Large private plantations, originating from the colonial period (from the second half of the 19th century) are major actors in this hegemony. Their production units cover areas that range from 20,000 to 30,000 hectares, and are part of Indonesian, Malaysian or European capitalist multinational firms. A firm may farm up to 400,000 hectares of plantations, which involves employing more than 40,000 permanent workers and providing support to their families.
>
> The organization of large plantations has characteristics specific to the second spirit of capitalism as Boltanski and Chiapello (1999) define it, i.e., capitalist organizations in which the hierarchy prevails as a mode of imposition and regulation of the involvement of employees. Orders flow vertically down the long lines of authority, from executives to managers, assistants, chief foremen, foremen, and workers.
>
> Permanent workers, unlike temporary workers, receive benefits in kind in the form of paternalistic protection in a country with a very liberal welfare state: housing, medical care, schools, savings schemes, sports grounds, places of worship are all services available to those with permanent worker status. However, these protections, even when combined with incentive pay, are not sufficient for a secure retirement, for which the financial allowance is low. Thus, workers need to secure their future through savings and investment strategies such as small businesses. The main objective of these strategies is the accession to land property (Barral, 2013). In regions newly colonized by plantation companies where much land is available, the vast majority of workers manage to acquire and farm an agricultural plot. On Indonesian pioneer fronts, the development of firm agriculture stimulates the emergence of family farming, even family business farming, at its periphery.

5.2.1.5 Social and Development Actions

When capitalist firms set up operations in sparsely populated areas with few roads, they sometimes participate with the State in the development and maintenance of road infrastructure which can prove vital for isolated agro-industries in Africa or Southeast Asia (Box 5.2). Remoteness of the location sometimes also leads agro-industries to undertake health care and education activities at varying scales. These social actions can be formalized through contracts between former landowners, agro-industry and the State, as is the case for the Cameroon Development Company Ltd (CDC) in the southwest of that country.

Public agro-industries have supervised development projects for smallholdings or *plantations villageoises*,[4] serving as the techno-financial interface between a development bank and the project beneficiaries. This was, for instance, the case of Sodepalm in Côte d'Ivoire or Socapalm in Cameroon, which developed village palm groves in the supply areas of their oil mills. These public agro-industries were

---

[4] These *plantations villageoises* are the French counterparts to the English nucleus estates and smallholders development schemes. They mainly differ in their origin (public or private) and their modes of intervention for territorial development, which can even impact the whole territory, including the financing of road infrastructure, or providing support to employees and their communities.

later privatized at the request of international donors, thus being freed from any obligation for development action for benefitting smallholders.

5.2.2 Managerial Enterprises

Managerial enterprises are generally smaller than capitalist firms. They are often created by domestic investors, with varying profiles (private, public officials, politicians, etc.). All of them are endeavoring to diversify their investments in agriculture, and mobilize their social and financial capital to this end.

5.2.2.1 Capital and Its Management

Capital is held by one or more investors away from the production unit, whose daily operational management they delegate. There are no links between agricultural production and the family sphere. Even if there is a single investor, he or she is involved as an individual and not as a family member. On the one hand, capital is generally less mobile than in the capitalist form, since the owners are normally individuals who are less likely to be driven by financialization rationales. On the other hand, however, the lack of attention often paid to environmental sustainability, especially if it is coupled with a pattern of management of human resources which is unfavorable to employees, can strain the sustainability of the investment, and therefore of the farms.

5.2.2.2 Access to Land

Managerial enterprises buy land when it is the only way to obtain access to it. Otherwise, they lease land (tenant farming) or obtain rights to use common lands from local authorities (customary or not). In this case, the acquisition of rights of land use is usually accompanied by compensation for the community, all the more important when the investors are not directly related to the community concerned. These compensations can take the form of providing employment to locals, social investments or production of public goods.

A new form is appearing in emerging countries, where there are more employment opportunities outside agriculture: agricultural contractors working on lands of farming families (planting pools called *"pools de siembra"* in Argentina, for example). Productivity increases through mechanization that family farmers cannot acquire on their own. They become "rentiers" with a status different than of lessors of land since the contracts allow substantial flexibility.

5.2.2.3 Technical Management and Performance

Through their relationships, these domestic investors mobilize the necessary technical information to create their enterprise: access to selected planting material, field recommendations, etc. The qualities of the recruited farm manager and the proportion of gross income reinvested by the enterprise's owner back into the farm determine the technical performance of the fields. Managerial enterprises supplying industrial processing units are of particular interest to capitalist firms, who can reduce the number of interlocutors by dealing with them. Capitalist firms can then focus their technical support and commercial facilities on them. This partly explains the generally good yields of managerial enterprises. Substandard agronomic performances are often associated with frequent changes of underpaid farm managers and difficulties the farm manager encounters when he does not have access to funds to pay for inputs and labor at the right time.

5.2.2.4 Human Resources

In managerial enterprises, it is often a manager who runs the organization and supervises all the tasks. There are no real villages of workers, but worker camps with more or less comfortable wooden huts. These managerial firms either employ permanent workers from neighboring villages or rely on a workforce that originates elsewhere and moves into the camp to stay or to work there for a few months.

If the managerial enterprise is located on a remote site or if local people refuse to work for it, the managers then bring in the workforce from known emigration regions. Some of these workers settle down, sometimes by marrying a native. They then obtain rights to use common lands or to buy land to set up their own farms. If this phenomenon becomes widespread and when land resources are exhausted, serious land conflicts can arise between the peoples of the region and those who arrived later.

Since living and working conditions in the camps of managerial enterprises are not easy, workers typically attempt to save enough to invest in their own businesses.

5.2.2.5 Social and Development Actions

In areas where pressure on the land is already high, the setting up of a managerial enterprise can have negative results in the form of expropriation of village lands by a single influential person who has connections and capital. It can also lead to positive results through the creation of jobs, visits by technicians of extension services, the setting up of a processing facility near the managerial enterprise or purchase of production by resellers or processors who organize the collection of the production. In the latter case, the setting up of the managerial enterprise extends the supply area of a processing unit to the area concerned by providing sufficient production volume to initiate its collection by the processor.

5.2.3 Family Business Farms

Two types of trajectories make up the category of family business farms. The first consists of rural or urban dwellers, with some capital, a pension, salary, income from some small commercial or craft activity, who decide to set up a farm they will manage themselves, using one or more permanent employees. The same strategy as managerial farms is adopted, though at a much reduced scale, with more modest means and with hands-on involvement of the farm's owner in its management. The other trajectory involves family farms which, on expanding, find themselves short of family labor and resort to the use of permanent employees. We find this trajectory on every continent, from the French *Beauce* to the Brazilian *Cerrados*. It is also common on oil palm family farms in Cameroon: the farmer establishes a family farm producing cash food crops and develops a palm grove with or without the assistance of an external project and at least partially funded by the sale of food crops. Once the entire palm grove is in production and the surface area allows it (10 ha), he hires a permanent employee and reduces food crop surface areas to all that is necessary for home consumption. These 10 ha of palm trees correspond to the farmer's "retirement."

5.2.3.1 Capital and Its Management

The appropriation of productive capital and principles for managing it are similar to those of family farms, but the separation between family and farm is greater. Management methods are geared more towards objectives of productive accumulation, partly due to the recourse to permanent hired employees.

5.2.3.2 Access to Land

This type of farm has access to land similar to the way the family farm does, with differentiation depending on particular land-tenure backgrounds. In all cases, access to land implies recognition by the communities concerned. This may explain the difficulties immigrant farm owners have as compared to native owners. Sometimes family business farms can also call upon administrative, union or political networks to obtain or secure access.

5.2.3.3 Technical Management and Performance

Technical management and agronomic performance of family business farms are largely dependent on the abilities of the manager. They do not differ from those of family farms, except by the more general use of equipment or motorized machinery based on financial capacity.

5.2.3.4 Human Resources

The permanent employees of family business farms need to be versatile in order to perform all or part of the farm work. They may even need to supervise the occasionally necessary temporary labor (clearing or pruning at the beginning of the rainy season) or some more technical tasks (harvesting, phytosanitary applications).

5.2.3.5 Social and Development Actions

Family business farms often join hands with neighboring farmers to purchase improved seeds and inputs from a semi-wholesaler and to establish contacts with distant extension services. When a family business farm invests in processing equipment which it cannot use at full capacity, it initiates an action of local development by offering the facility to others. This is the case in Africa outside the supply areas of palm oil mills: a small farmer buys a small-scale mill which he uses only a few days every fortnight.

We have seen how these three forms of non-family agriculture differ. To grow, they mobilize networks of different actors – mainly private and international, including senior officials, for capitalist firms; national and diversified for managerial firms; mainly local for family business farms. It should be noted that their impacts on growth, jobs and environmental externalities vary widely within each type. Their contribution to a balanced development is stronger in a clear regulatory (legal, political) environment and when asymmetries between actors are limited. Finally, we have seen that these types of productions are not disconnected from forms of family farms. We will explore this in the next section.

5.3 Links Between Family Farming and Other Forms of Agriculture

The coexistence of different forms of agriculture varies depending on agrarian histories. We can roughly distinguish three types – complementary coexistence, interlinked coexistence and competitive/conflicting coexistence – which can each evolve over time.[5]

5.3.1 Complementary Coexistence

The main scenario of complementary juxtaposition is that of the nucleus-plasma and smallholders development scheme. It corresponds to smallholdings set up around an agro-industry consisting of industrial plantations and an industrial processing unit. Clear land tenure situations then allow the synergistic coexistence of family and entrepreneurial forms of production. Industrial production units may process not only their own agricultural raw material but also that purchased from family farms in the area (often the case in rubber and oil palm smallholdings, for example). This coexistence is governed by more or less formal local contractual agreements between managerial farms (agro-industries in the sectors of rubber, palm, export pineapple, tea, export banana, etc.) and farmers in their supply area to increase the agro-industry's production (Box 5.3).

> **Box 5.3. The experiment of *alianzas* in Colombia.**
>
> Sylvain Rafflegeau
>
> The *alianzas* (or inclusive alliances) represent a new development model in which an investor takes the initiative of setting up an oil mill and a managerial enterprise specialized in palm groves. He partners with family, family-business and managerial farms to create smallholdings which supply his oil mill (Cano *et al.*, 2006; Lizarralde *et al.*, 2012). Smallholders who are associated with this scheme gain access to bank loans guaranteed by the *alianza*, while alone they would not be eligible to receive loans because of the lack of sufficient collateral. They also benefit from technical support disseminated by the State's extension services. This pattern of development which includes an industrial oil mill is different from the nucleus-plasma and smallholders development scheme in that it is not necessary to have a large concession for industrial plantations because each farmer plants in his own grounds, thus avoiding the risk of land expropriation. It also differs from situations of spontaneous development where production and processing capacities are not planned and where the purchase price of the palm bunches is not negotiated with smallholders. In fact, smallholders who are members of an *alianza* participate as associates in strategic decision making.

After the privatization of public development companies, capitalist agro-industrial farms have sometimes taken over from development companies in their supply areas, thus helping family farms, family business farms and managerial

---

[5] Chapter 7 explores the territorial implications of these configurations of complementarity and competition between family farming and enterprise farming.

firms that supply them. Generally, the more stable and direct the relationship between an agro-industry and smallholders is, more the contract is mutually beneficial: accessible services for smallholders, supply security for the agro-industry.

5.3.2 Dependent Interlinking

Other forms of complementarity/dependence are based on the existence, within a large farm, of family sub-units having a certain autonomy in the organization of work. Such is the case, for example, of the *colonato* system which existed in Brazilian coffee *fazendas*. In this system, each family was responsible for a plot with the possibility, in the early days of the plantation, of cultivating food crops (Stolcke 1986). We find the same dynamic of family gardens in some industrial plantations in Africa or the experiments of the Soviet kolkhoz and other State farms (USSR, Cuba).

We can also mention certain situations of *latifundio/minifundio* combinations in Latin America (outside of the agricultural frontier, when land tenure is stabilized). The *minifundio* plays a role in this context of the labor pool from which the *latifundio* can "draw" from depending on cropping calendars. This selling of labor brings in cash income for minifundist family farms, in situations where the marketing of domestic production is often difficult or not profitable. This dependence in situations of asymmetry makes the minifundists particularly vulnerable and sensitive to climatic or market hazards.

5.3.3 Territorial Competition and Processes of Eviction

However, in most situations, forms of production compete for access to land. These situations can be brought under control by regulations concerning farm ownership and rental, when regulatory and governance institutions are established and respected. Three of these situations are common.

The first situation is that of agricultural frontiers. The speed of their forward movement depends on the active role of family farming in clearing (slash and burn agriculture, establishment of pastures or perennial crops), replaced, often very violently, by family business or entrepreneurial farming. The Amazonian agricultural frontier and Indonesia are an illustration of this situation.

The second situation concerns the financialization of entrepreneurial forms of production (Box 5.4), which sometimes goes as far as land expropriation. It takes place primarily in sparsely populated areas, with land available for purchase or rental (in Ukraine, for example). When it concerns cases in which land rights pertain to different domains (customary, positive, etc.), family farmers can find themselves evicted, with only a few of them having access to jobs created locally

(Ethiopia). Worse, these situations sometimes have restrictions on access to other natural resources, for example, in Office du Niger in Mali, where Libyan investors have negotiated preferential access to irrigation water, making them a priority when availability is low.

Box 5.4. The financialization of South African agriculture.

Ward Anseeuw

While South Africa has not been affected by a process of large-scale land appropriations as observed in many developing countries (Deininger and Byerlee, 2011), it has seen new agricultural investment and production models emerge. They have taken the form, in particular, of greater involvement of new actors, traditionally unknown to the agricultural sector: financial institutions (commercial banks, investment funds), (agricultural) engineering companies and asset management companies, all seeking to diversify their portfolios. These investors perceive agriculture as an investment for the future and engage in a "Malthusian-oriented speculation." Their interactions with the more "traditional" actors lead to new modes of action, investment and production. As such, a new agricultural development model now tends to prevail with the South African agricultural sector undergoing a process of industrialization and financialization (Ducastel and Anseeuw, 2011). These dynamics are not underpinned as much by mechanization as they are by the transformation of agricultural and production structures. They are, in fact, based on shareholder structures, operating according to purely financial tools and strategies. The organization of agricultural production tends towards a highly integrated system, similar to Taylorist industrial chains which in some cases are incorporating all segments of the value-chain ranging from upstream to primary production and even to the downstream marketing segments (Ducastel and Anseeuw, 2011), but with actual agricultural activities usually being subcontracted. This is particularly the case of the EmVest investment fund, whose capital originates with the Emergent Asset Management company in London. Emergent's major investors include American universities Vanderbilt and Harvard (which has since left the consortium), which have invested over 500 million US dollars.

The third situation concerns access to loans or certain state subsidies, only available on formal control of the land. This requirement leads to a veritable land rush, in which the entrepreneurial forms (especially firms), whose fixed capital is more readily mobilizable (company capital, investment funds, traditional bank loans) have an advantage over family forms, whose access to the national banking system is more limited.

Similarly, the setting up of an agro-industry on an isolated site whose processing unit is supplied, partially or fully, by the production of smallholders transforms the region to an area of high agricultural value (Toulmin 2009). The opportunity to supply the processing unit by establishing a farm in its supply area results in a growing pressure on land that pits family farming against family business farming and managerial companies for access to land.

This quick review of forms of linkages between family and non-family farming shows a wide variety of situations which precludes any simplistic segmentation. However, one should note, on the one hand, the importance of path dependency vis-à-vis agrarian history. Indeed, contemporary agrarian structures are marked by land polarization, type of colonization or the history of more or less extroverted modes of land planning. On the other hand, we should note the importance of political choices that lead to bifurcations. These choices affect the development of one or other form of production, either abruptly through agrarian reforms, or gradually through recourse to different public policy instruments (modes of accessing credit or land, land taxation, helping the organizing of family farmers, etc.).

5.4 What Forms of Agriculture for Tomorrow? Society Has to Choose

This chapter has shown that non-family forms of agriculture have developed under some specific conditions (need for tight upstream/downstream integration to meet quality standards of the final product, specific sparsely populated farming systems, political choices, etc.). However, contrary to dominant nineteenth-century thought, family farming has demonstrated its flexibility, its resilience and its ability to compensate for supposed structural disadvantages (economies of scale, quality standards, etc.) thorough collective action.

Increases in urbanization and changing diets will accelerate and diversify modes of market integration of agriculture. Food systems limited to niches or to short-supply locations will develop alongside the globalization of trade of "industrial" and standardized agricultural products (Rastoin and Ghersi 2010).

Political choices on the functions to assign to agriculture will be critical in defining the future configurations, and the mobilization of civil society will play a decisive role in shaping agricultural policy. The quantity and quality of agricultural production, equilibrium of trade balance, creation of agricultural and urban jobs, environmental management of natural resources and territorial planning: what exactly should we expect from agriculture? The priority accorded to distant markets, the standardization of production and the industrialization of the downstream processes of agricultural production all favor non-family forms of agricultural production. In contrast, the importance paid to local markets, the desire to maintain rural employment during the demo-economic transition and the willingness for a holistic territorial development favor family farming (Chap. 7).

Part II
Helping to Feed the World and Territories to Live

Coordinated by François Affholder, Laurène Feintrenie, Bruno Losch

This book's first part defines family farming systems by what they have in common and by the characteristics that distinguish them from other forms of agriculture, while emphasizing and illustrating their diversity. This diversity of family farming systems and of the relationships between family, labour and the farm are explained by the presence of family farming systems in all regions of the globe, in diverse ecological environments and extremely varied economic and social contexts. They interact everywhere with other economic agents by establishing partnerships or competing with them for access to resources or markets. But if family farming systems are so pervasive in today's world, what are their actual contributions to the myriad needs of the planet's seven billion people? How do they contribute to the management of natural resources and territories? To what extent do they meet the global demand for food and raw materials for industry (fibers, resins, oils, alcohols and others)?

This second part of the book aims to provide the keys to analyzing family farming systems and, wherever available, provide the information on their contributions, whether positive or negative, intended or forced. This part is organized according to a progression across spatial scales, from the plot to international markets. Thus measured, these contributions of family farming systems will be put into perspective in relation to the major challenges agriculture will face in the twenty-first century, which are addressed in the book's third part.

Chapter 6 recalls the similarities in the histories of agrarian regions very distant one from another. The first known forms of agriculture were of shifting practices in which a field is cultivated for a short period (1–3 years) and then abandoned to forest regrowth for a much longer fallow period (over 15 years). These practices involve high mobility of families and put low pressure on land. Nomadic pastoral systems are also based on the temporary use of natural resources (pasture, grass and water) that are allowed to regenerate before being exploited again. Changes in agricultural practices and technical innovations allowed a family to produce more and better (healthier and more varied products). This led to the settling down of agricultural populations and to population growth. The same techniques have been implemented by agricultural families in different continents, for example, agroforestry. They originate from the observation of natural phenomena by farmers and the acquisition of empirical knowledge on the functioning of ecosystems and on the needs of domesticated and exploited plants and animals. Human migration and trade have gradually built other bridges between widely separated agricultural populations, with the adoption of technologies introduced from abroad and adapted to new ecological, cultural and socio-political contexts. Colonization and the development of international trade led to the expansion of crops for export such as cocoa, coffee, tea and cotton. The Green Revolution has spread the use of industrial inputs (fertilizers, pesticides, herbicides) and improved plant breeding material used in crops and monoculture plantations, leading to a certain homogenization of cultivation techniques and crop varieties. However, specific agricultural practices have been maintained or even developed within family farming systems, whether due to constraints, lack of access to these intensive techniques or through conscious choices by farmers.

While Chap. 6 focuses on the practices implemented by family farms from the plot scale to the farming system scale – in particular the methods of managing and enhancing natural resources –, Chap. 7 is devoted to the more encompassing level of local territories (ranging from marginal zones and pioneer fronts to urban centers). It explores the combination of family farm practices at this scale, especially with respect to their relationships with other forms of production. It also emphasizes the activation and mobilization of specific territorial resources, which depend on each territory's specific economic and social formation.

The expansion of international trade over the last two centuries has resulted in sustained growth of agricultural commerce. National markets however still retain a major role, most notably with regard to basic food products (cereals and tubers in particular). Chapter 8 examines the contribution of family farming systems to their production and to international markets, with the caveat that the analysis is complicated since family farming is not a statistical category. It thoroughly examines available data on world production and international trade, focusing on the characteristics of production and on economic development models. This review highlights the importance of family farming systems despite the role played by entrepreneurial forms (mainly agro-industry) for certain products. It uses as illustration some major value chains of tropical products, for which CIRAD's expertise is mobilized, to better understand the links between forms of production and types of markets.

In this context of multiple external influences and increased exposure to competition, family farmers have often limited capacity for action. Their banding together into agricultural or rural professional organizations (APO/RPO) and their contribution to local democracy allows them to defend their economic, social and political interests. These organizational dynamics are discussed in Chap. 9, which focuses on collective action and on the translation and expression of family farming practices and their motivations by representative bodies that carry their interests into the political sphere. The scale of intervention of these organizations, which remain firmly rooted to their productive and territorial dimensions, ranges from the very local to prominent international forums.

Chapter 6
Contributing to Social and Ecological Systems

Laurène Feintrenie and François Affholder

Family farmers are rooted to the land and territories – according to processes covered in Chap. 7 – that they pass on to future generations. Their activities at the scale of the plot and at the scale of the agrarian system are determined not only by the resources available but also by the security and continuity of access to these resources. Whatever the types of crops grown or livestock reared, the biophysical environment and natural resources remain the essential means of production of family farming (Tchayanov 1972).

The wide diversity of family farming is thus linked to the diversity of natural ecosystems. Farmers are compelled by necessity to be innovative and to adapt to the constraints and opportunities of local social and ecological systems. They do so by adopting a variety of strategies ranging from the mimicking of natural ecosystems – by reproducing ecological dynamics and functions to increase the production of ecosystem goods and services – to the extreme artificialization of the environment. This process is usually accompanied by an intensification of labour to overcome the economic and environmental constraints and, more rarely, by an increase in physical capital (land, farm equipment, and infrastructure). Regardless of the strategy chosen, family farming systems participate actively in the management of natural resources, either successfully or causing negative impacts, but always with a strong attachment to the location where they produce. Complementing the exploration of the complex social issues that arise and unfold within family structures (Chap. 4), this chapter uses a technical perspective to examine the various, and often contrasting, contributions of family farming in the management of social and ecological systems.

L. Feintrenie (✉)
ES, CIRAD, Montpellier, France
e-mail: laurene.feintrenie@cirad.fr

F. Affholder
Persyst, CIRAD, Montpellier, France
e-mail: francois.affholder@cirad.fr

6.1 Innovative and Efficient Production Systems

Family farming systems are primarily intended to produce food and other agricultural goods for the direct use of the family and, as far as possible, for generating income through the supply of local or distant markets. While human populations have sought to settle in the most suitable environments for farming, they do not always have the opportunity to do so due to political, social or cultural reasons having little to do with the environment's initial productivity and fertility. But even in the most infertile locations, farming populations, usually organized in families, have developed techniques adapted to local conditions by harnessing available resources as best as possible.

6.1.1 Production Systems Modeled on Natural Ecosystems

In humid tropical environments, the intensity and amount of rainfall damage the soil, most of whose value is stored in its superficial layers. When soils are not covered by vegetation which protects them from the direct impact of rain and which extracts large amounts of water through transpiration, rainfall erodes the surface horizons and drains large amounts of minerals needed by plants out of the rooting zone of the soil. To cultivate such environments, the farmer must be fully aware of their fragility in order to implement conservative and protective practices. A strategy common to several humid tropical agroecosystems consists of cultivating only plots of small surface areas and for short periods. The forest clearance is thus limited to the minimum necessary, as much due to the unavailability of sufficient labor as by the wish not to expose large areas to rainfall and to benefit from the regulation of water and minerals flows by the neighboring ecosystem. After a few years of cultivation, the farmer lets the forest cover regenerate during a long fallow period. This replenishes the stock of soil organic matter and reduces the seed stock of weeds by depriving weeds of sunlight, hence reducing the risk of weeds competing with crops during the following cultivation cycle (Rouw 1995). Such farming often takes the form of shifting cultivation where two or three years of food crops alternate with long fallow. These shifting systems can be sustainable as long as fallows are maintained long enough to allow the full regrowth of the forest. In fact, these are the most sustainable agricultural systems ever implemented on the planet, to the extent that they have ensured the survival of human populations for millennia (Griffon 2006). Moreover, they do so with such a low impact on the ecosystem that it is sometimes difficult to discern a primary forest from a forest restored by a very long fallow. But for these systems to be sustainable, there has to be relatively low population densities of the people dependent on them.

Some forms of agroforestry (association of trees and crops in a same plot) in humid tropical climates are the result of an intensification of shifting cultivation practices in response to increasing land pressure (Griffon 2006) or new business opportunities (Feintrenie and Levang 2009). These agroforestry plantations are capable of supporting higher human population densities and of producing not only food for families but also marketable products. Typical examples are the agroforests in Indonesia but others exist in the humid equatorial zone on all continents. These are smallholdings combining perennial cash crops such as rubber with other useful plants such as timber trees, fruit trees, food crops, craft materials (palm, rattan, bamboo and others) and medicinal plants. Species of pioneer, post-pioneer[1] and fire-climax[2] plants which appear in sequence in the establishment of a natural forest are replaced by a succession of crops with similar sunlight needs (Gouyon et al. 1993). The first step after slash and burn is to replace the pioneer plants by a stage of heliophilous[3] crops which grow rapidly and have short production cycles (rice, vegetables, bananas, papayas). They occupy the space in a few weeks and inhibit the growth of spontaneous pioneer species which are now considered weeds. This first stage of cultivation creates a shady and moist microclimate at the ground level, favorable for the germination and development of forest species (rubber, fruit trees, palm trees and timber trees). The post-pioneer phase is dominated by fast-growing crops with short immaturity periods, between 4 and 8 years, such as coffee, pepper, cinnamon and clove. This phase maintains a biophysical environment conducive to the growth of young trees. It also benefits from maintenance operations undertaken for annual crops (fertilization, weeding). After 15–20 years, agroforests present a complex plant pattern comparable to that of a secondary forest of the same age, with a high and closed canopy in which many forest post-pioneer species have spontaneously found their position (Gouyon et al. 1993). The renewal of the agroforest then depends on the death and falling of trees, creating clearings in which a new generation of plants can grow, either spontaneously or with the help of the farmer. Due to the continuous and spontaneous regeneration of many species, agroforests present a structure wherein forest trees of all ages are found. By pushing the mimicry of the forest ecosystem to its pinnacle, Indonesian farmers have developed a very detailed understanding of plant successions that constitute it. This knowledge has then been profitably utilized to cultivate native species in this relatively poor environment, very closed (little light, little space) and highly competitive. Competition between plant species must also fully be mastered to promote the production desired by the farmer at any given time. Temporality here is an important factor and choices often have to be made between crop quantity, frequency and quality.

---

[1] Pioneer vegetation is the first type of vegetation colonizing an area. It is characterized by species requiring significant exposure to the sun. It grows rapidly but has little resistance to competition from other plant species. The post-pioneer vegetation is the type of vegetation that succeeds the pioneer species.

[2] Specific plant species in a mature forest, not prone to natural disturbance. The fire-climax plant community is the final state of a succession of vegetation in the same place over time, and the most stable state for given soil and climate conditions.

[3] Plant species that need sunlight for growth.

The coffee plants grown in agroforestry systems, under shade trees, provide one of the most spectacular examples of improvement over time of agricultural production. A shade-tolerant plant in its natural environment – the forests of Ethiopia – the coffee plant can also be grown in full sun. In this latter case, the farmer must help the plant adapt to its new conditions of growth by providing it with a richly fertilized soil, watering it, increasing phytosanitary treatments, weeding and practicing rigorous pruning. The agroforestry coffee plant, on the other hand, requires little maintenance and few inputs. In the short term and if significant production facilities are available, coffee in the sun wins. Over the long term, the agroforestry coffee plantation is unbeatable: the plants produce less but they also have a longer life, expenses on inputs and labor are reduced and organoleptic quality improves. Other parameters such as mean temperature (reduced), the soil carbon content, nitrogen mineralization and microbial activity in the soil are more favorable in plots with shade trees (Nonato de Souza et al. 2012). Finally, agroforestry coffee plots, in much the same way as agroforestry cocoa ones, are havens of biodiversity (Deheuvels et al. 2012). Many family coffee or cocoa plantations in Mexico, Kenya, Indonesia, Brazil, Cameroon, Ghana and Costa Rica are agroforestry plantations.

Agroforestry is also a technique of adaptation to arid and semi-arid environments, used since time immemorial by African family farmers to create a favorable microclimate for annual and perennial crops by shading, and to optimize water and mineral resource cycles in the service of sustainable agricultural production (Box 6.1).

Box 6.1. The densification of trees in Sahelian landscapes.

Régis Peltier

The traits and functions of *Faidherbia* (*Faidherbia albida* (Del.) Chev.), a symbolic species of Sahelian agroforestry systems, are well known to agropastoralists, farmers and scientists. Its features include a deep root system to reach the water table on alluvial soils, an inverted phenology, with leaves present in the dry season and absent during the rainy season, and a capacity for vegetative propagation (suckers, coppiced stems and branches). Its positive impacts on associated crops, fodder (leaves and fruits) production and firewood production are also widely recognized.

Yet, *Faidherbia* agroforestry parks are nowhere near as widespread as they could be, despite the efforts of many NGOs and extension services. The example of northern Cameroon shows that research on crop productivity under *Faidherbia* led, in the 1990s, to a change in perception of this tree by agricultural development actors and services (Peltier, 1996). It then became possible to promote the restoration of these parks on a large scale, by mobilizing public funds and involving farmer associations and organizations to supervise and support the assisted natural regeneration of young *Faidherbia* (Smektala *et al.,* 2005).

Moreover, results of socio-economic surveys and pruning trials conducted in 2012 confirm the interest farmers have in pruning trees and show that it is possible to produce fodder and firewood sustainably by trimming the trees every six to eight years (Peltier *et al.,* 2013). The farmers' demand on the right to prune trees and freely use the harvested wood was taken into account in the proposed draft to reform the law on the forestry regime, introduced in 2012 in the Cameroonian parliament.

The agroforestry park is probably the most prevalent land use system in Africa (Von Maydell 1983). A wide range of native trees or shrubs are preserved in the fields by methods of assisted natural regeneration to protect spontaneously germinating plants. These practices are typical of family farms and help maintain cultivation and livestock rearing activities in locations where water is a very limited resource and soil fertility is low (reduced organic matter and nutrients).

Agroforestry parks have also contributed to the preservation of a rich agrobiodiversity. Thanks to these parks, trees such as shea, baobab, tamarind or marula (jelly plum tree) have been preserved in all their genetic diversity over successive generations of farmers.

Agroforestry and more generally the species associations practiced in all tropical environments enable farmers to optimize the use of nutrients and water, and thus constitute the foundations of a sustainable agriculture for human populations of relatively low density.

In the case of cereal crops grown without any use of synthetic inputs (such as pesticides and mineral fertilizers) nor to mechanization, grain and total biomass yields are remarkably stable in space and time within a cultivated ecosystem but – more surprisingly – also across different ecosystems. This holds true even when comparing situations as different as upland rice cultivation subject to annual rainfall higher than 2,000 mm in the mountains of Vietnam and millet in Senegal with less than 500 mm (Affholder et al. 2013). This remarkable stability of yields contributes to the adaptation of agricultural systems to climatic hazards (Baldy and Stigter 1993). Another constant of these systems is that their grain yields are very low, around 800 kg per hectare per year, or 5–30 % of the theoretical potential allowed by solar radiation, temperature and precipitation, while total biomass produced is only slightly less distant from its "climatic" potential (in the range of 20–50 %; Affholder et al. 2013). Therefore, over time, these family systems are likely to fail in meeting human needs due to increasing populations. They eventually start to draw excessively from natural resources, which leads to conflicts among communities to access these dwindling resources (Box 6.2).

> **Box 6.2. Family farming systems in forested and periforested Central Africa: the legacy of shifting slash-and-burn cultivation.**
>
> Jean-Noël Marien
>
> Family farming systems in Central Africa are based on practices originating directly from traditional techniques. These involve rainfed crops, with the most dominant being the tubers (cassava, sweet potatoes, yams, etc.), plantain, maize, rice, groundnuts, beans and oil palm. In forest areas, shifting slash-and-burn cultivation is predominant. In general, after clearing and burning the cultivation area, a mixed cropping system is implemented. After the harvest, the field is simply abandoned to fallow for up to 20 years. This farming system is not very productive and the quantities produced are proportional to the surface area. In some regions, especially when the overall household wealth and education levels rise, family agriculture, which is usually subsistence-oriented, changes rapidly and becomes more productive, adopting the use of inputs and more efficient equipment, while still retaining the family nature of the farm (family labor, secured access to land) and without changing the farm's main products. This system is based on an almost complete lack of inputs and durations of tree fallow sufficiently long to maintain and restore a level of adequate soil fertility (Marien *et al.*, 2013).
>
> But shifting subsistence farming constitutes, often in conjunction with the exploitation of firewood energy, the first cause of forest degradation and deforestation in Central Africa, far more than mining projects or agro-industrial plantations (Marien and Bassaler, 2013). This has serious consequences for forest degradation, soil depletion and water availability. A rapid increase in food needs due to great increases in populations, especially urban and periurban populations, alters the balance of this system.
>
> The increased pressure on land due to population growth and the development of other competing industrial activities such as agro-industry can also lead farmers to turn to forest lands where the soil is still rich. But the land will have to be completely cleared and conflicts can arise when it is relatively scarce (Chapter 12). Access to land is thus a source of conflict; it is a major issue and can even lead to serious geopolitical risks. This situation is aggravated by the fact that the major part of the Central African forest area is not part of any land use plan (e.g., forest concession or protected area) (Marien *et al.*, 2013).

Where a growing population happened to increase the agricultural demand, family farms have been urged to innovate. Thus new solutions have been emerging, with technical innovation leading to an ever-increasing artificialization of the environment.

6.1.2 An Artificialization of Environments by Intensive Production Systems

An emblematic example of such an artificialization of the environment is the rice terraces of Southeast Asia. They are known for their scenic beauty as well as for their high productivity. But they are the result of an extreme artificialization of the ecosystem by generations of family farmers, combining large-scale modifications of the topography to create flat cultivation terraces, building and deviations of water by irrigation and drainage canals, and sometimes the use of animal traction. This system has several advantages. First, the flooding of crops reduces weed pressure due to the limited number of species that can survive in this aquatic environment. Second, rice transplanting allows the farmer to start the crop cycle in the reduced space of the nursery while the previous rice crop on the field is not yet fully harvested. Thus, the intercropping period is minimized; crop cycles may even overlap. Another advantage of transplanting is to give a "lead" to the crops over the weeds likely to compete with them in the field. The techniques used here often include the use of chemical fertilizers and pesticides. Family farmers share the use of fertilizers and pesticides as spearheads of the Green Revolution with family business and entrepreneurial farming. Motorized mechanization, however, has made no inroads here given the small surface areas and the difficult access to the terraces. The fragmentation of rice fields into small units and the daily maintenance required to maintain the terraces are only possible in family units where labor is available.

These cropping systems are very labor intensive but have high per-hectare productivity, with grain yields of as much as 80 % of the maximum theoretical potential under this climate (Lobell et al. 2009; Van Ittersum et al. 2013). Thus paddy rice yields in China have tripled between 1960 (around 2 t per hectare) and 1995 (over 6 t per hectare), and increased from 2 t per hectare to about 4.5 t per hectare in Indonesia and Vietnam (Griffon 2006). Water has to be managed collectively by families growing rice in the same watershed. Cropping calendars have to be synchronized and the work done at the same time so that flooding or drainage can take place for everyone. Coordinating cropping calendars also helps guard against pests, particularly birds. Self-help and work-exchange systems are often implemented to undertake large cultivation work as a team, with remuneration for such work often having no relation to wages prevalent at the time and being strongly influenced by the social relationships involved.

Other examples of the artificialization of environments by intensive production systems can be mentioned, such as vegetable farming systems in peri-urban areas or those described in the first part of the book. Highly artificialized systems are found wherever the rural population density is very high. In strictly rainfed systems, soil fertility management and weed control are generally amongst the practices that are the most intensive in labour and input use. The use of the latter is usually only possible under certain market-access conditions, rarely met for family farming systems found in poor countries. Obviously, the use of inputs is only feasible if the farmer makes a profit after taking into account not only the income from product sales and the cost of inputs, but also the transaction costs of accessing the products and inputs markets. If the use of inputs is not feasible, fertility is usually managed via biomass flows through the territory and between crops and livestock, aiming at recycling as much as possible the resources extracted from the soil. Agricultural communities developing such agroecosystems are generally well structured. Thus, family units are linked to production areas, watershed basins or, more generally, to agricultural systems. Decisions of family farmers depend strongly on this context, which provides them with a physical, social, economic, organizational and institutional environment.

6.2 Do Agrarian Systems Help Manage Ecosystems?

The sustainability of the family farm unit depends on a defined and delimited territory and the natural resources that are attached to it, even if this territorial attachment is often combined with forms of outsourcing by the mobility of some members of the family or their pursuit of non-agricultural activities (Chap. 4). This anchoring of family farming systems may seem to tie them to sustainable practices since it is in the producers' interest to maintain and grow their means of production, intrinsically linked to the natural resources available to them. However, this state of things is still very dependent on the family farmers' perception of the degree of security and continuity of access to land and resources attached thereto (trees, water resources, soils and sub-soils), as well as on the social dynamics that underpin the territories (Chap. 7). Other conditions necessary for sustainable use of natural resources by farmers are the knowledge of and the ability to control their impact on the ecosystem.

6.2.1 Sustainable Forms of Family Farming, but Only Under Certain Conditions

We have seen that family farming production systems may remain respectful of the environment over long periods. But they may no longer remain sustainable if the balance on which they are based is disturbed: increased land pressure, a decrease in labor supply, an economic crisis, easier access to and improper use of mechanization or inputs. This is how, at several places in the great tropical and equatorial forests of the world, especially during the twentieth century, shifting cultivation systems lost their balance due to the increase in demand for land – for agriculture or for other purposes – or by use of forest biomass as firewood, construction timber or paper pulp. Slash-and-burn systems are thus stigmatized as being the cause of disastrous deforestations (see Box 6.2) where they had worked for millennia. In savanna regions, changes in equilibrium of agroecosystems based on fallow or agroforestry parks have been observed. Despite their recognized qualities and great adaptability to the arid Sahelian environment, agroforestry parks are often abandoned in favor of temporary agricultural activities. The *Faidherbia* park is emblematic of the debates between traditional family farming that has been proven sustainable in the field (see Box 6.1) and the agricultural model of the Green Revolution, still taught in some universities as the best technical model irrespective of the local conditions.

The impact of production systems based on the ability of the ecosystem to regenerate its fertility is closely tied to the balance between products exported from the field and the time that it is granted to restore its fertility[4]. There is always a

---

[4] From the recycling of what is extracted, and thanks to flows from the atmosphere, rocks, neighboring ecosystems and stocks of exported materials.

productivity threshold that cannot be exceeded without external supply of matter. This threshold is determined by ecosystem services such as those regulating water and nutrients flows (Box 6.3). It varies depending on the distribution of solar radiation and water, and the nature of soils and their relationships with the underlying rock. The spatial distribution of plants extracted (by humans or livestock) in relation to resources also influences this threshold at the plot scale. Moreover, if an increase in land productivity is not sufficiently compensated by imported fertilization, it can result in a fragilization of the ecosystem such as excessive erosion of depleted and impoverished soils, savannization or desertification. In this, family farms find themselves in the same boat as other agricultural forms (Chap. 5). They can be the cause of the degradation of the ecosystem if they do not have the means, the will or the knowledge necessary for a sustainable management of fertility and for employing suitable soil protection techniques.

Box 6.3. Ecosystem services and payments for ecosystem services.

Denis Pesche

The concept of ecosystem services originated in ecology and conservation biology scientific circles in the United States in the 1980s (Mooney, 1983). In 2005, the Millennium Ecosystem Assessment (MEA) brought this concept to the international stage. It identified 24 services, grouped into four broad heads: provisioning services (food, fresh water, wood and fiber, fuel, etc.), regulating services (regulating climate, flood and diseases; water purification, etc.), cultural services (aesthetic, spiritual, educational, recreational, etc.) and supporting services (primary production, nutrient cycling, soil formation, etc.).

In 1996, Costa Rica established a pioneering program of payments for environmental services (PES). In the continuity of forest policies implemented in the 1980s, this new program recognized four major services provided by forests (carbon, water, biodiversity and scenic beauty) and compensated forest owners for good forest management practices. This program has been funded by a fuel tax and also, since 2001, by loans from international donors. The excitement generated by this first experience quickly led to other similar experiments in Latin America, primarily under the leadership of the World Bank. This type of intervention, though innovative, is based on a simple principle: payment for a service is made directly to its provider. By introducing the idea of the market, instruments such as PES marginalize more traditional conservation approaches which rely on integrated development and are based on local coordination, with the State playing an important role.

MEA has indirectly contributed to accelerating the spread of the PES model in tropical areas on all continents. Spurred on by donors, major environmental NGOs and a growing number of enthusiastic forestry experts, the model is being tried out in various forms in many countries. It was not until the 2008-2010 period that criticism started emerging based on experiences of projects claiming to be PES. In recent years, criticism is also increasingly being heard of the risk of commodification of nature.

The idea of an Intergovernmental Platform on Biodiversity and Ecosystem Services (IPBES) fructified in 2008 with the ambition of providing policy makers, such as the Intergovernmental Panel on Climate Change (IPCC), with validated knowledge on matters relating to biodiversity and ecosystem services.

Moreover, some forms of agriculture, whether family or entrepreneurial, are only harmful to the environment when their installation is at the expense of an ecosystem of high conservation value, such as a forest or swamp with remarkable biodiversity or substantial carbon stock. In such situations, family farming still retains the advantages of its fragmentation into small landscape units (fields, meadows, plantations), which maintain hedges, woods and grassy edges producing several ecosystem goods and services (regulation of flows, attraction of natural predators of pests, windbreaks, etc.). But when it starts abutting on a pioneer frontier, as in the Brazilian Amazon, family farming becomes another tool of territorial colonization (Box 6.4).

Box 6.4. The farmer's cow in the Amazon. What a program!

Soraya Abreu de Carvalho, René Poccard-Chapuis, Amaury Burlamaqui Bendahan, Jonas Bastos da Veiga, Jean-François Tourrand

When we talk about livestock farming in the Brazilian Amazon, we most often refer, on the one hand, to the driver of deforestation that it has been throughout the last half-century and, on the other, to large cattle herds grazing in the vast land areas originating from successive phases of colonization since the 1960s and later transformed to grazing lands (Piketty *et al.*, 2005). Established alongside this livestock farming is another form of farming, similar enough in its structure, its management and its representation to almost every peasantry in the world. This livestock farming fulfills different functions depending on the season and on the period in the life cycle of peasant families.

The primary function of livestock farming in the Amazon is, of course, to provide food security. And following on from this security is the secured income for the family generated by this activity, even though the earnings are relatively low compared to other activities such as the cultivation of annual or perennial cash crops. This security is to be compared with the efficiency of the Amazon cattle sector, which also offers many different job opportunities, especially in the meat, milk and leather agro-industries, thus fulfilling a very significant and important territorial function. At the family level, one of the major features of Amazon peasant farming is its contribution to food security, as much as in the supply of milk and cheese as through income. This function is essential during the initial phase of new settlements when access routes are still unsatisfactory and the flow of productions to markets irregular (Chapter 6).

From an agronomist point of view, livestock farming has, through grazing, an important agricultural function in societies originating from colonization in the Amazon. After the slash and burn, the pasture is planted directly after the cultivation of the first crop. It covers the ground after the crop is harvested and limits the progress of forest regrowth. In addition, if well managed, it is able to provide, with low levels of input use, production of one to two calves per hectare per year, as well as two hundred to three hundred liters of milk or cheese in the case of family consumption. The result is a prevalence of the pasture in the colonized Amazon landscape, to the detriment of the original forest ecosystem (Sayago *et al.*, 2010). Family livestock farming also leverages by-products and animal residues: livestock can be used for traction and, to a lesser degreee, as pack animals. Moreover the manure produced is used in home gardens and orchards for growing vegetables and fruits.

The concept of ecological intensification was proposed by agronomists as expressing both the challenge of reconciling high productivity and respect for the environment over the long term, as well as the way of meeting this challenge by leveraging knowledge of the inner mechanisms of agroecosystems to manage them judiciously and sustainably. This ecological intensification is starting to find its place in the North and in entrepreneurial farms of the South, but so far remains marginal in the family farms of the South (Chap. 18).

While some agricultural practices are specifically designed to preserve the environment or strengthen the provision of ecosystem services, others provide the same environmental benefits without actually attempting to do so.

Family farming systems are often affected by a lack of means or by material or physical constraints which determine some of the practices implemented.

Therefore, organic farming, without necessarily being so labeled[5], is sometimes the result of a lack of access to chemical inputs. Similarly, the preservation of biodiversity is not always a goal of the farmers and, in some cases, can even be considered as a constraint, as in Indonesian agroforests (Box 6.5).

Box 6.5. Indonesian agroforests, a biodiversity not always desired.

Laurène Feintrenie

Family farmers in Sumatra moved from hunting and gathering to shifting cultivation of rice, then to agroforests. They adopted technical innovations – and even developed some on their own – to take advantage of new economic opportunities or overcome technical constraints. They converted primary forests into secondary forests and then into agroforests, at each step maintaining a high level of biodiversity and most of the forests' ecological functions (Feintrenie and Levang, 2009). For this reason, agroforests are often presented as the result of local peasants' wisdom. It is a mistake, however, to consider that the conservation of biodiversity was intentional on the part of the farmers.

The initial studies of Indonesian agroforests, conducted by ecologists, emphasize their remarkable ability to preserve the ecosystem services of nearby forests. The conversion of forests into agroforests allows the conservation of many plant and animal species found in the neighboring natural forests. However, it is primarily the scarcity of labor that allows forest regrowth in rubber plantations. Farmers do not explicitly favor this vegetation, they just let it grow. In addition, quite a lot of biodiversity is far from being appreciated by the inhabitants, especially mammals such as wild boar, tigers, elephants and monkeys. These major pests are considered inevitable because of the dietary restrictions of Muslim local populations, the protected status of certain animal species or simply the lack of means to control them (Feintrenie *et al.*, 2010).

However, in many cases, the maintenance of the landscape is carried out by family farmers, whose work in the matter – intentional or not – is not always perceived or recognized. Farmers are often responsible for the maintenance of watercourses, roads, embankments and ditches, all helping to maintain the road network in the area. Some ecosystems exist only in the presence of an ancestral agricultural activity, such as grasslands and savannas, peat moors or Mediterranean scrublands. Yet, the role of agricultural activities is not always well known, and controversies continue to surround the true nature of the environmental impacts of familial forms of cultivation and livestock farming (Box 6.6).

---

[5] More than 1.6 million farms were certified in the world in 2010, of which more than 34 % were in Africa (mainly in Uganda, Tunisia and Ethiopia), 29 % in Asia (mainly in India) and 17 % in Latin America (mainly in Mexico) (Agence Bio 2012).

> **Box 6.6. Herdsmen in the Sahel.**
>
> Abdrahmane Wane and Christian Corniaux
>
> Pastoral systems in sub-Saharan Africa are dominated by the Fulani, Moorish and Tuareg family production systems. They are characterized in particular by different mobility schemes developed in response to the constraints of the arid environments. However, forms of mobility are changing and local and global dynamics that affect land use planning and access to resources now influence these mobilities (Dedieu *et al*., 2010). If the use of space remains extensive, intensification increases on the basis of family labor. This is especially the case in the dry season when several family nuclei are formed. Thus, the transhumant departure to agricultural areas in the south of a part of the encampment with the majority of the herds, leaving a family group at the base camp with a few goats and some dairy cows, depends on the intensity of the drought and the proximity of dairy markets. There is also an increasing exodus of young people looking for seasonal or permanent employment in cities or in irrigated areas. Labor wages also go up and increasingly illustrate the monetization of pastoral practices. The human pressure on pastoral areas thus reduces accordingly (CIRAD-FAO, 2012).
>
> However, controversies on livestock farming's impact on the environment continue to swirl. It is repeatedly accused of participating in the environment's desertification or of contributing to the degradation of ecosystems. On the other hand, pastoralism is often the only productive activity that adds any value to the arid and semi-arid zones even if the share of value added remains very unequal. It is now widely accepted that over the entire area of the transhumant routes and with safe conditions of mobility, and therefore a controlled pressure on resources, livestock farming preserves reservoirs of biodiversity while remaining profitable and competitive (Vaissyères *et al.*, 2012).
>
> More generally, increasing population pressure and urbanization, mainly on the coasts, have altered the balance between food needs and resources in Sahelian and coastal areas. There are fewer opportunities for Sahelian populations to generate livelihoods locally as compared to external opportunities, even given the latter's risks. This has led to profound changes in family work and the division of encampments (Chapter 3). It is in the context of these global changes that the debate over the impact of livestock farmers on the ecosystem must be considered.

Whatever their impact on the ecosystem and natural resources, it is certain that family farming systems are experiencing changes linked to regional, national and even global contexts. Their ability to adapt and innovate often determines their very survival.

6.2.2 Flexible Farming Systems: Innovative Relationships with Ecosystems

Family farming systems are sometimes established in isolation. A remote community far from any other will develop production systems enabling it to achieve self-sufficiency and provide for all its needs. When this isolation is ended, the farming system is disturbed and must find a new balance. The settling down of Amerindians in Guyana (Box 6.7) illustrates how the modalities of family farm production can evolve to allow farmers to benefit from the facilities that are newly made available to them, without requiring them to abandon their farming practices. It is a form of resilience of the farming system, which integrates new elements while still maintaining its essential character.

Box 6.7. Territorial changes and adaptation of the shifting cultivation system by the Amerindians of Guyana.

Isabelle Tritsch, Valéry Gond, Philippe Karpe

The Wayampi and Teko Amerindians traditionally practice a shifting slash-and-burn agriculture, based on the cultivation of bitter cassava (*Manihot esculenta*). Their territory is managed under customary law with collective ownership of resources. Their livelihoods are today strongly oriented towards subsistence activities, but they have undergone significant changes since the 1960s due to various factors such as population growth, regrouping and settling down around public infrastructure, and monetarization. The practice of shifting cultivation has become difficult, and around towns the fallow periods have been reduced, undermining the sustainability of the system.

As a result, Amerindians are adapting their systems for the exploitation of natural resources by redeploying them spatially. Instead of intensifying farming systems through labor and capital, they adapt their practices mainly through spatial and temporal organization. On the one hand, locations of primary residences have been split with the creation of many villages close to towns, allowing an expansion of the surface areas of resource exploitation. On the other hand, production systems are being reorganized in a multisite strategy combining slash-and-burn with short fallow near villages – these are managed with intensive labor and have the purpose of ensuring a food reserve in proximity of houses – with remoter slash-and-burn sites, further from the villages (families need to make long journeys by canoe to reach them), with long fallows. These sites are structured around temporary habitations – such as camps – in which the families live during the seasons of agriculture work.

The spatial organization of these habitations is based on family networks and alliances, which allow a transposition of the customary mutual support traditions found within extended families. Such an arrangement also permits the sharing of transport costs. These costs are offset by the multifunctionality of the locations: in addition to agricultural activities, families also go to these temporary camps to hunt, fish and relax with family. These habitations thus become a second territorial base for extended families and allow easier access to forest resources, while a residence is maintained in the towns and villages to benefit from access to infrastructure and public services (Tritsch *et al.*, 2012). In this respect, the traditional cyclic displacement of Amerindian villages is replaced by multiple places of residence and places of production and by the introduction of very frequent functional mobility linking towns and forests.

Technical innovation and adaptation of the production system of family farms are sometimes required by law or the evolution of the political context. In Brazil, all types of farms – from family farms to agribusiness companies – are required to participate in the effort to preserve the Amazonian forest (Box 6.8).

Box 6.8. Family farming, forest conservation and transition to a green economy in the Brazilian Amazon.

Marie-Gabrielle Piketty, Isabel Drigo, Emilie Coudel, Joice Ferreira, Plinio Sist

In the Brazilian Amazon, family farming could play a major role in the sustainable conservation of forest resources. According to official estimates, family farmers and communities are responsible for the conservation of 40 million hectares of forests. This is because of the Brazilian Forest Code, which requires the conservation of 50% to 90% of all forest property (large or small). In some states, the current wood demand can only be met through an increase in the logging of family-farming forest reserves.

Since the conservation of these forests is required by law, the prospects for REDD+ type of payments are limited (Ezzine de Blas *et al.*, 2011) and this conservation is rarely the subject of financial compensation, although "green grants" are gradually being introduced since 2011 for the poorest families of conservation units. They can also be exploited for wood or non-wood forest products. The exploitation of wood from these forest reserves requires the submission of individual or community sustainable forest management plans.

Initiatives for community forest management for timber logging have increased since the early 1990s through government grants or support from NGOs (WWF, IUCN, etc.). This type of management is seen as one way to protect forests while generating direct income for family farming. However, studies assessing costs of and revenues from various schemes show that most of them struggle to provide a real alternative for family farming and are only a limited source of additional monetary income (Drigo *et al.*, 2013; Piketty *et al.*, 2013). Besides the cumbersome regulatory process, access to remunerative markets remains highly uncertain without any support structure for marketing or any guaranteed minimum prices.

Public policies are therefore necessary to encourage a change in production practices that can reconcile forest conservation with economic returns for family farming. This is all the more urgent now because, since 2005, the Brazilian government has been able to reduce deforestation in the Amazon by more than 80% by targeting large landowners and it is now targeting family farmers, as small areas account for over 60% of the current deforestation. However, as of now, very few family farmers have benefited from land and environmental regularization underway in the Amazon, excluding them from green economic sectors that are being established today (Coudel *et al.*, 2012). Consultation processes are emerging within the " Green Municipalities " project but representatives of family farming are often marginalized.

6.3 Sustainable Management of Natural Resources?

Family farming systems are very diverse and, like the ambiguities and debates on the social dimensions of their organization (Chaps. 4 and 5), they affect ecosystems in various ways. The challenge for any agricultural community is to overcome the limiting factors of production. Industrial approaches to production that triumphed during several decades of the twentieth century have sidelined some technical innovations based on an empirical understanding of the ecological mechanisms that regulate agroecosystems. The low cost of energy and easy access to transport infrastructure allowed farmers to be provided with synthetic material and products to manage fertility and eliminate pests. Today, integrated management practices for plots and for complete agricultural systems are increasingly being recognized for their relevance. However, the overall population is increasing in such a manner that it is not certain these integrated practices will be sufficient in themselves to lead to a sustainable agriculture able to face the enormous challenges that this human pressure is imposing on agricultural production models (Chap. 13).

Some forms of family farming, such as agroforestry and conservation agriculture, are inspired by natural ecosystems to manage resources at the micro-local level and are thus able to maintain the provision of ecosystem goods and services at high levels. In contrast, other family farming systems lay no claim to sustainability. Farmers intentionally exploit those resources intensely whenever they think they may not be in a position to benefit from them over the long term or if they think they cannot leave them to their children. The continuity of access to land and associated resources appears as an essential condition, but one that is not sufficient, to any hope for sustainable production. But family farmers are not alone in their territories. They are confronted by collective and individual public and private actors who have interests which may diverge from their own. These territorial dynamics are analyzed in Chap. 7.

Chapter 7
Contributing to Territorial Dynamics

Stéphanie Barral, Marc Piraux, Jean-Michel Sourisseau, and Élodie Valette

Because of the sheer extent of the geographical dimension of rural areas and the dominant role that agriculture plays there, family farming is at the heart of territorial dynamics. It involves the mobilization of territorial resources by individual or collective actors (Gumuchian and Pecqueur 2007). These resources can either be physical – abundant natural resources, favorable climatic conditions or very good market access, etc. – or intangible – traditional and ancestral know-how, political resources, cultural heritage, etc. The appropriation and use of these resources reveal strategies of action of various stakeholders. Thus, territorial dynamics can be defined as changes and translations, in a given area, of individual or collective actions planned and undertaken for the appropriation and use of limited resources, in specific institutional and political contexts (Piraux 2009). This chapter aims at evaluating the contribution of family farming models, as the sum of individual and collective actors, to these territorial dynamics.

The analysis is based on the consideration that territorial dynamics, as viewed from the perspective of human intervention on the environment, differ significantly depending on the distance to major urban centers. This distance determines different levels of pressure on resources (deforestation, desertification, landscape transformation, rehabilitation or artificialization of land, pollution, etc.). Furthermore, the share of agriculture in economic activity also tends to grow along this same gradient. Thus we propose analytical categories of rural spaces in terms of the

---

S. Barral (✉)
SAD, INRA, Champs-sur-Marne, Paris, France
e-mail: phanette.barral@gmail.com

M. Piraux
ES, CIRAD, Belém, Brésil
e-mail: marc.piraux@cirad.fr

J.-M. Sourisseau • É. Valette
ES, CIRAD, Montpellier, France
e-mail: jean-michel.sourisseau@cirad.fr; elodie.valette@cirad.fr

distance to major urban centers by differentiating isolated rural areas and marginal zones from typical rural areas and urban areas or those under urban influence. The border areas between these types of spaces, naturally porous and changing, demonstrate phenomena of transitions between these spaces.

This chapter's underlying assumption is that the expression of territorial dynamics specifically outlined by family farming is differentiated according to these spatial categories. Flows of people and products and the pressure on resources (water, soil, forest cover) in each category appear to be specific. Competition with other types of activities (residential, industrial, mining, forestry) and other forms of agriculture for the use and control of resources varies greatly in form and in terms of how it is managed. These factors directly influence the nature and functioning of family farming systems.

Of course, the forms taken by these various spaces as well as the transitions between them can encompass a wide range of situations. In particular, transitions of marginal areas to typical rural ones sometimes take the form of agricultural frontiers. We emphasize here these transitional spaces not because of their significant spatial extent, but because they are undergoing rapid and radical changes that show the processes of evolution of family farming systems within territories in sharp relief.

The chapter is structured according to four broad analytical categories, shaped by a gradient of environment's anthropization. After characterizing these categories' specificities, we will focus on showing the contribution of family farming in the economic and social construction of these spaces, through its own dynamics, but also through its linkages or competitions with other forms of agriculture.

7.1 Family Farming in Marginal Zones

Marginal zones are sparsely populated regions, such as forest areas of the Amazon, Congo and Indonesia or the steppe and desert areas like those found in Central Asia and Saharan Africa. The term may also refer to more localized zones with difficult agricultural conditions, almost exclusively developed by family farming, and whose management may often also ensure the ecological and economic balance of larger areas. Natural resources are used very extensively, the main goal being to meet the needs of local populations. The economy is barely monetized and social division of work is limited, as is the presence of government services.

In these poorly developed regions, the social structure of agriculture is based on the nuclear or extended family, which is usually part of community groups controlling all or part of the means of production. Shifting slash-and-burn agriculture, hunting and fishing predominate in forest regions. In transhumant livestock farming regions, the living place and its resources are collectively managed, the production is extensive and integration into marketing or technical advisory networks is limited. Infrastructure facilities are rudimentary at best and public authorities have a minimal presence. Nomad populations move several times a year, depending on the availability of fodder resources.

The natural environment of these marginal zones can be precarious which implies that it requires limited human pressure on resources. Therefore, any change in forms of exploitation – arising from any reason – can imperil their transition to agricultural frontiers (in the case of forest areas) or to more typical rural zones due to the significant risks of overexploitation.

In Mongolia, for example, the change of political regime in 1991 (from a tutelary supervision by the USSR to the sudden liberalization of the economy) resulted in an impoverishment of urban populations, who then turned to livestock farming activities. The resulting increase in grazing pressure jeopardized the food and health equilibriums of herds and, by consequence, the food and monetary resources of families (Devienne 2013).

Agricultural firms may also cause changes in the way value is added to marginal areas. The changes that result from their arrival in forested areas can show complementarities with the rationales of family farming. This is the case, for example, of artisanal and industrial sectors of palm oil production in Cameroon or large Indonesian private palm oil plantations that generate agriculture in the form of small plantations at their periphery (Chap. 5). On the other hand, the introduction of corporate activities can be very disruptive and cause the breakdown of local production systems. This is the case of mining or of large Indonesian forestry firms, in the 1970–1990 period, whose activities have resulted in depriving local people of their production systems (Durand and Pirard 2008). The research work of Marshall (2011) on the setting up of agro-industrial firms on the Peruvian coastal desert clearly shows the ambiguity of encounters between capitalist organizations and local communities. While they can lead to conflicts or the exodus of inhabitants, they can also open up economic opportunities. There is also an increased risk of depletion of natural resources. Since the 2000s, the increase in land expropriation by agro-industrial and mining companies (with approval by – or in connivance with – public authorities) or by sovereign wealth funds further threatens these marginal areas.

7.2 Family Farming on Agricultural Frontiers

Agricultural frontiers constitute an initial phase of construction of new territories (Monbeig 1996). This phase unfolds over time in a dynamic process resulting from the decisions of migrant families on taking up agricultural activities. These frontiers represent a transition of marginal forested zones to forms of typically rural agriculture by substituting shifting cultivation with subsistence agriculture such as plantations or livestock rearing undertaken by migrants in search of agricultural land. Even though not constituting a permanent space but a process of transition, this phase of territorial genesis remains fundamental, if not critical, in order to determine the future trajectories of these regions and the conditions under which they will follow sustainable forms of development (Poccard-Chapuis 2004).

Research work has highlighted the diversity of practices of colonization and of their contributions to a specific historical process (Dufumier 2010). The role of family farming systems in the structuring of frontiers has been analyzed from three angles: the migrations which are responsible for land colonization, the structuring of social relationships underpinning economic and territorial development, and the evolution of agricultural frontiers due to increasing pressure on land resources.

Ruf (1995) shows how information about the possibility of developing an agricultural activity on a frontier circulates within communities, transmitted from early migrants to family members back in the villages, and helps initiate spontaneous migration flows. Migration and pioneer processes can also be organized by the State in order to conquer and control new territories, as was the case in the Brazilian Amazon, in Laos and in Indonesia, with the well-known programs of *kolonisatie* (1905 to 1945) and *transmigratie* (from 1945) in which the colonial government – and later the national government – organized the displacement of hundreds of thousands of people (Levang and Sevin 1989). In the case of agro-industrial firms setting up activities on an agricultural frontier, the lack of local labor impels them to recruit migrant workers who become potential land colonization actors (Barral 2012). These three drivers of migration can be found together in the same place.

The gradual settling down of families on agricultural frontiers is accompanied by the development of social ties in several ways: family and community ties are used as important information and mutual aid vectors; land colonization is structured around commercial relations between migrants; finally, relationships between farmers and firms also help define the territorial structure of frontiers.

Family ties impact the way land is appropriated and more broadly the way the whole territory is colonized. Moreover, even though land grabbing is organized around the family unit, collective dynamics are also observed, mainly in the form of mutual exchanges of labor (particularly for clearing) and circulation of seeds. In some cases, the first settlers who were able to capitalize on farming activity recruit new migrants to colonize new lands. In these cases, land development is based on these associations, with one party providing the capital and the other the labor force.

The presence of agro-industrial firms can be decisive in determining the forms of production adopted. For example, in the case of palm oil production (in Asia in regions where there is no small-scale processing), small farmers depend on the firms' oil mills, as well as on the work and wages provided by these firms. Family farming, however, still remains an important driver of local development. Strengthening and consolidating such family farming on these pioneer territories is a strong and recurring challenge for social sustainability.

As the agricultural frontiers move forward, colonization evolves into situations where – with increasing population densities and decreasing yields and land availability – a movement towards intensification of production is observed and new institutional arrangements rise. Conflict is a constant fact: mainly conflicts between local people and migrants, conflicts amongst migrants, and conflicts between local people and firms. It repeatedly takes place during the process of appropriation of resources – a process which is nearly always poorly regulated by legislation – and increases as resources grow scarcer.

On these evolving territories, access to natural resources is especially critical to the viability of the first production systems that settlers may be able to establish. Natural resources are initially managed through a process akin to mining: first deforestation, then the exploitation of the temporary fertility of forest soils, which then rapidly deteriorate. If farms do not adopt new techniques, the farmer may be compelled to sell his land holdings to neighboring farmers or to a large farm. He then moves to an urban periphery or advances further forward on the front, looking for new land to clear. This explains the main features of Amazonian frontiers: creation of small towns, advance of the deforestation line, but also violence and poverty (Poccard-Chapuis 2004).

In these regions, it becomes necessary to change the basis of production systems in order to stop deforestation sustainably and permanently. This requires the insertion of family farming in territorial dynamics. One typical feature of the frontier is the difficult emergence of collective action due to the diverse origins of migrants and their projects and to the instability of initial stages of territorial development.

The evolution of these agricultural frontiers also depends on legislative and normative contexts of each State. Support to the integration of families tends to lower their dependence on agricultural income. Local identities are consolidated with the emergence of second and third generations of migrants. Meanwhile, local institutions are created and infrastructure is developed for agricultural diversification. These transitions lead to "typical rural zones," as per our analytical categories.

7.3 Typical Rural Zones

On the anthropization gradient between the urban and marginal zones, the "typical rural" zones encompass a wide variety of situations. Conventionally designated as the countryside, these spaces can include small or medium towns. A typical rural space is characterized by three essential criteria. First the average population density is lower than in urban areas, which is linked with lower availability of equipment, infrastructure, goods and services. Secondly activities are most of the time dominated by agriculture, both for employment and land use. Croplands, natural areas and small towns together shape landscapes. The final criterion is a socio-cultural and political one. Cultural values and political systems reflect the influence of the agricultural world and of its historical hierarchies on territorial representations and governance institutions.

7.3.1 A Typical Rural Space Where Family Farming Influences Territorial Dynamics

The most common representation of typical rural territories is a land which has been put to use by a more or less dense pattern of family farms, accounting for most of the territory's workforce. In these agricultural landscapes, there is not much scope for land concentration, and industrial and family business farming has historically struggled to gain a foothold. This has resulted in varying population densities depending on national trends, but generally higher than in marginal zones, agricultural frontiers and areas with corporate farming and family business farming (see Raton 2013).

These areas are found in most parts of the world, but they are especially widespread in rural areas of West Africa and Central Asia, where the vast majority of the world's agricultural workers are found and where agriculture retains an important weightage in wealth production and employment (Chap. 3).

This category encompasses a very wide diversity of agrarian configurations and systems, which stem not only from eco-climatic conditions but also from the history of human settlements, social and cultural structures, access to markets, etc. (Chaps. 2 and 5). However, regardless of the configuration, the prevalence of family forms of agricultural production ensures that the drivers of territorial construction are anchored locally. This is, of course, not to say that external dynamics are irrelevant. They do have an impact, especially the economic ones, but it is how these dynamics are perceived and accepted by the rural community incorporated around family farming – which, in fact, is where community leaders and sometimes even elected officials come from – which determines the path taken.

In areas dominated by family farming, the link between the domestic unit and the production at the level of families and farms is related to the structure of the local level. Organization of settlements (more or less dispersed depending on the availability of resources); patterns of movement of goods; trade-offs between resource management, climate risk and performance of agricultural systems (Chap. 6); and management of common goods (institutional, physical or natural) are all territorial results of the direct links between the family unit and the farm (Box 7.1).

> **Box 7.1. The place and roles of family farms in the structuring of rural territories in the Sudano-Sahelian zone of West Africa.**
>
> Jean-François Bélières
>
> Unlike more humid zones where industrial plantations were developed during and after colonization, the savannas of Sudano-Sahelian West Africa are largely dominated by family farming, which has helped shape the landscape and economies in the region for decades.
>
> The territories are organized around medium-sized villages (800–2,000 inhabitants) which have links to secondary cities where a significant part of trade takes place. Territorial configurations of a countryside connected to a city seem to strengthen with the increased mobility of people and goods. In this respect, we note an increased supply of food products to these secondary cities (Raton 2013). The farming family is the actor of change in these agrarian structures and their economic and social impacts.
>
> Family farms are shaping the architecture of villages. The homestead, huts and granaries express the social and economic organization. The head of the farm decides what to produce (crops) and the means to devote to it, using as reference the habits and customs which govern village rural life with its taboos, its days devote to work on the farm, its free days for dependent members, its obligations for collective work dedicated to the village community.
>
> The responsibility for managing land is gradually transferred from the community to the family, with strong appropriation at the family farm level. Common property – in particular the grazing areas – however remain extremely important as much due to their surface area as their economic and social role. At the same time, the densification of rural spaces requires incremental adaptations, still under the control of family farms. Stopping or shortening of fallow periods, or the settlement of pastoralists, are examples of challenges for territorial governance that family farming systems and their organizations have to face.

Family farming developed in these rural areas contributes, as elsewhere, to fulfilling multiple, social, economic and environmental functions. These functions are underlied / shaped by local identities, due to an often longer historical perspective – which are less rapid than in urban areas and agricultural frontiers – in conditions of access to means of production. Despite their importance, they are often poorly understood and therefore inadequately valued (Box 7.2).

> **Box 7.2. Multifunctionality in New Caledonia.**
>
> Jean-Michel Sourisseau
>
> In New Caledonia, Kanak people practice family farming on plots of very small size (90 % are smaller than 25 acres) based on complex production systems (22 plants grown on average per family) which, though intensive, have a very low consumption of mineral inputs (Guyard et al. 2013). This agriculture involves 96 % of families living in tribes, even though it contributes only 6 % of their monetary income and offers much lower labor remuneration than found in other sectors of a rapidly growing economy with almost full employment. The attachment the Kanak have to the land is to be found elsewhere. Agriculture is practiced because of reasons of food, of course, but also because of its social and identity functions. Closely linked to hunting and fishing, agriculture brings with it affiliation to a community in that it maintains knowledge and relationship with its environment. Products are as often given away as consumed and, if valued at market prices, contribute to 28 % of family resources and significantly reduce inequalities between families. Through donations, hierarchies and alliances are recognized and maintained; through food, knowledge and relationships with their ecosystem are activated.

The concept of localized agri-food system (LAS, Box 7.3) offers an insight into the principles of multifunctionality in which agriculture is embedded. In particular, it highlights the significance of local knowledge, largely hybridized but anchored to agroecosystems, not only as a base, but also as a driver of activation and leveraging of local resources offered by the massive presence of family farming.

> **Box 7.3. The localized agri-food system (LAS): a key to understanding the dynamics of family farming in rural zones.**
>
> Claire Cerdan
>
> Localized agri-food systems (LAS) were originally defined as real organizations of actors providing agri-food products and services (agricultural production units, agri-food enterprises, commercial undertakings providing service, catering, etc.). LAS were associated through their characteristics and functioning to a specific territory (Muchnik *et al.*, 2008). This definition tended to focus on rural areas, where family farms and small-scale enterprises were engaged in a strategy of differentiation of agri-food production. Building relationships between production processes and a geographical space can however take different economic, social and agricultural dimensions and is not limited only to mechanisms protected by a certificate of quality or origin.
>
> Thus, more recent work takes the territorial transformations of agri-food activities and the evolution of relationships between society and the agricultural sector into account. It is no longer a matter of leveraging and building on the specific local know-how of family farms, but also of trying to meet the challenges of managing spaces and implementing new forms of governance of resources and territories.

7.3.2 A Typical Rural Zone with Spaces Dominated by Agro-industrial Firms

In contrast with typical rural spaces built by family farming systems is another type of rural world, dominated by corporate farming and family business farming. This is especially the case in some South American, South African, Australian and Indonesian regions. These areas are often sparsely populated. Small- and medium-sized towns host secondary and tertiary sectors related to production. In transiting from being dominated by family farming via varied and singular trajectories, these areas have lost the traditional character we describe above. Local knowledge and specificities go into sharp retreat and can even die out altogether.

In fact, firms in rural environments treat the land as a mere mean of production which has to be made efficient in order to produce agricultural commodities and to ensure the proper functioning of agro-industries. Family farming finds little, if any, place here. When it is present, it is in a very dependent way: tightly integrated with industry, typical examples being the poultry or milk sectors. There is a push for integration and financialization of agri-food systems (Box 7.4). Social relationships between firms and family farming structures remain often tenuous and tense because of the proletarianization of local population (Chap. 5). This is the case in large palm oil or banana plantations. These are also areas where the processes of temporary or permanent immigration for work are significant.

> **Box 7.4. Private enterprise of former South African cooperatives.**
>
> Ward Anseeuw
>
> To promote the efficiency of a market economy in South Africa, the State support system for agriculture was dismantled. Through the privatization of their physical and financial assets accumulated through grants and subsidies during apartheid, existing cooperatives were transformed into powerful agribusinesses. These changes were accompanied by processes of vertical and horizontal integration of agrifood systems, mainly through mergers and acquisitions. From being institutional intermediaries, cooperatives were transformed into technical and financial intermediaries, integrating upstream and downstream segments, providing services both to large-scale white-owned farms which developed during the apartheid era as well as to small-scale black-owned farms mainly resulting from land reform. This new form of agribusiness has gradually come to control more and more of the primary production not only through contractual arrangements but also by the way of land acquisition through a process of integration. In this way, it now enjoys strong territorial and political control.
>
> The best illustration of this process is without doubt AFRGRI. A former cooperative, now listed on the Johannesburg Stock Exchange, it is currently one of South Africa's largest grain traders. It is the fourth largest wheat and maize miller; the third biggest actor in the poultry feed sector; and has major interests in the seed and pesticide production industries, agricultural extension services, etc. Reaching beyond the structure of the South African agricultural sector, AFRGRI, like many other former cooperatives, has now ventured into the rest of Africa (Boche and Anseeuw 2013).

Productive specialization of land through agricultural concentration should have a favorable effect on other sectors of the local economy and on the regional development, rather than just on family production itself. If competitiveness diminishes, territorial trajectories can change dramatically. Companies tend to move to other areas, usually without retaining responsibility of the negative externalities their activities generated. This is the case, for example, in Brazil, in Goiás, where poultry industries relocate further north in Mato Grosso rather than deal with land pressures generated by the expansion of sugarcane cultivation. In other words, major agribusiness groups often employ a pioneer strategy to serve primarily their own interests, with the result that the territory as a whole later suffers.

7.3.3 Areas Where Family Farming Co-exists with Corporate Farming and Family Business Farming

Between these two contrasting situations exists a continuum in which complementarities are built or, in contrast, processes of domination are exercised by corporate farming or family business farming. Territorial structuring is then mainly dictated by the nature of the relationship between these two types of agriculture, depending on processes of juxtaposition and competition – more rarely of interdependence –, in ways described in Chap. 5. Such examples are numerous: the stabilized Amazonian frontier, irrigated areas, situations around Indonesian palm oil plantations, etc.

Any complementarities between the different forms of agricultural production often depend on the weightage of family farming (in quantity but also from the institutional and political point of view). It is also and especially dependent on social conditions of production, primarily the security of land tenure. Very often, if land rights of family farmers are not secured, tensions emerge between family and corporate or family business farming. This is the case of Office du Niger in Mali, where hybrid forms of agriculture have developed and where modes of access to farmland (ownership or rental) have diversified (Box 7.5).

Box 7.5. Family farmers and foreign investors: going beyond clichés in Office du Niger.

Amandine Adamczewski

In Mali, the Office du Niger area has enjoyed irrigated agriculture since the 1930s, practiced by family farmers moved there by the colonial State. Some hundred thousand hectares are currently irrigated, but developable land could reach an extent of as much as one million hectares. Since the 1990s, family farms in Office du Niger have increased their production ten-fold and now meet the majority of the country's rice needs. But the farm situation remains worrisome: 63 % are below the poverty line, while farm profitability has been compromised by the reduction of their sizes, 7.8 ha per family in 1982 to less than 2 ha in 2006.

Hard hit by the decline in development assistance, Mali launched an appeal to domestic and foreign investors, public and private. Consequently, since 2005, it has tentatively allocated more than 870,000 ha to these investors. Family farms are worried by this development, fearing unequal competition in the rice sector and concerned about access to irrigated and dewatered plots. The latter, destined for the cultivation of dry crops, livestock rearing and fuel wood, lie within the domain of the State and can be allocated to investors without notice. It is no surprise then that relations between these two forms of agriculture, family and entrepreneurial, have become strained. However, private investors are struggling to develop their concessions. We observe a stopping of some investments and the emergence of hybrid solutions for development (Adamczewski et al. 2013). Thus, by subletting on the perimeters of land allocated to macro-investors and intermediate investors (domestic and foreign investors with less than 50 ha), family farming systems have access to irrigated land. Family farms can also become investors, for example in the form of collective organizations which can bring together from 50 to 300 family farmers.

The boundary between farming methods is therefore porous, and highly subject to change. Open crises may occasionally themselves erupt when investments take the form of sudden land expropriation, but most of the time actors adapt by entering into land tenure arrangements which enable them to continue their agricultural activities.

Other examples perfectly illustrate the various links between small family farms and corporate or family business farms.

In areas of stabilized frontiers in the Brazilian Amazon, on the outer boundaries of land reform, traditional communities benefit from increased land security. Complementarities between family farming and corporate or family business farming can then prove significant. They can take the form of lending of farm equipment to work the land, animal transfers between family farmers and landowner farmers, as well as seasonal or continuous employment. In the case of palm oil production, this coexistence can take the form of relationships of complementarity, domination or conflict depending on the situation. In Indonesia, for example, around the large plantations, forest land is suitable for the setting up of small oil palm plantations by farm workers. Corporate farming helps the emergence of family and family business agriculture and boosts the income of farmers which is often otherwise insufficient (Chap. 5).

This coexistence – which is also found in the Brazilian Amazon – is different in case of contract farming. Independant households / planters are not employees of the plantations, they work farm plots whose initial investments for cultivation were made by the firms. To repay the latter, the independant households / planters deliver their production to the firm at prices set by the firm. These situations of dependency can induce very significant territorial reconfigurations (land, prices of food crops) and can sometimes even lead to open conflict (Timone 2013). This results in a quasi-proletarization of family farmers. Such situations also concern millions of hectares for sugar in southern Brazil, or soybeans in Uruguay and Argentina, where large agro-industrial firms lease the land of family farms as well as the services of its owners. In this way, some family farmers become service providers. These family farming situations are the ones which become depleted most often (Clasadonte et al. 2013).

Finally, there exist situations – increasingly frequent, especially in the Indonesian landscape – where investors set up oil mills and purchase all their requirements from local producers. This configuration exists in most African palm-oil producing countries, where artisanal and industrial red oil have distinct commercial outlets. Households using improved seeds can also sell their products to industrial mills. This arrangement provides a bargaining advantage for them; their production is not lost if they refuse the terms offered by any one firm (Rafflegeau 2008).

All these examples illustrate that once we go beyond the clichés, we perceive that complementarities can build social relationships and are often the basis for strengthening territories.

7.3.4 *Rural Territories Where Agriculture Has Turned Marginal*

Finally, the share of agricultural population continues to decline in some countries, in particular in the rural areas of developed countries which have experienced decades of structural transition in the form of massive exoduses out of agriculture to secondary and tertiary sectors. Residential and tourism economies have come to dominate many rural areas, with an ensuing and obvious loss of power, recognition and freedom of action of the family and non-family farming worlds. In these territories, environmental requirements, quality of life, and even equity in access to cultural goods and landscape aesthetics, take center stage due to this population inversion. They constitute new standards for the positioning of agriculture in human life.

France is illustrative of these transformations. In its rural areas, fewer than 10 % of jobs are now connected to agriculture. This is explained by the combination of a trend towards concentration of agrarian structures by a gradual decline in the number of small farms and a more recent trend of urban migration. As notes Perrier-Cornet (IHEDATE 2011), only one quarter of the 1,700 rural living areas in the country are home to agri-food industries. The French government's "Territories 2030" foresight, launched in 2010, offers among its scenarios for rurality a configuration of a "generalized residential countryside" which is already at work but likely to develop further (IHEDATE 2011). In this scenario, the countryside populations continue to grow but intensive agriculture, which is no longer defended by politicians, is invited to invest elsewhere. National standards, which tend to be defined in relation with conditions of urban life, become the reference, as much from a social point of view as economic or in terms of population displacements. Agriculture, which can retain its family character (or encompass it), becomes part of a focus on quality by minimizing as much as possible its local nuances, but continues to lose its grip and its influence on territorial construction.

7.4 Urban and Peri-urban Agriculture

Family farming is finally present in and around cities. Areas under urban influence form a unique category, where the share of agricultural activity is minimal and being further reduced gradually with the advance of the urban front. Farmland tends to be converted to urban uses, mainly residential but also industrial and recreational. At the same time, the nearby urban market can be an important outlet for agricultural production. All agricultural activity – family farming as well as other forms of agriculture – is variously affected by these situations of competition between land uses. However, for the purpose of our analysis, we distinguish between urban and peri-urban agriculture. Urban agriculture persists in the intra-urban space, with a relatively stable spatial extent in the gaps left between the dense buildings. Peri-urban agriculture, on the other hand, sees its perimeter reduce and fluctuate with urban growth.

Urban agriculture is essentially a family one in nature and it is broken up across very small surface areas. In the peri-urban context, on the other hand, all forms of agriculture are to be found. Whether in the context of a developed, emerging or developing country, corporate farming can be found adjacent to small farms owned or rented by a family.

In 1996 there was an estimated 800 million urban and peri-urban farmers worldwide (Smit et al. 1996), a figure that accounts for, for example, a quarter of the urban population in some African capitals (Orsini et al. 2013). But this number is difficult to pin down with any accuracy, first at the level of statistical censuses, since informal and secondary agricultural activities are often excluded, second because the artificial dichotomy between urban and rural sectors often generates contradictions. The spatial extent of agriculture in towns and cities is also significant but is often underestimated. Smit et al. (1996) have compiled data from various sources for the percentage of space occupied by agriculture in urban areas. In Beira, the second largest city in Mozambique, 88 % of the urban green spaces are used for family farming. In Bangkok, 60 % of the metropolitan area is used for agricultural activity.

7.4.1 Growing Weightage of Urban and Peri-urban Agriculture Despite High Land-Tenure Insecurity

Even though the share of farmers in the general labor force will ultimately decline, the number of family farmers or workers practicing farming in the city will remain stable or even increase. While urban growth forecasts show that by 2030, 60 % of the population in low- and middle-income countries will reside in the city, in many countries, particularly in sub-Saharan Africa, income from agriculture will still be the mainstay of a majority of the population (for example, 60 % in Cameroon, 70 % in Benin). In developing countries, the lack of social security makes the practice of agriculture a vital necessity. This is also the case in other contexts where food security is undermined by political or economic upheavals. For example, in Moscow, the number of families growing fruits and vegetables increased from 20 % in 1965 to 70 % in 1990 (Smit et al. 1996).

The productive function of urban and peri-urban agriculture remains its most important one. In developing countries, as discussed above, this agriculture tends to develop in urban clearings and gaps (Bon et al. 2010), especially in situations of food or urban crises. It helps compensate for the lack of social safety nets, it is a source of employment and income without large upfront investments and it allows the poverty of the most vulnerable to be alleviated.

The general increase in population densities, however, impacts the maintenance and growth of agriculture, both urban and peri-urban. Family farming is especially affected by the loss of fertile lands traditionally located near large cities. In Meknes, a city in northern Morocco, urbanization has compelled family farmers to reconsider their strategies in the context of high land prices. They have a strong incentive to sell their land rather than change their agricultural practices (Valette et al. 2013). While criticism is often directed at the continuing loss of agricultural lands on city outskirts, land-market regulations to prevent such loss remain unimplemented most of the time (Box 7.6).

> **Box 7.6. Supplying milk to Greater Cairo.**
>
> Véronique Alary, Christian Corniaux, Salah Galal
>
> Milk is supplied to the Greater Cairo area (20 million inhabitants) by two sectors: 20 % from industry, divided between imported milk powder and large farms with generally 100–1,000 heads of cattle. The remaining 80 % come from the traditional sector, called "loose milk" by the industrial sector.
>
> Family farms that form this traditional sector are found partly on the outskirts of Cairo or in the Nile Valley and Delta. Besides these traditional farms, which manage a number of dairy cattle depending on the size of the land they have – on average, 1,000–2,000 m2/animal is required – some family farms rear cattle indoors. These latter are heavily dependent on the market. Due to the unstable socio-political context in Egypt and urban growth, these urban units are very vulnerable. They bear the full brunt of rising prices of feed due to the devaluation of the Egyptian pound and strong speculative land pressure in urban areas. Starting in the early 2010s, there has been a mass exodus of farmers to outlying areas or to newly developed areas in the desert and some have even abandoned farming altogether.
>
> Urban expansion is the main factor of change in the livestock farming systems of the Greater Cairo and Nile Delta. Since the revolution of 2011, building on more than 20,000 ha per year has taken place on agricultural land in the north of Cairo. The increased government ineffectiveness in regulating land use has fueled speculation and has thus led to an irreversible loss of rural spaces in urban areas. In addition, peri-urban family farms are experiencing increased constraints in animal rearing in urban areas (pollution, logistics for inputs and products).

7.4.2 An Unobtrusive but Dynamic Agriculture

One of the characteristics of urban and peri-urban family farming systems is its low recognition by public policies. In the Cameroonian capital city of Yaoundé, while peri-urban vegetable gardening is recognized and even encouraged (Temple et al. 2008), subsistence family farming meets indifference or outright hostility. Urban markets there do not suffer from supply problems thanks to highly productive rural areas especially on the western plateau. The social functions of subsistence family farming are not recognized and health problems caused by wastewater use on vegetable crops eaten raw help explain this negative attitude. This lack of institutional recognition is an important constraint and increases land-tenure insecurity. For example, cultivated areas or unauthorized constructions can be destroyed at any time and their populations evicted.

The environmental role of food crops is ignored even though they limit erosion on the slopes, contribute to flood control in the lower areas of cities, and allow the recycling of wastewater and solid waste safely since the main products, even if in direct contact with this water, are consumed after cooking.

Governmental attitude towards agriculture is thus paramount given the challenges of food security and safety, environmental protection, exclusion and poverty. This attitude varies greatly from one country to another. In developing countries, many initiatives have existed for several years. In Accra, Ghana, where market supply is precarious, local authorities have set up measures to protect agriculture, in particular by reserving land for farming. Also in Ghana, Operation Feed Yourself, launched in 1972, is intended to encourage the practice of food farming by urban residents, especially when they lack market access (Obosu-Mensah 2002). In Cuba, in the pursuit of food security, governments have allotted land to families in Havana. Family gardens were estimated to occupy more than 35,000 ha in the late 1990s (Moskow 1999).

7.4.3 A Multifunctional Agriculture

At one time clearly confined to supplying food (mainly vegetables) to urban markets, urban and peri-urban agriculture is now being commended for its social, recreational, environmental functions. In industrialized countries, it is increasingly being sought for its multifunctionality, adapting to societal demand. Family farming is at the forefront of these changes, in particular through the development of short supply chains and of leisure activities complementary to those of agricultural production. The revival of community gardens is based, in addition to their productive function, on their dimension of social integration.

To contrast food and economic functions in developing countries with recreational, landscape and environmental functions in industrialized countries would nevertheless be simplistic. The contribution of agriculture to urban food supply is being increasingly acknowledged and encouraged in Europe and North America in particular, while its landscape functions are starting to be recognized in Africa and South America, in a context of international urging towards sustainable urban development. The environmental function of agriculture, however, *is* recognized across all countries.

Is family farming the best form to maintain this agriculture in a context of high social mobility and land pressure? Here too the answer varies greatly depending on the context. The increasing search for housing space and the need to ensure the ability to feed cities by a reorganization of food systems – recognized at all political levels – promote, at least in the industrialized and emerging countries, very space- and labor-efficient solutions which are incompatible with the maintenance of family farming. This has led, for example, to the corporate development of vertical farming, often involving hydroponic cultivation in towers or on walls of limited spatial extent. These experiments are concurrent with very different initiatives, supported by some city authorities, which aim to relocate agriculture and its markets, and rely instead on the maintenance of family farmers within and near cities. In Africa, the issue of food security, particularly prominent since food riots of 2008, prompts hypotheses of the maintenance and development of subsistence family farming, as it already exists in the intra-urban space. At the same time, consumer demand for healthy and quality food is growing, a demand that family farmers may find difficult to satisfy. In Hanoi, for example, where a large majority of family farmers work on very small surface areas, difficulties are obvious despite proactive and favorable public policies (Moustier 2010).

As far as the maintenance of family farming as the primary provider of urban markets is concerned, the role of public policies is crucial in terms of preservation of agricultural land. It involves urban planning, financial support (input costs, market prices, price stabilization, access to credit), a research and extension systems for vegetable gardening and more profitable and sustainable livestock farming. More importantly, integrated agri-urban policies recognizing the role of family farming in urban systems need to be conceived and implemented (Box 7.7).

Box 7.7. Socio-spatial arrangements of intra-urban family farmers in Bobo-Dioulasso.

Ophélie Robineau

In Bobo-Dioulasso, Burkina Faso, three main types of farmers coexist: ethnic Bobo farmers native to the area, whose lands have been progressively urbanized and who have intensified their agricultural activities towards vegetable gardening on land along streams, where construction cannot take place; migrants with no employment opportunities in the city who undertake intensive rearing of pigs in their yards to support their families; and, finally, migrants with stable jobs (civil servants, doctors, merchants) investing in pig, poultry or cattle rearing as a secondary activity on land on the urban fringe.

A study on how vegetable gardeners access organic fertilizer shows how the production of vegetables in the city is maintained through social relations based on complementarities between spaces, actors and activities. In Bobo, gardeners use three main types of organic manure: pig manure, cattle manure and urban waste. The lack of available cash does not allow gardeners to stock up on manure, especially since, unlike chemical fertilizers, manure availability is variable in space and time, and transportation requires access to a cart.

To obtain organic manure, gardeners develop links with livestock farmers and carters who offer their services for the transport of various products. Family and neighborhood networks, favorable relations between ethnic groups (Peulh herders and Bobo farmers), frequent visits to the same social places and anchoring of social relations in time help construct relationships of trust. Livestock farmers, located in the neighborhoods surrounding the vegetable garden sites, provide animal manure. The link gardeners have with them allows them to build relationships with other spaces, actors and activities complementary to their own production activity. Once these links grow, livestock farmers reserve manure for gardeners they know well and reduce its price for them. In return, growers pay for the manure in a timely manner. On the other hand, the carters are sought for two tasks: bringing animal manure from the yards of livestock farmers to the vegetable garden sites and collecting municipal waste for the same destination. Here too, the establishment of relationships of trust is fundamental. These practices based on the mobilization of social and relational resources of proximity are involved in maintaining family intra-urban gardening.

7.5 What Role for Family Agricultures in The tomorrow's Territories?

Territories are shaped by a diversity of family farming models, whose impact at the landscape level also echoes economic and social effects. These depend on spatial categories structured by distance to large cities, which determines different pressure levels on resources. The relationship between family farming and other forms of agriculture ranges between complementarity and conflict, in relation with complex and unstable territorial constructions.

The diversified, multifunctional and pluriactive forms of family production contributes significantly to local dynamics, but it is expressed differently depending on the structural changes that rural areas are undergoing. In Africa in particular, pluriactivity is clearly promoted with very diverse patterns. This is partly the result of urbanization, marked by the emergence of small and medium towns and by the increased mobility of populations (Losch et al. 2013). All over the world, there is an increasing mixing of rural and urban areas, which questions common oppositions of their representations. However, this mixing does not seem to curb the problems of poverty (Chap. 10), and the multifunctionality of family farming systems, even if they become more oriented towards urbanity, still involves, at least in developing countries, defensive strategies and risk management.

Even though they are important to our understanding, all these dynamics are poorly captured by national statistics, which widely underestimate not only cultivated areas and volumes of production but the real contributions of this form of agriculture to territorial dynamics (Chap. 3). Due of this, territorial resources are not fully known, thus distorting local representations of development, particularly the hybridizations between commercial and non-commercial rationales, between family farming and other forms of agriculture, between the city and countryside. This lack of knowledge limits the opportunities for innovation that adequate territorial public policies may be able to promote. In fact, in many countries, public policies remain dual and sector-specific. Yet, to let family farming models express themselves and their strengths, it is imperative to come up with real territorial policies that can leverage these different dimensions and functions.

With the increasing urbanization of rural areas, and an ongoing financialization of agriculture, territorial competition between family farming and corporate and family business farming is likely to increase, as indeed it is between the overall agricultural and non-agricultural sectors. While this can be a source of tension, opportunities will also abound, provided that adequate public policies are implemented. To confront these challenges successfully, farmers will need to take collective action (Chap. 9). Such action is, in any case, necessary to facilitate the marketing of agricultural products given that the competition is no longer limited to the scale of the local territory. Indeed, with the growing liberalization of markets, farmers face increasing competition in selling their products. Prices can be driven down by competitors benefiting from lower production costs or better negotiation skills. Despite this more competitive environment, family farming models nevertheless continue to occupy a prominent place in the production and supply of domestic and international markets. This is the topic of the next chapter.

Chapter 8
Contributing to Production and to International Markets

Sylvain Rafflegeau, Bruno Losch, Benoît Daviron, Philippe Bastide,
Pierre Charmetant, Thierry Lescot, Alexia Prades,
and Jérôme Sainte-Beuve

The predominance of family farms in agrarian structures in various regions of the world (Chap. 3) and the size of their territorial extent (Chap. 7) ensures that their share in the production of food and agricultural commodities is substantial. Their dominance is also reflected in their contributions to international markets, which are largely supplied by family farms. However, there exist significant differences depending on food and non-food types of products: staples (cereals, tubers and plantains, legumes, oilseeds, sugar plants), animal products, fruits and vegetables, stimulants (coffee, cocoa, tea), textile fibers and rubber.

These differences stem from market configurations – especially due to the share of local consumption (domestic markets) relative to distant consumption (exports) –, intrinsic product characteristics (quality, perishability), processes of production and processing and, finally, models of agricultural development.

We will use these factors as a key to understanding the importance of family farming systems in global production and the supply of agricultural markets. A specific and detailed analysis of their effective contribution by product type is, however, not possible without suitable statistical systems. And since family farming systems do not form an analytical category of either agricultural or trade statistics (Chap. 3), we will limit our analysis to some major sectors of tropical products. We will draw on CIRAD's expertise to better understand the

---

S. Rafflegeau (✉) • P. Bastide • T. Lescot • A. Prades • J. Sainte-Beuve
Persyst, CIRAD, Montpellier, France
e-mail: sylvain.rafflegeau@cirad.fr; philippe.bastide@cirad.fr; thierry.lescot@cirad.fr; alexia.prades@cirad.fr; jerome.sainte-beuve@cirad.fr

B. Losch • B. Daviron
ES, CIRAD, Montpellier, France
e-mail: bruno.losch@cirad.fr; benoit.daviron@cirad.fr

P. Charmetant
Bios, CIRAD, Montpellier, France
e-mail: pierre.charmetant@cirad.fr

© Éditions Quæ, 2015
J.-M. Sourisseau (ed.), *Family Farming and the Worlds to Come*,
DOI 10.1007/978-94-017-9358-2_8

characteristics of products and their markets and the respective shares of family farming systems and other types of agriculture in global production.

8.1 Overview of Global Production and Agricultural Markets

Global agrarian structures are characterized by an overwhelming proportion of family farms and small-sized of production units, of which 70 % are smaller than 1 ha, 85 % smaller than 2 ha and 95 % smaller than 5 ha (Bélières et al. 2013; Chap. 3). These are the structures which supply agricultural product markets.

These figures reflect the significance of Asia, and partially of Africa, in agricultural demographics, since agrarian structures in Europe, North America and South America are more diversified in size, with 10 %, 30 % and 40 % respectively of their farms being larger than 20 ha. In these areas, agriculture predominantly takes the familial form, although for larger production units, the scaling up involved usually implies a family-business orientation characterized by the use of hired employees.

These differences in agrarian structures and technical systems are not directly related to the regional distribution of global production, which pertains rather to the natural environment and the history of development of the various productions. They have, however, a greater impact on the role of exports in regional productions, depending on the size of the domestic market and exportable surpluses.

8.1.1 Market Configurations and Regional Distribution of Production

It is not possible to review all agricultural production, especially since most of it is exchanged in a processed form: not only primary processing in the form of husking, shelling, drying or conversion into oil, but also secondary or tertiary processing by the agrifood industry. These changes of state result in a multiplicity of products which considerably complicates any statistical analysis.

Two approaches to the regionalization of production are possible: an aggregated approach, by converting volumes into a single measurement unit, or an approach by the main types of products, which can provide an overview of global and regional agricultural production. In this exercise, it is not so much the actual regional distribution of production that is of interest to us, but rather what it reveals about the place of family farming with regard to each region's agrarian systems and development models.

The Agrimonde foresight study presents an overview of the production areas by aggregating large volumes of different products after converting them into food

Table 8.1 Share of the world's major regions in the global production of calories of plant origin in 1961 and 2007[a] (in %, calculated on the basis of Gkcal/day)

	1961	2007	Total production (multiplication factor 1961–2007)
Africa (sub-Saharan)	6.2	6.7	3.07
Latin America and the Caribbean	8.0	13.1	4.68
Asia	29.9	40.8	3.89
Former USSR	13.5	5.9	1.26
Middle East and North Africa	3.1	3.7	3.45
OCDE	39.3	29.8	2.16
World	100.0	100.0	2.86

[a]The Agrimonde analysis covers the 1961–2003 period and was updated for 2007 (Dorin 2012). The regional breakdown follows that of the Millennium Ecosystem Assessment. The OECD countries include Europe (including the former Eastern Europe), North America, Australia, New Zealand and Japan, but Mexico, South Korea and Turkey, who are OECD members, are instead included respectively in Latin America, Asia and the Middle East

calories (Paillard et al. 2011). Due to the difficulty of analyzing animal products, only crops – which, in 2003, accounted for 88 % of total agricultural production in terms of calories – are represented here (Table 8.1).

Because of their demographic weightage, it is Asian countries that contribute the most to global crop production. They have taken the lead over the OECD countries, where levels of labor productivity, however, remain higher and where there is greater specialization in animal production. Latin America is the third largest crop producing region and accounts for more than double the production of sub-Saharan Africa.

Table 8.2 shows the regional distribution of production for 16 plant products that illustrate the range of dietary needs in terms of energy and nutrients. Cereals alone cover approximately 45 % of the global population's caloric needs.

Asian countries have a significant presence in the global production of rice, palm oil, tea, coconut, rubber and cotton fiber. The most notable regional exceptions to this Asian dominance are soybeans (North and South America), maize (North America), coffee (South America) and cocoa and plantain (Africa). The regional distribution is more balanced for the other products. Among the 16 products analyzed, wheat is the only crop whose production is relatively strong in Europe, but this continent has a significant presence in animal products and fruits and vegetables, which are not included here.

The analysis of the structure of product markets still shows the importance of national markets and domestic consumption, despite the development of international trade over the last two centuries. Nearly half the world's population is engaged in agriculture, and home consumption remains high, especially for food staples (cereals, tubers, some legumes). The supply of domestic markets also remains the priority for large numbers of non-OECD countries.

Table 8.2 Share of the world's major regions in the global production of 16 major crops in 2009 (in % of volumes)

	Rice (eq. paddy)	Maize	Wheat	Cassava	Soybeans (seeds)	Palm Oil	Coffee (green)	Cocoa (beans)	Tea	Coconut (nut)	Banana	Plantain	Pineapple	Sugarcane	Rubber	Cotton fiber
Africa	3.3	7.1	3.8	47.8	0.7	5.5	12.0	64.8	12.7	3.2	13.0	72.0	10.8	5.4	5.2	5.8
America (North)	1.5	41.9	12.8	0.0	42.5	0.0	0.0	0.0	0.0	0.0	0.0	0.0	1.1	1.7	0.0	12.7
America (Latin) and the Caribbean	4.1	12.1	3.3	15.1	42.7	5.3	58.7	14.4	2.0	8.5	27.7	24.3	35.3	53.9	2.7	6.1
Asia	90.5	28.6	43.2	37.1	12.5	89.1	29.2	20.7	85.1	86.2	58.6	3.6	51.9	37.1	92.1	72.4
East Asia	31.7	20.3	17.1	2.0	7.0	0.5	0.3	0.0	34.4	0.5	9.4	0.0	8.0	7.1	6.4	30.5
South Asia	29.6	3.0	17.7	4.5	4.6	0.0	3.3	0.3	35.3	20.3	28.9	1.5	8.7	20.8	10.0	29.9
Southeast Asia	28.9	4.5	0.0	30.6	0.8	88.6	25.4	20.4	10.4	65.4	19.6	2.2	35.2	9.2	75.7	0.4
Europe	0.6	10.3	33.6	0.0	1.5	0.0	0.0	0.0	0.0	0.0	0.4	0.0	0.0	0.0	0.0	1.4

Source: FAOSTAT

International markets for agricultural products have developed relatively recently (nineteenth century), with an acceleration in the past few decades. Until the nineteenth century, trade was mainly focused on meeting European demands for a few tropical products (stimulating plants, sugar and spices). It was the transportation revolution (steamships, refrigerated transport, railways, aviation) and the development of the futures markets, made possible by progress in communications (undersea cables, telex), which resulted in the elimination of distance as a constraint.

Apart from a strong integration between Europe, North America and the Southern Cone countries of Latin America, international trade is very significant for only a few products – stimulating plants, palm oil, rubber – and significant for pineapple, soybean, cotton fiber, wheat, sugar and animal products, with the exception of fresh milk. For food staples, except wheat and sugar, the share of production that is traded internationally is marginal. However, these data on raw materials lead to an underestimation of the importance of trade insofar as it is sometimes the products resulting from primary processing that are exported. This is the case of soybean oil and meal, copra oil, cocoa butter or even cotton thread (Table 8.3).

The share of exports in total production, however, reveals significant regional differences for major food commodities. Unlike Asia and Africa, the Americas and Europe confirm their status as agro-exporters. As regards market shares, the situations are much contrasted with different "regional champions": North America for maize, wheat for Europe,[1] the Americas for soybean, Southeast Asia for rice and cassava (Table 8.4).

The relatively small share of international markets in agricultural supply should not however lead us to underestimate their economic importance. They accounted

---

[1] The importance of Europe in wheat exports must be tempered by the large volumes of intra-EU trades.

8 Contributing to Production and to International Markets   133

Table 8.3 International trade for major crops in 2009 (in % of volumes)

	Rice (eq. paddy)	Maize	Wheat	Cassava	Soybeans (seeds)	Palm Oil	Coffee (green)	Cocoa (beans)	Tea	Coconut (nut)	Banana	Plantain	Pineapple	Sugarcane	Rubber	Cotton fiber
Source	1	1	1	2	1	1	1	2	2	2	2	2	2	1	2	1
Exports/ production	8	11	21	10.2	36.0	74.0	75.0	176.7	42.9	4.2	20.8	1.3	33.7	28.0	66.9	36.0

Sources: (1) USDA; (2) FAOSTAT
NB: the level observed for cocoa is explained by the large amounts of re-exports

Table 8.4 Share of major regions in global exports and share of exports in regional production (food staples and soybean in 2009) (in % of volumes)

	Rice		Maize		Wheat		Cassava		Soybean (seeds)	
	Region/world	Export/production	Region/world	Export/production	Region/world	Export/production	Region/world	Export/production	Region/world	Export/production
Africa	3.0	5.9	2.0	3.7	0.5	3.4	0.2	0.1	0.2	11.9
North America	11.5	52.3	46.8	14.3	24.0	49.7	0.0		51.9	45.1
Latin America and the Caribbean	8.9	14.3	18.3	19.5	6.3	50.3	2.2	1.5	43.9	38.1
Asia	69.9	5.2	5.6	2.5	7.0	4.3	97.0	26.7	0.7	2.0
East Asia	2.8	0.6	0.6	0.4	0.6	0.9	0.9	4.2	0.5	2.7
South Asia	16.2	3.7	2.7	11.6	0.3	0.5	0.1	0.2	0.1	0.4
Southeast Asia	50.6	11.7	1.8	5.1	0.3	NA	96.0	32.1	0.1	5.3
Europe	6.7	74.4	27.2	33.9	52.3	41.4	0.5		3.2	79.7

Source: FAOSTAT
NA not available

for 1,700 billion USD in 2012 (UNCTAD data), almost half of the value of petroleum exports. Thus, the international rice market, which accounts for only 8 % of global rice production, is worth 25 billion USD. And for products with high added value such as fresh fruits and vegetables, international trade touched 154 billion USD.

8.1.2 Characteristics of Products and of the Production

Agricultural production and the various products derived from it have intrinsic characteristics that result in very different levels of technical limitations in their use and marketing. Perishability is a prominent example of these product-specific features – it is a function most notably of water content (fresh and dried products) –, and results in very different allowable post-harvest periods for processing and storage or for consumption. Types of quality and heterogeneity of products (such as shape, color, sugar content, fiber resistance or length, or bacterial content) are other intrinsic characteristics whose marketability is determined by the requirements of industries or consumers or by health standards.

Production processes and methods of transformation of products also have very different characteristics. Thus the production cycle can be divided into three main categories: annual crops; multiannual crops with cycles of 2–5 years; and perennial crops (fruit trees, plantation crops, shrubby herbs), whose life cycle is much longer (often decades) and consists of a juvenile phase followed by a production phase. These differences have implications not only for capital investments and returns on investment, but also on the work required (specialization and quantity, depending on the rapidity of various tasks, itself dependent on requirements of seasonality or perishability, for example).

These various characteristics affect the organization of production and markets because the types of qualities, types of requirements and the need for specific investments generate uncertainties, risks and high transaction costs (pertaining to the obtaining of and mastery over information, negotiation between economic agents, quality control, etc.). Organizations and institutions can help minimize these uncertainties and costs, hence the importance of producer and professional organizations, particularly for family farmers. They result in modalities of specific functioning of markets, often structured in value chains or sectors around a product at its different stages between production and end consumption. Contracts between economic agents are another possible solution and their sophistication often echoes the degree of uncertainty and risk that they must manage.

Thus the situations of totally free markets where product supply and demand meet directly (spot) can be divided into two main types: very localized trade of uncomplicated products and futures markets which determine the reference price of certain agricultural commodities. For other products, the setting up of organizations at various stages of the supply chain which operate on the basis of contracts is the most common solution. In some specific situations, these organizations may even go on to incorporate some degree of vertical integration by incorporating different stages of production and processing of products.

These considerations have often determined the structuring of sectors and influenced the modalities of the production. Nevertheless, we must remember the "constructed" nature of these different product characteristics, production processes and markets. Few natural characteristics are immutable, except perhaps perishability and seasonality of production. Indeed, the types of qualities and complexity available are primarily the result of the demands of industrial or end consumers. These demands evolve over time based on multiple criteria, such as requirements of taste, shape or color, or of health, environmental or ethical considerations (organic-agriculture products and fair-trade products). The length of the production-processing-marketing cycle also changes according to technical innovations and their adoption. Examples include short-, early- or late-cycle varieties; new methods of preservation, storage and processing; the speed and quality of transportation; and type and performance of tools and infrastructure at distribution and marketing locations.

All these features are the result of the different histories of each sector and each market, the balances of power between the various categories of economic agents, the dynamics of technical and institutional innovations, and the quality of the overall economic and institutional environment. They have a considerable impact on family farmers, who have to adapt to them by organizing to meet the norms and standards of different markets or by negotiating for the necessary support from other operators. Producer organizations can also influence the definition of some criteria of differentiation, by participating, for example, in the development of labels, or through other marketing or collaboration methods (short supply chains or solidarity networks).

8.1.3 Specific Characteristics of the Main Models of Agricultural Development

The history of agriculture and the patterns of development of the various sectors, stemming in part from the characteristics mentioned above, have strongly influenced current forms of agriculture.

It is possible to distinguish two main models of development of production: one from endogenous processes, subject of course to many external influences, and the other resulting from external interventions juxtaposed on existing production systems. Disruptive situations stemming from collectivist experiments (espousing the elimination of family farms and private agricultural property) have historically concerned only a few countries and exist today only in a few specific situations. And yet, they have often had lasting impacts on the capacities of initiative of various actors, including farmers, and on the organization of economic chains, and thus have slowed down processes of change. These situations have sometimes been exploited by land-investment capitalist and managerial entrepreneurs, as can be observed by developments in several East European countries and those of the former USSR.

In models of endogenous development, the original farming systems (Chap. 2) have gradually evolved to meet the growing demand for food caused by the triple phenomenon of population growth, urbanization and market expansions – responding first to local demand for staples, then to domestic demand[2] through increasing integration of markets, and finally to international demand. These requirements were met thorough technical progress, the organization of production and sectors, and the development of new varieties or types of production. Former settlement colonies (North America, the Southern Cone of Latin America, Australia and New Zealand), where the original agrarian systems were wiped out by land expropriation by European settlers, witnessed the same trends.

In countries which experienced the technological revolutions of the last century, especially motorized mechanization and chemical fertilization, which led to a dramatic surge in yields and labor productivity, the endogenous development followed by family farms often resulted in a specialization in certain products, accompanied by processes of integration into sectors. This trend led to a differentiation of farms, with a shift towards family-business and even managerial forms. In

---

[2] In colonial times, this response to domestic demand included the demand from the "home" country, especially in the context of the "imperial preference" system of trade agreements and customs tariffs.

countries where production technology advanced less, diversification of production often remained the rule for farms, with the continuation of basic food production. This did not, however, prevent them from developing export crops.[3] Farm types remained more homogeneous and anchored in family forms.

In the case of "exogenous" development, value is added to existing crops or new ones planted in ways different from those of existing farming systems – involving new operators, most of them foreign, new techniques and organizations, and often new export crops. This configuration usually takes the form of an enclave, such as found for agro-industrial plantations. This model is prevalent in non-settlement colonial plantations in tropical regions, where large estates and agro-industrial enterprises have inserted themselves into existing farming systems without challenging them (at least not locally). Major development projects undertaken after independence and often involving public or semi-public companies belong to this model. This model has permanently shaped many regional situations worldwide and persists today in a renewed manner with large-scale land acquisitions by foreign public and private operators (Chaps. 5, 7, and 13). But various dynamics of appropriation of this model are at play since local investors also invest in agro-industrial plantations (the case of palm oil in Southeast Asia), thus contributing to a hybridization of local development processes.

8.2 The Share of Family Farming for a Few Emblematic Products

Because of the statistical difficulties we have repeatedly referred to, it is not possible to draw up an overview of the contribution of family farming for all types of production. However, in order to present some practical situations, we reviewed a few main characteristics of several sectors of tropical products by attempting to measure the shares of family farming systems in their overall productions.

Table 8.5 shows the proportions of the production of ten crops for each of three different types of farming. Experts arrived at these figures by combining and cross-referencing existing statistical information with sectoral data, information originating with the sectoral companies concerned and field survey results. Hence, these figures are estimates. For some sectors, the distribution between the different types of farming was calculated on the basis of volumes produced, for others it could be calculated only on the basis of planted surface areas. In this latter case, the effective weightage of entrepreneurial agriculture is underestimated insofar as yields from agro-industrial plantations are, on average, higher than those of family farmers

---

[3] In colonized territories but without European settlement, the introduction of new crops into existing farming systems has resulted in a hybridization and development of local agriculture, such as cocoa cultivation in the countries of the Gulf of Guinea or rubber in Southeast Asia.

Table 8.5 Experts' estimates[a] of family farming's contribution to the production of some major crops (in %)

Sector	Entrepreneurial farming models - Family farming models		
	Enterprise farming	Family business farm	Family farm
Rice	2	4	94
Cotton	3	8	89
Bananas (all)	13	18	69
plantains	2	16	82
export dessert	78	13	9

Sector	Entrepreneurial farming models - Family farming models	
	Industrial plantations	Small holdings
Coconut (ha)	4	96
Coffee	5	95
Cocoa	5	95
Rubber (ha)	24	76
Oil palm (ha)	59	41
Sugarcane (ha)	60	40

[a]Estimates calculated by F. Lançon (rice), M. Fok (cotton), T. Lescot (bananas), A. Prades (coconut), P. Charmetant (coffee), P. Bastide (cocoa), J. Sainte-Beuve (rubber), S. Rafflegeau and C. Jannot (oil palm) and R. Goebel (sugarcane). The table indicates the contribution in percentage of production volume or, in some cases, of planted surface areas (ha)

(Feintrenie and Rafflegeau 2012), as illustrated by the case of oil palm (presented later in this section).

The lack of systematic information on production structures compelled us to use two different classifications for the various sectors. For some of them, a distinction could be made between family farms, family business farms and enterprise farms (rice, cotton, bananas). In other cases, only two broad types were distinguishable: industrial plantations, owned by capitalist enterprises, and smallholder plantations, a more composite category that includes family farms, run entirely by the farmer and his family, family business farms – very common – with the use of permanent salaried employees and, finally, managerial farms where the management and cultivation are fully delegated to employees (managerial staff and laborers, Chap. 5).

The sectors presented in the table reveal a predominant place for family farming for rice, cotton, coffee, cocoa and coconut, with a production share of around or exceeding 90 %, while family business farms, enterprises and industrial plantations range from significant (rubber) to very significant for other sectors (oil palm, export dessert banana, sugarcane).

For highly perishable produce such as some fruits and vegetables, family farmers – when their financial capabilities are limited – usually have an interest in marketing their products on local markets on the morning of the harvest itself, eschewing intermediaries. In contrast, corporate farms and family business farms usually have the financial wherewithal to invest in a primary processing unit or a cold chain for transport or even for export. The collective organization of farmers can overcome some funding constraints, as can contracts with agricultural enterprises or specialized operators, which can then manage the packaging or processing.

The technical constraints of processing often pertain to industrial choices which stem from the sector's development. These choices depend on the various interests involved (equipment vendors, financial operators, types of shareholders) and modes of organization used (decentralized or highly integrated).[4] Quality criteria and the volumes of consistent quality required for export are another example since quality standards reflect a given supply/demand state.

8.2.1 Productions for which Familial Forms of Agriculture Predominate

8.2.1.1 Basic Food Staples

Represented here by rice and plantains, most of the volumes of food staples are produced by family farming systems. This dominance results from the significant amounts of these crops that are home-consumed, their place in the structure of farm activity, and the importance of domestic markets. In regions worldwide, family farming has responded to the growth of global and urban demand. Family business farms have come up, especially in areas where motorized mechanization is possible, with a consequent increase in cultivated surface areas (this is particularly true for wheat and maize). Some farms located near cities have specialized in the marketing of food staples with the help of hired labor.

8.2.1.2 Cocoa

Cocoa is essentially produced by smallholders. The global surface area in production is estimated at ten million hectares and the annual harvest is about four million tonnes of commercial cocoa. Eight countries account for 90 % of production, of which 65 % comes from countries on the Gulf of Guinea, with two of them dominating production: Côte d'Ivoire with 38 % and Ghana with 20 %. Family farms and family business farms are the rule (95 % of smallholdings). Most farms cultivate from 0.5 to 30 ha. The use of improved varieties – some certified, some not – is widespread but yields vary widely, from 80 to 4,000 kg/ha/year. The global average yield is estimated to be around 400 kg/ha/year. Smallholder plantations are often old (sometimes very old, over 30 years) and require rehabilitation, a process

---

[4] The example of industrial hulling of coffee in Côte d'Ivoire in the 1970s and 1980s – which ended in fiasco – is an illustration. It was based on factories with capacities of tens of thousands of tonnes rather than on the maintenance of small artisanal hullers.

complicated by problems of access to quality planting material and technical advice. Rehabilitation is sometimes funded by private operators from the downstream sector. In areas where the cost of labor is higher, as in Latin America, the emergence of commercial cocoa farming based on hired labor with surface areas ranging from 100 to 2,500 ha relies on the intensification of production, with yields of over 1,500 kg/ha/year. In Ecuador, for example, 30 % of production now originates with entrepreneurial agriculture.

8.2.1.3 Coffee

Small family farmers produce most of the world's coffee, a production that provides livelihood to some 25 million people. Eleven million hectares of coffee plantations produce 7 million tonnes per year, of which 5.6 million are exported, representing a turnover of around 20 billion euros. The continuing increase in production in Brazil and Vietnam is more than compensating for the decline in African and Colombian production. Thus, in the last decade, global supply grew faster than demand, leading to a stagnation of prices. Coffee represents only 2 % of export earnings for most of the major exporting countries, but this figure is 59 % for Burundi. Improved varieties account for, at best, 10 % of coffee planted worldwide. Plantations are characterized by a high average age and a wide range of yields. Family farms often grow coffee without recourse to inputs and with a minimum of labor, which explains their lower agronomic performance. These systems which occupy large surface areas offer low returns per hectare, but they allow small farmers to adapt to market fluctuations by deciding to harvest or not, and then to devote more or less labor for post-harvest processing. Low world prices have thus caused some farmers to replace coffee with other crops and have also resulted in a decrease in the quality of processing operations, thus creating a downstream supply problem. The production of "certified" coffee, a label which has been coordinated by downstream operators, although it represents only 8 % of the coffee exported, has helped the coffee sector escape this downward spiral and ensure sustainability in some cultivation regions. Certified coffee production has led to the modification of processing practices and has strengthened producer organizations and their business strategies (promoting the use of labels). Industrial operators sometimes also intervene in upstream activities by contributing to the improvement of planting material through distribution of plants. Large associations of producer organizations also take similar initiatives and furthermore provide technical advice to their members.

8.2.1.4 Coconut

Coconut production is mainly concentrated in Asia and the Pacific, which together account for 87 % of planted areas. In addition to their primary food function, coconuts are also being used, starting from the late nineteenth century, for the production of copra (dried kernel) and its oil. Copra is a source of occasional and additional income to many farms in subtropical coastal areas. Farms there are mainly small family farms (96 % of the surface area) with generally 0.5–4 ha of coconut grove per farm. Some industrial farms are present in West Africa (Côte d'Ivoire, Ghana) and East Africa (Mozambique) where coconut cultivation is clearly a legacy of colonial times, grown in monospecific plantations extending for hundreds of hectares and whose main product remains copra. On plots of family farms in Asia and Oceania – as in India, Sri Lanka, the Philippines and island areas –, coconuts are seldom grown alone, but rather in agroforestry plots, in association with many other food crops such as bananas, roots and tubers, spices (pepper, vanilla) or sometimes cocoa. India, the Philippines and Indonesia are the three major producers of coconut. In each of these countries, the supply chain is based on a network of cooperatives or groups of well-organized producers. They supply not only the long-established and powerful industrial base but also small artisanal and innovative processing units offering products with very high added value: drinks based on coconut water, virgin oil, products derived from the shell (xylose), coconut sugar, etc. Since the early 2000s, the sector is undergoing a comprehensive revival. In some countries such as Brazil, industrial plantations of thousands of hectares of coconut trees are being set up to supply the international "health" beverage market. The demand for vegetable biofuels has also upset the oilseed market since 2006 and has created increases and instability in the price of copra oil, thus affecting small producers. Therefore, coconut smallholders, long dependent on the sluggish copra sector, today find themselves in a position to obtain improved valuations for their products – as long as they modernize their practices and are supported in this change of orientation.

8.2.2 Productions for which Non-family Forms of Agriculture Predominate

8.2.2.1 Rubber

With industrial plantations being responsible for 24 % of its world cultivation, rubber finds itself in an intermediate position among perennial tropical crops. At one end is oil palm cultivation, where capitalist agriculture dominates, and at the other are coffee, cocoa and coconut cultivation which are dominated by forms of family farming. Rubber's intermediate position stems from the historical structuring of the market, with the major tire manufacturers (Firestone and Michelin) initially integrating the sector vertically by establishing their own plantations. The flexibility of family farming, however, has long allowed it to respond to the dramatic growth in demand for standard rubber, whether in the form of artisanal latex processed into smoked sheets or of cup lump rubber. Three quarters of the surface areas under cultivation are found in smallholder plantations, with or without combination of food crops in the juvenile phase (tapping of rubber can take place only 5–7 years after planting), or in the form of rubber agroforests that are relatively common in Southeast Asia (Chap. 6). For example, Thailand with 12 M ha of which 8 M are in production, is the world's biggest rubber producer, accounting for 33 % of global production in 2013. Nearly all of its production (99.6 %) is by smallholders.

8.2.2.2 Oil Palm

The main two producing countries in the world, Indonesia and Malaysia, produce 46 million tonnes of crude palm oil from a world total of 54 million tonnes. To meet the explosion in demand for vegetable oil, production surface areas have been increased by 54 % in 7 years, from about 9.2 million hectares in 2005 to 14.2 million in 2012, of which 10.9 million hectares are in Indonesia and Malaysia (Oil World 2013). The dominant role of agro-industries in this sector arises from the industrial rationale that has dominated the development of the market, driven by global demand for fats. The cost of an efficient oil mill requires a certain level of minimum supply for a profitable investment. The cultivation system has therefore taken the form of plantation-factory combination in these two countries. Despite these global trends, there also exist smallholder plantations, small independent growers or those under contract with an oil mill, and very traditional forms of production within diversified production systems (Box 8.1) in the three production areas of Southeast Asia, Central and West Africa, and Latin America.

Box 8.1. "Natural" palm groves: a particularity of family farming in West Africa.

Sylvain Rafflegeau

The most commonly grown oil palm in the world, *Elaeis guineensis*, is native to the Gulf of Guinea. The existence of wild palm groves, regionally called "natural" palm groves, is at the origin of a traditional family farming system: use of traditional methods of oil extraction – treading and washing with water to extract the red oil from the pulp and heating of the kernel to extract kernel oil – and the use of other parts of the plant – palm wine and alcohol from sap, decoration products, mats and brooms from the stalks (Cheyns and Rafflegeau, 2005).

These palm groves coexist with monospecific agro-industrial and smallholder plantations, sometimes associated with food crop production during the juvenile phase. In Benin, in the region of Pobe, for example, many family farms select the best palm trees for selling bunches or for oil. They use the other trees, in plots essentially used for growing food, mainly for the production of palm wine. This extensive form of family farming now accounts for the majority of the oil palms in Nigeria, where 80% of the production comes from smallholders (Vermeulen and Goad, 2006). Nigeria has not really increased its production in the last 50 years, dropping from the rank of the world's leading producer to being fifth now – trailing Indonesia, Malaysia, Thailand and Colombia. While it is clear that this type of family farming system cannot satisfy the world's booming demand, artisanal red oil it produces does meet local demand for use in cooking specific dishes of the region (Cheyns and Rafflegeau, 2005). This traditional red oil production is not exportable beyond Africa since, with a few exceptions, the rest of the world consumes only refined palm oil, after having removed its carotenes, vitamins, etc.

8.2.2.3 Bananas

Like many tropical food and horticultural crops, bananas are an important part of the daily diet. In the tropics across the world, the vast majority of banana is grown on small family farms or family business farms (between 0.3 and 10 ha) whether on a monospecific plot or in combination with other crops. Banana fruit production – mainly for export to temperate countries or consumption in cities in producing countries – is, however, dominated by the industrial plantation sector, which has existed for a very long time but has shown its ability to grow rapidly to meet global demand. Economies of scale are necessary to satisfy growing volume and quality requirements. The major capitalist plantations date back many years and are found in Latin America, the Philippines, West Africa (Côte d'Ivoire and Ghana) and Central Africa (Cameroon). They have grown steadily and are complemented by the gradual emergence of specialized medium-sized family business farms (50–100 ha) (Box 8.2).

> **Box 8.2. Small African, Haitian and Malagasy family orchards supply international markets with products of recognized quality.**
>
> Magalie Lesueur-Jannoyer, Michel Jahiel, Jean-Yves Rey, Éric Malézieux
>
> Tropical family orchards exhibit a wide diversity in surface area size and in the number of crops grown. While they are largely oriented towards domestic sectors, they still manage to contribute significantly to international markets for certain products, particularly in Europe and North America: examples are the mango orchards in Senegal (Grechi *et al.*, 2014) or Haiti and lychee and clove orchards in Madagascar. In fact, Madagascar provides 65% of European imports of lychees, with 20,000 tonnes exported annually (*FruiTrop*, 2010). In 2012, Senegal became the tenth largest supplier of mangoes to the European market and became the second largest African supplier (after Côte d'Ivoire) (source: Eurostat).
>
> These orchards allow farmers to manage local resources in a sustainable and adaptive manner, while still inserting themselves into international standardized supply chains. They play an important role in food security both through their contribution in vitamin and caloric intake as well as through their contribution to incomes and employment. Indeed, these orchards are used to adjust household incomes during critical periods or to meet certain requirements for goods or services, especially education and health. The quality of fruits and products from these orchards is recognized on international markets and much sought after. It allows organizations with adequate supply chains to meet the expectations of specialty markets and certification requirements (biological, GlobalGAP, etc.) (Subervie and Vagneron, 2013).

8.3 A Need for Improved Statistical Knowledge

Family farming systems occupy a dominant position in agricultural product markets. An improved statistical knowledge of their economic weightage in the various sectors is an important issue that should warrant coordinated action among research and development actors.

Family farming's share of the overall production is massive, but it also faces strong competition from most other forms of agriculture, which sometimes occupy a dominant position for certain products due to specific models of development or special technical conditions. This competition is also reflected in access to natural resources, especially land and water. Here too, improved information would lead to a better understanding of this competition and its consequences.

Increasing vertical integration of certain sectors and the growing interest expressed by large global processing firms in production – in order to ensure their supplies and consolidate their market shares – are reflected in the development of different contractual forms that often place family farmers in asymmetrical bargaining positions. In order to defend their interests and rights, either with respect to large firms or within inter-professional organizations, and to benefit from economies of scale in production and access to technical support, family farmers have to strengthen their organizations. These collective structures will also allow them to participate in public debates at various scales, ranging from the local to the international, as discussed in Chap. 9.

Chapter 9
Contributing to Innovation, Policies and Local Democracy Through Collective Action

Pierre-Marie Bosc, Marc Piraux, and Michel Dulcire

This chapter[1] deals specifically with the many forms of collective action that family farmers undertake. It focuses, on the one hand, on the diversity of functions performed through collective action and, on the other, on the deeply evolutionary nature of organizational forms developed – even though some are part of long-term trends – and concentrates on three specific issues: innovation, insertion of agricultural products into markets, and the formulation and orientation of public policy. These multiple commitments have contributed in a more general way to the consolidation of local democracy where organizations of family farmers play an important role. Farmers have been coming together in organizations for a very long time in Western European countries and in those which experienced high levels of European emigration in the nineteenth and twentieth centuries. The movement is slower and more recent in developing countries; it follows on from the independence of former colonies and the democratization movements of the 1990s. And yet, collective dynamics in these countries often play today a central role in local and national political arenas.

---

[1] Part of this chapter relies on earlier collective works cited in the references (Bosc et al. 2001, 2004). The authors thank Eric Sabourin for his written contribution and Denis Pesche for his critical and constructive inputs.

P.-M. Bosc (✉) • M. Dulcire
ES, CIRAD, Montpellier, France
e-mail: pierre-marie.bosc@cirad.fr; michel.dulcire@cirad.fr

M. Piraux
ES, CIRAD, Belem, Brésil
e-mail: marc.piraux@cirad.fr

9.1 Definition and General Positioning

The organization can be seen as an instrument of collective action that family farmers use to reach the minimum level of cooperation necessary in order to achieve shared goals for purposes of internal and external coordination and to strengthen their negotiating capacities with other stakeholders. In this chapter, we will use the term "peasant and rural organization" (PRO) in order to emphasize the organization's local anchoring through the term "peasant" (in the sense of "which is tied to a region or a community") as well as enlarge the perspective to beyond just the agricultural sector through the use of the adjective "rural."

The organization has always been a historical constant in agriculture. It finds its origin in the need to respond collectively to the drudgery of work or to the unequal distribution of labor demand on family farms during the growing season. The organization is also necessary to make certain major investments for land improvements (terraces, polders, irrigation) for the purpose of obtaining more favorable agricultural and livestock production conditions. However, collective action pertains to many more objectives. They range from access to natural resources as common property to either tangible (equipment, infrastructure) or intangible (information, innovation, markets) common goods to the production of public goods (umbrella organizations, networks, product quality standards, etc.). Collective action's ambit also includes the formulation of rules and regulations affecting the conditions of production and exchange.

Collective action often intrudes into the public sphere because, through some of their functions, farmer organizations produce or help to produce common or public goods. Collective action may also sometimes have the goal of influencing the government to obtain decisions favorable to individuals or family groups who are part of said action.

Nevertheless, the forms that organizations take depend heavily on economic and institutional environments in which family farming has to operate. This translates into several nested levels of organization, from the family farm and the local level (local-development organizations) to the national level (national representations such as ANOPACI in Côte d'Ivoire) to the continental level (such as Asian Farmers Association) to the international level (Via Campesina). We broadly distinguish producer organizations (family or family-business) by products within sectors (such as Fedepalma in Colombia for palm oil), organizations for artisanal processing of family farming products (such as Fedepanela in Colombia for processing of artisanal sugarcane), and territorial organizations (local and territorial development associations).[2]

We present an overview of the modalities of collective action developed by family farmers, and illustrate the types of functions – economic, social, and environmental – that such action expects to develop. We also examine

---

[2] http://www.erails.net/CI/anopaci/anopaci/; http://asianfarmers.org; http://viacampesina.org/en/; http://web.fedepalma.org/; http://www.fedepanela.org.co/

organizations and collective action as individual and collective learning processes which facilitate participation in decision making and the establishment – or even the imposition – of new regulatory frameworks at various geographical levels and within sectors (ranging from local markets to international markets). Conversely, we will show how these normative and institutional frameworks "condition" the dynamics related to PROs and family farming.

9.1.1 Putting Forms of Collective Action in Perspective

Family farming has historically developed by relying also on cooperation between family farms mainly because of occasional labor shortages for particular farming operations.

Mutual assistance and exchanges of labor (sometimes accompanied by agricultural equipment) at the farm level often largely relied – and continue to rely – on non-commercial and non-monetary mechanisms of reciprocity. They are governed by family or proximity ties between similar family farms.

In many cases, natural resources have been improved through work organized collectively at the level of communities based on kinship and proximity. These efforts have various objectives, such as controlling the flow of water and preventing erosion, improving soil fertility through inputs of organic matter, limiting flood risks, colonizing new lands in the sea or in brackish areas.

Organized on relatively localized areas and following the rules defined by the communities, these systems have often functioned for centuries and have, at certain times and in certain areas, encompassed considerable amounts of space. Mention can be made of polders in the Netherlands; the arrangement of paddy fields in the Casamance region of Senegal; the rice terraces in Asia, in mountain areas in France and in the Andean countries; the irrigation systems of the *huertas* in Valencia in Spain; etc. The outcome of these collective actions was not only a natural environment rendered more productive (controlling of erosion, management of the water resource, improved soil fertility, etc.) but also the establishment of collective rules for the preservation and maintenance of the infrastructure developed (Ostrom 1990). These rules and mechanisms have not been designed in isolation but take into account the inherent uncertainty, risks of various kinds and opportunistic behaviors through mechanisms of collective sanctions and mobilization in order to respond to changes in the environment natural and to accommodate human behavior that is not "naturally inclined towards cooperation."[3]

However, these modalities of collective action run up against their limitations when family farming attempts to control and regulate its relationships with the

---

[3] This is a universal trait that has been observed across societies and their histories, and is aptly underlined by the subtitle of the work of Crozier and Friedberg (1977): "Constraints to collective action."

society that "encompasses" it, in which it has historically occupied a dominated position (Wolf 1966; Shanin 1986). "New" organizations have emerged to tackle issues with a scope that exceeds the farm's level and for which the capacity for action and possible responses based on groups founded solely on reciprocity are no longer effective. Such organizations have historically been created as part of the deepening of commercial relationships between family farmers and actors of agricultural trade.

Collective action is thus shaped by legal, institutional and cultural conditions and the policies in force. Legislation is not essential to the emergence of these organizations, but the political climate and freedoms of expression and association are factors favoring their development. We can refer to the French laws of 1880 on the right to associate, political liberalization in Africa in the 1990s, and the transition to democracy after dictatorships like in Brazil, for example. However, some organizations have emerged under authoritarian regimes and have then survived and even consolidated themselves subsequent to democratic transitions.

Conversely, when organizations are brought under the ambit of coercive laws of authoritarian regimes or when cooperative movements are hijacked by political and administrative machinery, quite the opposite effect takes place as far as family farming is concerned. Even after the regime is replaced and a more propitious environment prevails, there is continued aversion towards – and rejection of even the idea of –cooperation or collective action. And these strong feelings towards cooperative movements are similar regardless of the type of political regime. In these situations, the idea of associations or of cooperatives brings back memories of a past or of practices that family farmers would rather not see revived (Box 9.1).

Box 9.1. Cooperatives in independent Africa.

Pierre-Marie Bosc

Cooperatives have, on the whole, remained "foreign" to African peasantry, to the extent that one could even talk of "cooperatives without cooperators" (Gentil 1986). Imported forms of organization, inspired directly by cooperative organizations created in the industrialized countries at the turn of the twentieth century, found themselves imposed against the realities whose propensity for "cooperation" had been grossly overestimated on the basis of faulty analyses enthusing over collective values of African societies (Augé 1973). Rarely involved at the beginning, and generally excluded from decision-making and control of the cooperative mechanism, the producers never found their place in a highly controlled movement. In fact, these cooperatives set up by officials and rural public figures have also been used to serve the interests of political parties as a source of funding or as propaganda tools. We find these attitudes in some Latin American countries, such as Peru, with the "dismantling" of cooperatives (Bonilla 2008) and in the former planned economies which are currently in transition, as in Vietnam (Dao The Anh et al. 2008), China and the countries of Central and Eastern Europe. In all these situations, the term "cooperative" acts as a disincentive to collective action.

9.1.2 Functions Assumed by the Peasant and Rural Organizations (PROs)

The PROs can fulfill five major types of functions whose importance vary according to particular contexts and issues. We are thus led to distinguish economic functions; social functions; representation and advocacy functions; capacity building and access to information functions; and finally coordination functions. The latter two functions are transversal in nature and necessary for the proper fulfillment of the first three.

- The economic functions include procurement, production, processing and marketing of inputs and products, as well as services useful to farmers in improving the management of production factors, such as water for irrigation, access to land, labor or collective management of agricultural equipment. These economic functions pertain to the management of natural resources, access to production factors, including access to credit, and the processing and marketing of products.
- Social functions concern the domestic dimensions of the farm. Depending on context, they may be more or less important or have greater or lesser scope. They pertain to, amongst others, social protection, education, access to basic services such as health and sanitation, forms of mutual assistance and sharing, and culture.
- The functions of representation and defense of vested interests are exercised at different scales according to the forms of governance in force with the perspective of participating in the "crafting" of policies (Mercoiret 2006).
- The functions of capacity building include the sharing of information, communication – both internally and with other actors –, and the development of training (managers, executives, members).
- Finally, coordination is a key function, because PROs are able to establish organic links between the level of territories where family farms operate and other levels – regional and national – where decisions are taken that affect the incentivizing of the environment of the production units.

9.1.3 Producing and Managing Goods and Services

Organizations produce various types of economic goods and services, ranging from private goods to collective and public goods. The collective goods – equipment for processing, storage and mechanized farming, joint purchases of inputs, access to subsidized credit, etc. – contribute to the production of private goods at the individual farm level. They lead, for example, to an increase in production and income because of access to technology, inputs and technical advice. The goods produced can also be collective, and tangible, such as buildings or collective equipment for storage or processing of products, or intangible, such as access to information.

When there is a failure of the State or local authorities in the provision of public goods (health, education, road maintenance, etc.), organizations seek to mitigate the situation by partially picking up the slack. This is the case, for example, of basic health services, literacy or training. This corresponds to the transposition at the collective level of the family farm's domestic and social dimension. These situations are dynamic and correspond each time to specific configurations that are not expected to persist if the environment improves (Box 9.2).

Box 9.2. The contribution of the National Federation of Coffee Growers of Colombia in providing public goods.

Pierre-Marie Bosc (Based on Bentley and Baker, 2000)

In 1927, the National Federation of Coffee Growers of Colombia and the Colombian government agreed to the funding of a portion of public investment through a tax on coffee exports to be used to benefit communities in coffee growing regions.

Thus, the Federation uses a portion of the funds to build roads (12,882 km of roads built and 50,672 km improved), clinics, water supply systems, schools (16,923), and to recruits teachers. The proportion of public investment financed by the Federation in the "central belt" area of production is significant, and even isolated family farms are connected to the electricity grid.

Thus, to support these organizations, it appears counter-productive to want – from the outside – to limit their scope to strictly sectoral issues. Organizations do not systematically respond to social or domestic requirements, but when they do, it is important to identify the reasons why. PROs intervening in the management of natural resources which are considered common or collective property must take into account existing arrangements, which pertain in many situations to customary authorities and local mechanisms regulating rights and obligations. They correspond to social networks that govern the relationships between families and their farms within a locality. Finally, the PROs' participation on a larger scale also helps to produce public goods through their many contributions to policy formulation at different levels, ranging from local policies to international debates.

9.1.4 Between Multipurpose and Specialized Organizations: An Assessment of the State of the Environment at Different Scales

PROs therefore oscillate between being multipurpose and specialized, between sectoral anchoring and an ambition to meet multisectoral social demands, between local action and a desire to influence decision levels hierarchically and politically beyond the local context. Not reducible to mere family and social solidarity, they are not completely detached from it either, and their capacity of action depends on the operational and professional alliances they manage to forge outside of local social networks. Furthermore, far from being an anomaly, the multisectoral and multifunctional nature of these organizations is rooted in their members' complex livelihood strategies. There exists therefore no standard on what could or should such an organization do in such a condition. But there does exist a relationship between the complexity of the economic and institutional environments in which organizations operate and their own level of specialization or diversification of activities. Anchored to the local level, they incorporate different territorial levels either through sectoral activities or through representative mechanisms (federations). And this is how they gradually build links with economic, political and institutional actors at these various levels.

9.1.5 Family Character as Collective Identity

According to the agrarian histories, different types of agricultural organizations are often created for different types of agriculture, whether entrepreneurial farms – sometimes in association with international financial operators –, or small or large family farms, either specialized or diversified. Some organizations claim the family nature of production as part of their identity, such as the NFU (National Farmers Union)[4] in the United States, whose slogan is "United to grow family agriculture." In addition to its sector-specific actions, the NFU had been active, for example, in the promotion and implementation of a rural health system in the 1970s. In other contexts, the family farm represents the social and economic model championed by the PROs. This is particularly the case for the Network of Peasant and Producers Organizations in West Africa (Roppa). Here too, the family model is a strong assertion of identity, going beyond the agricultural sector since it pertains to the family as pluriactive structure. It is the desire to highlight the family character of agriculture that led the World Rural Forum[5] to propose – successfully – to the United Nations to declare 2014 the International Year of Family Farming. It is a matter not only of defending but also of promoting forms of production that are specific and distinct from entrepreneurial forms.

9.2 Organizations, Family Farming and Innovation Processes

Schumpeter's (1935) work on economic development dates back to the first half of the twentieth century. He clearly defines in it the different types of innovation: new goods, a new production method, the opening up of a new market or outlet, the

---

[4] http://www.nfu.org/
[5] http://www.ruralforum.net/

mastery over a new source of raw material and, finally, establishment of a new organization of production. Furthermore, the OECD also explicitly recognizes that innovation exists also – beyond merely technical practices – "in matters of organization" (OECD 2005).

The "detour" via the collective dimension, via common construction is therefore understood as an extension or expansion of individual action (Commons 1934) centered on the family farm, and which allows a more favorable insertion of individuals into markets with changing requirements and for which access costs are such that collective action (product adaptation, processing, bargaining power) is justified.

As far as the wide diversity of family farming models is concerned, collective action in matters pertaining to technical innovation is justified because of the very high degree of standardization of the technical models proposed, or even imposed. Agricultural research has long ignored the diversity of family farms, whose mobilizable capital varies greatly from one family unit to the next (Jamin 1994). Collective production networks oriented towards innovation (Faysse et al. 2012) are formed gradually according to their own needs and conditions, through exchanges and dialogues. Thus, family farmers engage on technical topics in order to adapt standard proposals to their specific conditions (Chiffoleau 2005) through a "technical dialogue" (Layadi et al. 2011), but which soon expands to encompass economic and organizational conditions.

The general framework of collective action is that of modalities of exchanges and interaction between farmers, *de campesino a campesino* – from farmer to farmer – (Hocdé and Miranda 2000), in order to renew and maintain the support mechanisms negotiated with other actors. These peer exchange dynamics produce structuring effects and train leaders through collective learning mechanisms. However, it would be unrealistic to believe that exchanges without external inputs[6] (technical, economic or institutional) will be sufficient to meet the challenges of the technical adaptation of production systems. The emergence of new skills to meet current challenges requires the help of technicians (agronomists, animal scientists, environmentalists, etc.) who are in a position to combine approaches of farm diversity and specialized technical skills within a holistic vision. This refers to the general issue of the conditions of taking into account the diversity not only within the family farming category but also of collaborative arrangements between various types of farms.

Thus family farmer organizations build within them a space for exchange, characterized by three features: consolidating and sharing their know-how and innovations developed by the farmers themselves; establishing specific support mechanisms, often with external funding; and participating in the definition and monitoring of activities of research and development organizations.

---

[6] The distinction between what is internal and what is external is often difficult to pin down. Not only do we consider organized interactions with actors of formal support mechanisms, we also take into account the intrinsic learning capacities of individuals and communities through their own informal networking (opportunities for geographical mobility and openings to new ideas, new worlds, new ways of living, etc.).

Innovation also pertains to the modes of relationships with markets. Historically, family farmers in the North organized themselves for better negotiating ability, which gave birth to the powerful cooperative movement in Western Europe and in "white settlement" countries (United States, New Zealand, etc.). Today, new constraints in the form of commercial standards are forcing family farmers in the South to come together to improve their effective bargaining power and decision-making abilities in view of the conditions imposed (Ton 2012). Organizing therefore remains more vital than ever, even though collective action and its modalities have to adapt to market and end-consumer demands. This can pertain today as much to the manner of production (organic, agroecological, ethical, etc.) as to new business models (direct sales, short supply chains, etc.), with or without value added to the raw product. There is always a search for complementarity between the collective dimension and the family dimension. The former allows the funding and maintenance of equipment of significant size, corresponding to the requirements of one or more particular standards, which would normally be outside the scope of an individual investment (artisanal canning workshop, for example). Family dynamics can thus incorporate a processing business, supplementing family employment through the use of this collective equipment. Consequently, a goal of many organizations today is to increase their share of the wealth produced in their sector(s). Organizing together gives family farmers a chance not only to capture markets but also a greater number of steps in the value chains (Box 9.3).

Box 9.3. Fall and rise of cooperatives: the revival of Pima cotton in Peru.

Michel Dulcire

The cradle of cotton cultivation in Latin America is to be found in Peru, where pre-Columbian cultures were already using it for weaving cloth. The Pima variety consists of exceptional elements in Peru: fineness, strength and fiber length (40–45 mm). The demand for this type of quality cotton accounts for 2 % of world production, and it is therefore a niche market with relatively attractive prices. In the past, the "white gold of Piura" was even listed separately on the Peruvian stock exchange. In the 1970s, land reforms gave rise to production cooperatives. But since the 1980s, they have found themselves economically and institutionally constricted due to a process of individual land reallocation, and consequently providers of State services destined for small family farmers gradually disappeared. Furthermore, the reduction of import tariff had an adverse effect on the market. The Peruvian textile industry stopped paying cotton farmers a fair price and took advantage of low import prices to buy cotton from the United States of the same variety, but of lower quality and whose cultivation was subsidized. Moreover, as a peasant leader said, "They make clothes with imported ordinary cotton and label it Peruvian," thus exporting as if they were made from Peruvian Pima cotton. No wonder, as a result, 60,000 ha of Pima of the 1960–1980 period dropped to only 1,500 ha in 2010.

The multiple-services cooperative Tallán-Chusis (COSTACH Ltd.) is an association of Pima cotton producers in the region of Piura in northern Peru. Currently, its membership includes 5,600 small family cotton farmers, each with 3–5 ha of land. It has helped in the revival of Pima cotton. In 2011, the association took charge of collecting the cotton and processing it for export by contracting a ginning factory to produce fiber and vegetable oil. COSTACH's activities and of its partners helped to "paint the fields white," covering 12,000 ha for the 2012 season. The organization is now recognized by leading entities such as the Ministry of Agriculture, investment banks and municipalities, and is going to set up its own cotton gin. It has also begun the process of creating an appellation of origin for Peruvian Pima cotton to curb unequal competition with low quality imported textiles.

"We are in the process of being reborn," affirms COSTACH. However, the organization must find the strength and resources to confront private companies which currently have a stranglehold over the economics of Peruvian cotton production. For COSTACH, this means "building relationships," both domestically and internationally, to improve the activities of farmers for whom it is regaining part of the value addition in the Pima cotton sector.

The diversity is thus immense, both in terms of socio-economic conditions of families as of their modalities of integration into highly segmented markets whose conditions of access constitute real entry barriers. Given these observations, mechanisms for providing support to family farmers must guard against offering standard or one-size-fits-all models. They should rather provide support that is flexible enough to adapt to contexts and states of organizations (Bosc et al. 2002). Discovering, communicating, sharing, working in groups to set one or more common objectives, managing risks and uncertainties, acquiring analytical skills and critical thinking, and "freeing" creative capacities to innovate together and share with others are all skills which need to be acquired by family farmer organizations. Such organizational learning (Argyris and Schön 1978) translates into improved abilities to respond individually *and* collectively to changes and continuous transformations of environments. More broadly, this is a process of gradual empowerment (Rondot and Collion 2001) or even of emancipation (Dulcire 2012). Finally, recognizing that learning is a necessity is also to accept that membership in an organization requires a commitment to more activities and responsibilities as compared to the usual relationship with other actors, in particular for tasks of collective monitoring and control.

It is important therefore to invest in order to support family farmers in organizing themselves especially so that they can respond better to changes in local, national and international markets. There exists no ideal form of organization to meet these requirements and a pragmatic posture, freed from standard models, is required. These observations lead us to emphasize two important challenges. The first is of identifying the modes of organization and coordination best suited to different family farming models so that they can better assert their rights, according to their location, the sector concerned, and the technical, economic and social environments. The second is a matter of determining the modalities of building alliances in order to assert choices, an endeavor made difficult in contexts where other types of agriculture exist and sometimes are in competition for accessing resources.

9.3 Contribution to the Construction of Normative and Institutional Frameworks

The government has the responsibility of defining public policies for rural areas. However, these policies are most effective when based on compromises negotiated with the various stakeholders. Such a process of co-construction reassures these stakeholders that the policies will indeed be more suited to the variety of contexts to which they apply. Policy formulation thus requires the mobilization of, and inputs from, a large number of actors defending different interests, with PROs playing a particularly important role because they lead to a collective – and therefore stronger – representation of issues and proposals for rural producers, who still largely predominate in rural areas. The contribution of family agriculture to the formulation of normative frameworks[7] also helps consolidate the PRO's local experiences and to strengthen them.

But the participation and co-construction cannot be decreed. It is a matter most often instead of long-term social and institutional innovation processes whose implementation facilitates, through the confrontation of multiple intra- and inter-group interests, the gradual incorporation of family farmers' concerns and issues into the political agenda. These changes also depend on the broader political and social contexts, including the democratic environment which determines the dynamics of the PROs' involvements and their representations. Participation requires, above all, the joint formulation of a strategy to defend causes which does not correspond to the strict sum of the members' interests (Pesche 2007). Finally, it requires that members and leaders of the organization be capable of analyzing the place and role of agriculture in society in order to better contextualize their proposals and thus increase their room for maneuver.

In fact, the various contributions of family farming mentioned in this book's second part (contributions to agrarian systems, to territorial dynamics, to global markets) are highly dependent on normative and legislative government frameworks that structure the institutional environments of family farming systems and, more generally, determine the expression of democratic representation, decentralization and the level of government involvement in some advisory support functions.

In all the countries of the South, the involvement of family farmers and their organizations in the negotiation or implementation of public rural development policies is becoming increasingly common. This is due to the more or less interrelated convergence of several factors:

– the existence of a political environment guaranteeing individual and collective freedoms of expression and organization at the sectoral level and more generally to all organizations of civil society;
– the emergence and proliferation of producers' organizations capable of defending their interests (at regional and national federative levels);
– the emergence of a coalition of actors, structured around the PROs in association with NGOs, universities, technical services, etc., whose alliances allow not only the adoption of influencing strategies or a diversification of the interests being defended (ecological, social, cultural, etc.) but also bestow some financial autonomy on PRO movements;
– changes in the level of state intervention for some functions, especially economic ones and for support to the rural sector (technical assistance, training,

---

[7] The normative framework is considered, on the one hand, as a set of general principles, legally formalized standards, laws and mechanisms that prescribe individual and organizational actions and, on the other, in a more cognitive approach, as the production of normative structures by actors in a position to give meaning to their actions (Lascoumes 2006).

commercialization and regulation of sectors, credit) subsequent to the implementation of structural adjustment plans;
- the processes of decentralization of State action, based on the extension of democratic principles to the local level.

The participation of PROs is characterized by institutional modes of representation of the interests of different rural populations, according to their particular level of articulation with the government. It is therefore highly dependent on political regimes (Dugué et al. 2012).

This participation also pertains to different types of national policies which may be sectoral, especially with regards to the level of support extended to production sectors; transversal, such as land policies, or those related to agricultural services; or more general, such as the drafting of laws concerning agriculture. As has already been emphasized, PROs prefer to work at the national scale so that they can influence agricultural and rural policies (Pesche 2007). More recently, however, networks of PROs from several countries are being formed to perform the same type of functions but at the supranational level. The Via Campesina movement is a good example of this trend. Such networks work at two levels: the sub-regional level, where agricultural policies are becoming increasingly more important given the transformations of agriculture and rural life in the countries of the South (WAEMU, ECOWAS, Mercosur[8]), and the international level, which involves trade negotiations or agreements on the environment, having repercussions on both international trade and local agriculture. Moreover, at this latter level, there is often a reconfiguration of alliances, often temporary, between a country's PROs and its government in order to defend common causes at the international level (cotton in West Africa, international trade).

For PROs thus involved, the multiplicity of levels of governance (local, national, global) manifests itself in a need to participate in numerous decision-making and opinion-forming forums. This fragmentation of public decision making also complicates the work and – since there is only so much that a PRO's managers and officials can do – makes new alliances necessary. This also renders the task of effectively assessing and monitoring the effects of policies extremely difficult (Dugué et al. 2012).

9.3.1 In Latin America, and in Particular in Brazil

The twentieth century in Latin America has been marked by strong contrasts between authoritarian regimes (military dictatorships in Argentina, Brazil, Chile) and populist regimes (Mexico with the Institutional Revolutionary Party, the Brazil of Vargas or the Argentina of Perón) who, however, shared the unfortunate trait of

---

[8] WAEMU: West African Economic and Monetary Union; ECOWAS: Economic Community of West African States; Mercosur: Southern Common Market.

hijacking democracy. The 1980s thus saw the region engaged in democratic processes that opened up spaces for dialogue that was openly welcomed by social organizations. The Washington Consensus also allowed the nascent democratic openness in these countries to be combined with the adoption of liberal policies (Ducatenzeiler and Itzcovitz 2011) under the influence of key multilateral agencies.

Thus the international injunctions of the 1990s and 2000s resulting from this movement of liberalization encouraged countries to restructure their public policies using new modalities and instruments based on participatory democracy and the territorialization of public policy. In Brazil, the transition to democracy opened up opportunities for the formulation of new policies, with the participation of civil society organizations born in the associative movement of resistance and struggle against the dictatorship.

This process of involving the target population in defining – or even implementing –public policies has been the subject of positivist discourse and of systematic support of international institutions and NGOs, regardless of the form and intensity, or even the legitimacy, of the consultation and participation modalities used. So there exists quite a body of work supporting the validity of these approaches (democracy, governance, social capital, empowerment, participation in development, etc.) as well as on the methods of supporting and encouraging political participation. The assessment of these policies requires specific approaches and methods which have been explored in publications[9] or books dealing with these issues using cases from the North and the South (Neveu 2007). In Latin America, two main processes, now associated with the periods in which they took place, can be identified regarding the drafting of public policies and the promotion of this logic of participation:

- a first phase of popular participation from the late 1980s to the late 1990s. The first participatory rural-development support programs were launched by the World Bank with its Community-Driven Development (CDD) approaches. This process has been linked to the creation of community associations and organizations of small rural producers across the sub-continent;
- a second phase of territorial development starting in the 2000s. This process, in addition to the emergence and "fashioning" of the "territorial" dimension, concerned three mechanisms: the decentralization of agricultural development organizations, the emergence of local-development initiatives, and the diversification and proliferation of organizations representing rural actors.

Within the framework of the 2004 federal policy to support rural territories through the National Program for Territorial Development (Pronat or PDSTR), CIRAD analyzed these new orientations by assessing their strengths and drawbacks in promoting family farming. Thus, Massardier et al. (2012) show that participatory

---

[9] In particular, see the journal *Participations*, http://www.cairn.info/revue-participations-2011-1.htm, retrieved 19 February 2014.

democracy has opened up new opportunities for actors previously excluded from public policies, most notably for the representatives of landless or family farmers. These new actors have demonstrated significant progress in their ability to learn and to participate in the drafting of public policies. In this process, local leaders and traditional representatives of farming communities have also benefited from these new forms of participation. They have exhibited a professional approach to negotiations on public policy projects for rural development. These participatory processes have strengthened their ability to negotiate with technicians of public services, enabling them to become mediators essential to the proper functioning of these public policies. The role played by these actors is crucial in the implementation of policy guidelines, sometimes to the detriment of real popular participation, thus undermining the principles of participatory democracy professed by territorial public policies.

The representation is also highly dependent on local conditions, especially on local trajectories, which determine the presence of state services and which have helped – or hindered – the consolidation of the movement towards unionization in alliance with NGOs (Piraux et al. 2013). Moreover, territorialization is hampered, especially by the administrative structure, which remains dependent on the federal system (Bonnal and Kato 2009). Bureaucratic procedures for disbursements and for project implementation are too restrictive. Despite these new policy directions, farmers are still dependent on expert systems of technicians and agronomists. These latter have organized in socio-professionals networks and retain the authority to approve or reject projects. PRO reactions have varied widely depending on local contexts, and especially according to their strengths and powers of influence, ranging from a passive attitude to a few confrontational situations.

In addition to – and separate from – these dynamics which are strongly promoted by the State, civil society in Brazil has directly created and implemented public policies, such as those related to the promotion of agroecology, particularly in semi-arid zones. This example illustrates the ability of the PROs, when they have the expertise and a sufficient level of organization, to enter into dialogue and coordination mechanisms with the State for programs specific to, and suitable for, family farming conditions. This experience emphasizes the importance of linkages between and across levels and the need for the institutionalization of appropriate mechanisms. It also stresses the need for concomitant development of government standards in matters of shared public action.

9.3.2 In Africa

In Africa, despite the political liberalization of the 1990s, very contrasting political regimes in terms of the democratization of public life today still strongly influence the mechanisms of representation and advocacy on behalf of family farmers.

Representing the interests of rural populations in Senegal and Mali, for example, is easier than of those in Chad or Togo, where political regimes are less open to dialogue with civil society (Dugué et al. 2012). In general, according to these authors, countries with a francophone tradition favor a form of representation and advocacy through a dialogue and confrontation with administrations and political authorities. Countries with an Anglo-Saxon background, on the other hand, seem to promote less exclusive and more open modalities and channels to defend interests, with a more important role accorded to parliamentarians and NGOs.

In any case, the level of influence on policy that producer organizations are able to bring to bear is often linked to their ability to forge alliances within the rural world (for example, the National Council for Rural Coordinated Action and Cooperation, CNCR, in Senegal), to become part of formal decision-making mechanisms (for example, the National Union of Burkinabe Cotton Producers, UNPCB), or to partner with other actors of civil society or of the private commercial sector. In this, the ability of PROs to use the resources of the processes of democratization often arms them with a decisive advantage (role of parliamentarians and the media to support the cause of rural populations). This demonstrates the importance of the national level – which is, as we have already noted, the preferred level of influence of PROs – to influence the decisions and the construction of public rural development policies (Dugué et al. 2012).

The recent involvement of PRO networks at sub-regional and international levels has led to undeniable successes in the drafting of public policies. For this, the PROs have had to adapt their structure to the scale of their interlocutors, which is indeed what African PROs did (Box 9.4).

Box 9.4. Establishment of supranational networks of producer organizations in Africa.

Dugué MJ, Pesche D and Le Coq J.-F

In the 1990s, several regional meetings allowed PROs from different countries of West Africa to establish contact each other. In 2000, the prospect of establishing a WAEMU-wide agricultural policy led to the creation of a network of West African peasant and producer organizations and farmers, known by its French acronym ROPPA. Since then, this network has been involved in negotiations to influence agricultural policies at various levels: WAEMU in 2001–2002, ECOWAS in 2005, and, in 2007, the Economic Partnership Agreements (EPAs) between the European Union and West African countries.

During this same period, several other regional networks of PROs were formed, often with the same goal of representing producers in the processes of defining regional agricultural policies. This is particularly the case of the Eastern Africa Farmers Federation (EAFF, established in 2001) and the Regional Platform of Peasant Organizations of Central Africa (Propac, in 2004). Furthermore, building on the momentum of the success at Cancun, specialized producers also organized at the continental level with the creation, in 2005, of the Association of African Cotton Producers (Aproca).

Finally, there exists, since 2008, a pan-African platform of peasant and producer organizations with the objective to speak with one voice with continental authorities, especially with NEPAD (New Partnership for Africa's Development).

9.4 Collective Dynamics Remain Inseparable from the Major Challenges of Family Farming

The dynamics of organization of family farmers, their emergence into the public debate and the consolidation of transnational social movements represent a major change at the beginning of the twenty-first century for developing countries. These movements, which usually exceed the framework of just the agricultural sector, have often emerged as a reaction to dominant models or those imposed from above, many of which date back to the colonial period and which have continued without any real breaks up to the 1990s.

Leaving behind the municipal or village level to structure themselves into national federations, or rebuilding the "empty shells" of State-imposed organizations, then engaging at supranational levels to defend their interests in appropriate forums, these movements have gained recognition and earned a prominent place in the public debate with the support of NGOs, international organizations, intellectuals and sympathizers. In developing countries – unlike in the richest nations because of their relatively longer history of organized social movements –, these movements are often more recognized in external forums than at the national level, even though family farming retains considerable demographic, economic and social importance. This paradox, which exhibits large variations depending on local situations (Cissokho 2009), is explained by the lack of representation in the political sphere which pertains to the history of the countries concerned.

PROs are now actors in their own right in rural development. They deserve support not on the basis of ideological community values but for more pragmatic reasons: their economic efficiency, their capacity to innovate by taking the needs of markets into account, and their ability to defend their interests. Because of their territorial anchoring, family farmers and their organizations find themselves struggling to deal with the agricultural challenges of the twenty-first century. These challenges – food security, poverty and employment opportunities, and even health risks – for which family farming and its organizations have a prominent role to play will be partially explored in the third part of the book. But there exist also broader challenges, such as the long-term management of resources and the environment and the need for new forms of regional development, for which too their contributions will be critical.

Part III
Meeting the Challenges of the Future

Coordinated by Philippe Bonnal, Ludovic Temple

This third part of the book focuses on the role and contributions of family farming in the countries of the South in the context of major modern-day challenges facing humanity: poverty, employment, energy, animal health and plant protection – and their implications on human health –, and the environment. This is a necessary starting point because family farming encompasses a significant segment of the planet's population. It provides the vast majority of the world's jobs, maintains a close and enduring relationship with natural resources, and produces a broad range of goods and services for the benefit of society (a range that can be expanded depending on societal demands) that affects or could affect a large percentage of the global population. Given its considerable socio-economic influence over populations in the South, family farming is one of the keys to problems faced by society: through its active participation in societal responses to challenges, by being a fallback sector to mitigate impacts of crises or, conversely, by being the primary victim of the negative effects of development processes.

The nature of these challenges urges us to use existing knowledge to determine whether and how family farming is capable of responding to them and to any consequent controversies, as well as to suggest ways of transforming this form of agriculture. An initial issue concerns the ability of family farming models to help alleviate poverty, particularly in developing and emerging countries. Features of family farming that help create and diversify income sources, land-use planning and intersectoral development can be partially explained by examining the lopsided extent of the demographic transition and the development of the agricultural workforce (Chap. 10). A second issue concerns the ability of farmers to fulfill expectations arising from globalized intersectoral development which has been imposed by developed and emerging countries. Family farming is thus called upon to offer possible solutions to meet demands from international, regional and local markets. Can family farming systems satisfy global food requirements (Chap. 10)? Can they facilitate the energy transition by producing biofuels

(Chap. 11)? Answers to these questions are subject to differing viewpoints which are discussed and explored in Chaps. 11 and 13.

In a transversal manner, strengthening the productive efficiency of a family labor-intensive agricultural system is a key element in helping family farming adapt to increasing health risks caused by climate change (Chap. 12), globalization of trade and an industrially induced homogenization of technological processes. Marked by its limitations, family farming is nevertheless seen as a repository of innovative processes that could help address such risks in the future, due largely to its ability to manage natural resources (Chaps. 12 and 13). Thus, by ensuring food security in rural areas, it helps decrease the industrialization of production and the centralization of agricultural production. By promoting local agriculture based on a higher use of labor, it promotes quality products, reduces environmental impacts and contributes towards a better management of biodiversity, regarded as a strategic resource for future innovation processes.

These issues, however, interact and generate new questions. Can family farming meet the dual challenges of increasing its productive efficiency while maintaining the increased use of family labor that characterizes it? Should family farming transform itself to be able to supply globalized food and energy world markets, or should it attempt first to cater to local demands of neighborhood markets, knowing that these choices generate different development trajectories? How can governments strengthen family farming's capacity to innovate?

Answers to these questions highlight the importance of understanding the existing dynamics of transformations that family farming systems are undergoing, and especially of their adaptability and limiting factors. These answers, however, also challenge interactions between family farming and other production models (family-business, entrepreneurial or agro-industrial). These interrelationships, depending on the context, reveal situations of conflict in access to resources but can, in some cases, be characterized by complementarities or juxtapositions.

In addition to contributions in understanding challenges and characterizing structural issues that determine the future of family farming, this third part of the book highlights, in a transversal manner, the need to contextualize debates. One must pay attention to differentiated development levels between Africa, Asia and Latin America (as characterized by intersectoral adjustments, demographic transitions, functionalities of public policies, democratization, etc.), where each of these specific contexts can consist of several contrasting regional realities.

In the emerging countries, family farming, whose demographic weight has decreased given the urbanization of lifestyles, appears increasingly like an economic and social factor that could help manage negative social and environmental externalities of economic growth. This is the reason why it is increasingly being recognized and sought after by governments in these countries. However, in less developed countries, family farming, which is regarded as the dominant socio-economic category, appears as a key resource in the management of demographic transitions, in a context where intersectoral development and migratory flows have not been able to solve the employment problems created by such transitions. The role of family farming in such a context enjoys, paradoxically, little recognition by these countries' governments.

Chapter 10
Challenges of Poverty, Employment and Food Security

Philippe Bonnal, Bruno Losch, Jacques Marzin, and Laurent Parrot

At the global level, family farming is the primary source of employment and the main supplier of food products (Chaps. 3 and 8). It is also, paradoxically, the sector which harbors the largest number of poor people, mainly due to its central place in the economies of many developing countries (IFAD 2011). There are many reasons for this paradox and they must be sought in the heterogeneity of agricultural performance. This heterogeneity results from radical differences between technical systems (Chap. 6) and the economic and institutional environments that result from power relationships built over time within each country's particular context. These varied histories have led to very different statuses and room for maneuver for family farming systems at the economic, social and political levels.

Expected population growth over the next few decades (median scenario of 9.6 billion at the end of 2050[1]) immediately places family farming at the core of the food challenge. Family farming also becomes central to the efforts undertaken to alleviate poverty and create jobs. These three challenges are closely interlinked because, while food security pertains to the issue of production, it is first a problem of income to be able to afford decent food, which requires activities and jobs which can generate this sufficient income. Poverty is also a vulnerability to risks, chief among them in many regions being that of food insecurity. These challenges are particularly daunting in sub-Saharan Africa and South Asia, both regions which are

---

[1] According to the United Nation's *World Population Prospects* (2012 revision), http://esa.un.org/wpp/Excel-Data/population.htm.

P. Bonnal (✉) • B. Losch • J. Marzin
ES, CIRAD, Montpellier, France
e-mail: philippe.bonnal@cirad.fr; bruno.losch@cirad.fr; Jacques.marzin@cirad.fr

L. Parrot
Persyst, CIRAD, Montpellier, France
e-mail: laurent.parrot@cirad.fr

the world's poorest and the most "rural" and where, in 2050, 75 % of the growth in world population will take place.

This chapter's aim is to provide a brief overview of the current state of poverty, in connection with employment and food security. It describes the current responses to these challenges, especially those in the form of the economic environment, public policies and support programs. It is also a matter of identifying the conditions necessary for a durable strengthening of the role of family farming models in reducing poverty and improving employment and food supply, especially in the world's poorest regions.[2]

10.1 Current Status and Trends

Rural poverty can come in extremely varied forms, stemming from a variety of contexts which makes comparisons all the more difficult. Although poverty has declined in absolute terms, due to changes in populous countries such as China, some regional situations remain critical and manifest as a strong pressure on employment and food security.

10.1.1 Types of Rural Poverty and their Manifestations

10.1.1.1 The Evolution of Concepts and Difficulty in Implementing them

Even though poverty had long been reduced solely to a lack of sufficient income, its multidimensional and systemic nature is now widely accepted. There are many factors of rural poverty. They mainly pertain to the constraints of life and work, which are most often expressed in the framework of the family farm, with the most significant constraints being the inadequate access to land and water and the lack of secure land rights. Other factors of poverty also exist, such as low wages for farm labor, lack of productive and commercial infrastructure, weak public services, especially those for education and health, lack or inadequacy of social policies, the lack of recognition of social and political rights and freedoms of choice (Sen 1999), the predominance of some social norms, particularly regarding the role of women and children (Sindzingre 2005), loss of status – whether social (Paugam 1986) or socio-territorial (Bourdieu 1993) – or, more directly, the acceptance of what marginalization means for society (Sélimanovski 2009).

---

[2] This chapter focuses on the situation in countries of the South, but the issue of agricultural employment also concerns countries of the North in a context marked by the crisis of productivist agricultural models and the extent of structural unemployment.

However, the lack of homogenous global data on this multidimensional aspect of poverty complicates international comparisons and usually obliges recourse to the most simplistic criterion: that of disposable income per day, generally expressed in USD at purchasing power parity (PPP) based on specific income thresholds: USD 1.25, 2, or 5 per day.[3] Comparisons also encounter the difficulty of reconciling different definitions of poverty, with each expressing the characteristics of a particular society and determining the types of public action and development policies. The structural adjustment plans of the 1980s and their economic and social impacts have led to the taking of social indicators such as life expectancy, literacy and health into account. Similarly, the worsening situations in many countries during the 1990s have led to further widening of the definition of poverty to include more encompassing indicators (risk, vulnerability, freedom of expression) – a concept formalized by the capabilities approach.[4] But while the concepts have evolved, income continues to be the main indicator of poverty because of its ease of use.

10.1.1.2 A Decrease in the Incidence of Poverty at the Global Level, but Marked Continental Differences Persist

In 2010, IFAD estimated that 1.4 billion people live in extreme poverty in the world, of which one billion suffer from hunger,[5] mainly children and adolescents. It also stated that 70 % of people living in extreme poverty were to be found in rural areas (IFAD 2011).

These numbers represent a net decline in the prevalence of extreme poverty in the world (threshold of USD 1.25 per day) over the past three decades, so much so that the first of the United Nations Millennium Development Goals (MDGs), to "reduce extreme poverty by 50 % between 1990 and 2015," was attained by 2010 itself (United Nations 2013). But this general positive trend should not obscure the fact that the situation is still grave and that there exists a marked differentiation in the distribution of poverty in the developing world (Chen and Ravallion 2012).

---

[3] The index of absolute poverty is based on the definition of a poverty line fixed at USD 1.25 per day that measures the ability of an individual to meet his or her minimum basic needs. This threshold is determined theoretically using a basket of goods in a manner recommended by the World Bank and has been adopted by many countries. It has been chosen by international financial institutions (IFIs) as the upper threshold of extreme poverty. This evaluation method differs from the measure of relative poverty, whose use is especially widespread in Europe, which is based on the median salary of the population (usually 50 % or 60 %).

[4] Sen (1999) defines capabilities as "the fundamental freedom to make alternative functional choices (or, less formally, the freedom to choose between several lifestyles)." This expanded definition calls upon the notion of freedom, broken down into four dimensions: political freedom, economic freedom, social opportunities, and fertility and coercion.

[5] An assessment by the FAO in 2012 estimated that 850 million people suffer from hunger (FAO 2013a).

The most dramatic reduction is in East and Southeast Asia, including China – which is the largest contributor to MDG1 –, where large numbers of rural workers have migrated from the countryside as a result of economic growth, urbanization and industrialization. At the same time, Latin America has halved its extreme poverty and has greatly reduced its incidence. The same can be said of Eastern Europe, the Middle East and North Africa. A peculiarity of all these regions is that the number of poor in rural areas is now lower than in urban ones.

It is in South Asia[6] and sub-Saharan Africa that the incidence of extreme poverty remains the highest. In South Asia, although halved since 1990, it is still very significant, since growth here resulted in fewer jobs and smaller increases in incomes for the poorest than it did in China. Jobs in South Asia continue to be concentrated mainly in rural areas. In sub-Saharan Africa, there was almost no reduction in extreme poverty between 1981 and 2010, and it still affects nearly one in two persons. This situation is explained by the impact of the recession caused by the structural adjustment plans of the 1980s which has not yet been offset by growth of the 2000s.

However, these two regions of high concentration of rural poverty exhibit very contrasting dynamics. Even though the greatest number of people living in extreme poverty in rural areas is to be found in South Asia, this number is decreasing there. In sub-Saharan Africa, on the other hand, the number of poor is increasing and the proportion of rural populations in situations of extreme poverty is higher (IFAD 2011). Incomes in rural African households are characterized by extremely low levels[7] (Fig. 10.1).

Choosing a higher poverty level (USD 5 per day), however, tempers the impression of a relative improvement (Fig. 10.2) and shows that, while extreme poverty is on the decline, limited income still marks the daily life of a large part of humanity.

Moreover, the pursuit of urbanization – the world's population is predominantly urban since 2008 – and the gathering pace of the processes of economic transition – with a shift of the workforce from agriculture to other sectors – are gradually changing the nature of the problem of poverty (Box 10.1).

---

[6] South Asia consists of the Indian subcontinent (India, Pakistan, Bangladesh, Sri Lanka, Nepal, Bhutan, and the Maldives) to which are added Afghanistan and Iran.

[7] In certain regions of Kenya, Senegal and Mali, the 5 % of the poorest households do not earn more than 30–50 USD PPP per year (Losch et al. 2012).

Fig. 10.1 Changes in poverty levels with a 1.25 USD per day threshold, in % of the total population

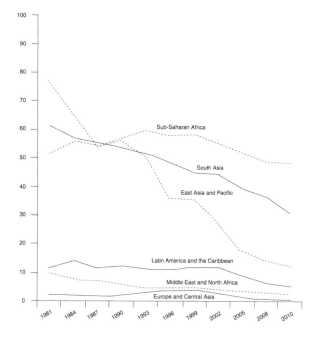

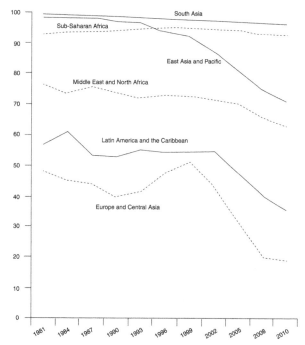

Fig. 10.2 Changes in poverty levels with a 5 USD per day threshold, in % of the total population (Source: http://databank.worldbank.org)

> **Box 10.1. Poverty, employment and agriculture: the case of Cameroon.**
>
> Laurent Parrot
>
> In Cameroon, reports from national statistics bodies state that poverty is higher in rural areas: 90% of farm households are poor and nearly 60% of the poor are farmers (INS, 2008). When activity is similar and depending on residence, income from the main activity of the head of the household is generally almost three times higher in urban areas than in rural ones. However, current reports do not provide information on purchasing power parity between urban and rural areas. Poverty, exacerbated by successive shocks of the CFA Franc crisis and the structural adjustment programs of the 1980s and 1990s, has affected almost all of Cameroonian society: public services were seriously weakened during this period. Life expectancy, which was growing steadily since 1960, even dropped from 55 in 1992 to 47 years in 2004 (United Nations, 2006).
>
> In a context of a total absence of any social safety nets, all workers, including those in agriculture, are particularly vulnerable: 90% of active workers do not have employment contracts, formal insurance, pensions or social rights. Underemployment affects 70% of the agricultural workforce and many workers (one in three) depend on multiple jobs to supplement their incomes.

The global changes in poverty in recent decades can be summarized in four points that redefine the challenges of employment and food security: a relative reduction, in which China, due to its demographic weight, played a predominant role; a difficult and specific situation in sub-Saharan Africa, where the number of poor people continues to grow because of insufficient job creation, and where three out of four poor people live in the countryside; growing inequalities in many countries that raise the issue of national cohesion; and, finally, a restructuring of geographical and sectoral characteristics of poverty.

10.1.2 Employment and Food Sufficiency: A Challenge for Family Farming

The issue of employment varies widely from country to country depending on each country's particular structural situation and the level of diversification of its economy. The main sectors providing employment vary quite dramatically between countries, and this is especially true for agricultural employment, which accounts for a wide range of 2–80 % of the total workforce in different regions of the world. And yet, despite these significant differences, it is agriculture, with 1.3 billion workers which occupies the global employment center stage, accounting as it does for 40 % of the total global workforce. The evolution of agricultural employment is therefore of strategic importance to many countries, especially those in Asia and sub-Saharan Africa which together account for 90 % of the global agricultural workforce (Chap. 3).

In Africa and Asia, agricultural employment must be put into perspective with the expected growth of the labor force and the rate of urbanization. While East Asia has experienced a rapid demographic and economic transition, South Asia and sub-Saharan Africa will still have a largely rural population up to the 2040s.[8] In the

---

[8] *World Urbanization Prospects*, 2010 Revision.

next 15 years, 560 million and 330 million young workers will enter the labor market in these regions respectively: 65 % and 60 % of them will reside in rural areas where agriculture will continue to remain the main sector of activity (Losch 2012b).

One of the key features of agricultural employment worldwide is that it essentially consists of direct jobs within family farms. The share of salaried employment is generally very low and it is significant only in a very limited number of countries.[9] However, agricultural employment has two specific characteristics. The first is its seasonality inherent to agricultural cycles, forcing family workers into partial inactivity, an inactivity sometimes aggravated in situations where access to productive resources (most notably land) is not secured. This relative underemployment in agriculture usually promotes the development of pluriactivity, should local opportunities exist (in services, handicrafts, non-agricultural employment), the recourse to mobility (temporary or permanent migration) or the emergence of pluriactive and multilocalized systems combining diverse activities in multiple locations. The second characteristic is that its level of labor remuneration depends on the situation of the agricultural sector and the national context – general economic and social environment, efficiency of technical systems, the existence of national or regional regulatory and support systems, etc. In low-income countries, there is a large gap between agricultural labor income and income from other activities. This income gap between sectors gradually decreases as the level of overall wealth rises.

This predominant role of family farms in agricultural employment extends also to food security, regardless of the level of economic development. In the poorest countries, the role of family farming in supplying food has increased. Formal and informal distribution channels depend on it for its products and it provides a significant part of urban and rural diets (HLPE 2013), either through the development of products sold or by satisfying a significant share of home consumption among farmers (Box 10.2).

---

[9] One salaried employee for every nine farms on the average, according to a partial breakdown by the FAO (Bélières et al. 2013).

Box 10.2. Food markets in West Africa.

Nicolas Bricas

Food consumption surveys conducted on large representative samples of urban and rural populations in virtually all countries of West Africa over the last ten years show that between 60% and 85% of the food for consumption is purchased. The domestic market in this region for agricultural products is considerably larger than the export market, and therefore constitutes a major outlet for regional agricultural production. The domestic market is divided almost equally between towns and rural areas. In recent decades, there has been an increase in rural towns where consumers can purchase some of their food and this expansion has also affected farmers who sell part of their production in order to be able to buy food. In terms of value, these markets can be divided as follows:

- A good third of expenditure on food consists of basic products which account for two-thirds of calories consumed (45% for cereals alone and 20% for roots, tubers and plantains). Within this category, imported rice and wheat dominate the market in the cities;

- About a third of food expenditure consists of animal products: meat, fish, eggs and dairy products. Consumed more in urban areas than in rural ones because of the higher purchasing power of urban residents, these products now mainly originate from local production, except for milk which is still imported in significant quantities;

- A final third of food expenditure consists of sauce ingredients (vegetables, legumes, oils, condiments), fruits, sugars and drinks. Here too, the majority of food is produced locally, except for sugar, which several countries in the region import.

Thus, urban markets are generally dominated by food imports from the point of view of nutritional intake. But from the point of view of their economic value, they cannot be considered to be predominantly supplied by imports. Instead, they have become important markets that have contributed to the transformation of what used to be called "subsistence" production, mainly destined for home consumption, into commercial food production (Bricas *et al.*, 2013).

But paradoxically, given this massive contribution of family farming to global food security, under-nutrition still affects a large proportion of rural families in several major world regions. As is the case with underemployment or low labor remuneration, under-nutrition is explained largely by the lack of access to the means of production, natural resources, especially land and water, and consequently to the lack of income to buy food items (IFAD 2011; Bélières et al. 2013). These handicaps result in an increased sensitivity to climatic, economic and political shocks (FAO 2013b).

The food situation in many countries is also constrained by the increasing volatility of prices of agricultural products. This volatility results from various factors related in part to the liberalization of trade, namely the growing share of imported food products, the greater rigidity of the demand due to changing diets and the emergence of a more affluent class of consumers. Also a factor, albeit controversial, is the political support for biofuels (HLPE 2011). Countries that suffer most from the increase in agricultural prices – demonstrated in dramatic fashion in 2007 and 2008, and again in 2011 – are insufficiently diversified and vulnerable to strong international competition, which is the case of most countries in sub-Saharan Africa.

As far as employment and food availability is concerned, family farming in agriculture-based countries of sub-Saharan Africa and South Asia suffers from a double paradox: it is the main source of employment but offers low labor wages; and it is the main supplier of food products but it is home to a large part of the population that suffers from malnutrition and hunger. In both cases, the reasons are due to the lack of stable access to productive resources, to the ever-present risks (natural and economic), competition from capital-intensive agricultural systems, and also due to poor infrastructure, public services, etc. The solutions to these recurring problems are largely exogenous to the family farm. They pertain to the improvement of the economic and institutional environment and refer to the content and modalities of public policy between the different categories of actors.

10.2 Existing and Future Solutions

Solutions already applied – and those yet to be applied – to support family farms in addressing challenges posed by poverty, the demand for jobs and food security cannot be considered in isolation. They have to be part of a comprehensive approach which takes the opportunities and constraints specific to the processes of structural change into account, in particular, the diversification of economic activities and urbanization. The pace of these transitions largely determines the scope of possible employment outside agriculture and increases the demand for agricultural products, which are themselves major drivers of change of agriculture.

Because of the structural income gap between rural and urban areas and between agriculture and other activities in the secondary and tertiary sectors – where labor productivity and remuneration tend to be higher –, the radical solution to rural poverty is to exit altogether from the agricultural sector. This is the path that was historically taken by the richest countries and is currently being followed by emerging ones. It has been part of public support policies that have encouraged individuals to choose to migrate to cities, other national regions or abroad. In poor rural areas, young people have similar aspirations and many would want, given the opportunity, to leave agriculture and move to cities in the hope of a better life (White 2012).

However, this solution leads to the question of this global development model's viability and its sustainability (urban concentrations, increased difficulties in managing natural resources and territories). At a more local and immediate level, the specific situation of countries with large agricultural populations becomes pertinent, as does their significant employment problems and limited room for maneuver. Their situation calls for massive support to be extended to family farming systems, either to facilitate their economic transition or to invent other developmental trajectories based on a more inclusive agriculture.

This debate culminates in two broad categories of responses: an improvement of the economic and political environment and a reinvestment in development strategies; and more specific actions targeted at agriculture and at rural and territorial development.

10.2.1 Improvement of the Economic and Political Environment

Since the 1990s, two major interconnected processes have helped create new opportunities to fight poverty in many developing countries: economic openness, on the one hand, and changes in scope and content of national public policies and international aid programs, on the other.

10.2.1.1 Economic Openness and Its Effects

Commercial deregulation of the 1980s, which contributed to making national economies of the North and South competitive, had very contrasting effects depending on the country. In most emerging countries, a phase of protectionism that allowed them to build an industrial base for the domestic market was followed by a period of new business opportunities, some stemming from industrial relocation and financial movements. These opportunities allowed some of these countries to strengthen their industrial and service sectors, develop urban employment and boost their domestic markets. The emergence or expansion of an urban middle class in these countries has structured demand for food goods, with the result that family farmers who have become part of market supply chains have benefited.

These effects were less clearly beneficial for the poorest countries, whose farmers and nascent industries suffered from international competition and where the economic and social situation often deteriorated as a result. The process was exacerbated by the brutality of the structural adjustment plans (SAPs) supervised by the IMF and the World Bank which imposed a trade liberalization – including in low-income countries – and dismantled regulatory mechanisms, whose benefits ultimately proved very relative in the richest countries (Stiglitz 2006). The SAPs have heavily penalized the purchasing power of the middle classes and have led to the contraction of the domestic market (see the example of Cameroon in Box 10.1). The deteriorating food situation in these countries, following the crises of 2007 and 2008, showed that they were very sensitive to price volatility and, more generally, to economic and climatic shocks.

Thus economic openness coupled with structural adjustment measures had varying effects depending on the countries and their place in the international balance of power: building up of commercial trade, development of the labor market and expanding consumption markets for some; contraction of formal employment, reduced purchasing power and shrinking consumer markets for others. The latter case corresponds to most countries of sub-Saharan Africa in the 1980s and 1990s.

Family farming played different roles in each of these situations: in the first, they largely contributed to the growth of and supply to the domestic market and sometimes to the export market as well, especially in Southeast Asia; in the second, they served as buffers to the crisis with employment devolving towards rural activities and the informal economy. They thus demonstrated their ability to adapt in an offensive manner in the first case and in a defensive manner in the second.

10.2.1.2 Changes in National Policies

Changes in the scope and content of public policies were facilitated by the very rapid growth of information networks and, in some countries, the spread of democratic values and standards. When the conditions were right, these processes reinforced the capacities of local actors to engage with – and even challenge – policy, facilitated popular expression as far as denouncing of poverty was concerned and led to demands for public action. Social-actor pressure groups, in alliance with NGOs and taking advantage of international solidarity networks, multiplied in number[10] and were able to make their voices heard during the food price crises of 2007 and 2008.

This social movement echoed the evolution of the international aid system which recognized the negative social effects of the adjustment period. The fight against poverty became a priority, as enshrined in the first MDG in 2000. Its eradication justified public action and opened the door to other types of interventions, particularly in the field of territorial development. These evolutions also led to changes in family farming's political positioning with its role now expanding to beyond the sole domain of production.

Thus, in Brazil, as in other emerging economies, redistributive policies have emerged (Box 10.3) that aim to correct, or at least reduce, the most glaring social and territorial disparities.[11] This development has been facilitated by economic growth, which released new resources for public intervention, and by the re-legitimization of the State's social and environmental regulatory function, necessary because of the negative externalities of the liberalization process.

---

[10] For example, in Thailand in 1997 with the Assembly of the Poor movement (Missingham 2003); in Brazil in 1994 with the *O grito da terra* movement; and in Mexico in 1996 with the *El campo no aguenta más* movement.

[11] These territorial development policies were supported by regional and international institutions and draw on the experience of industrialized countries. They have resulted in the transfer of public policies, for example, with Leader type European policies which have been promoted by the Inter-American Institute for Cooperation on Agriculture (IICA) in many Latin American countries.

Box 10.3. Income transfer programs.

Philippe Bonnal

Policies of redistribution have focused on the setting up of income transfer programs in one of two ways: with or without conditionalities.

The first type, strongly encouraged by international financial institutions, is based on making the disbursal of state aid to the poorest citizens conditional on the improvement of human capital, especially through child health and education. In this case, social safety nets mainly resort to "conditional cash transfer" (CCT) mechanisms.

In some countries, these mechanisms cover a large part of the poor population, as in Mexico's Oportunidades program (5.8 million families), in Brazil with the Bolsa Familia program (12 million households) and in the Philippines with the CCT program (700,000 families). CCT programs have also been established directly by international donors in countries such as Ethiopia, Bangladesh and Pakistan.

In the second type, income transfers are directed towards all the poor but this mass approach is fraught with difficulties of implementation, particularly because of the difficulty the poorest have in accessing existing institutional mechanisms.

The case of India is the most iconic since its welfare program serves 800 million people. The solution envisaged by the Indian government for distributing financial aid to the poorest – while avoiding corruption and clientelistic mechanisms – is to open a bank account for each person and issue a bank withdrawal card. This is a major undertaking, requiring the identification of each person and his or her place of residence, the installation of local ATMs and implementation of mobile banking mechanisms.

The Plano Brasil sem miseria (PBSM) scheme, implemented by the Brazilian government in 2011, takes the same approach and aims for an outright eradication of extreme poverty.

China, the country with the most impressive reduction of poverty, adopted very particular anti-poverty policies which alternated between umbrella policies directed at the masses and those which were specifically targeted on population categories, age classes, sectors or regions (Box 10.4).

Hence, for some emerging economies, the goal of eradication of extreme poverty is no longer seen as unachievable. Incorporated into the political agenda, its pursuit is even often based on the mobilization of the concerned country's own resources. This eradication is justified on both moral grounds – an argument widely espoused by the middle class, which is now the largest segment of the population in Latin America and which is now increasingly asserting its political clout (World Bank 2013) – and economic ones since incorporating the poor in consumer markets helps expand the domestic market, including for agricultural products. These measures, which mark an undeniable social progress, however, take the form most often of "social" approaches to the problem, rarely addressing the fundamental imbalances whose reduction would necessitate more fundamental structural reforms.

Box 10.4. Anti-poverty policies in China.

Jacques Marzin

The fight against poverty is a highly strategic issue in China given the historical and demographic scale of the phenomenon. Placed under the direct control of the State Council, the measures implemented over the last 30 years have had an exceptional impact on the reduction of absolute poverty (from 390 to 37 million people living on less than 1.25 PPP USD per day between 1981 and 2010), even though, at the same time, inequalities did increase (a rise in the Gini coefficient of income from 0.30 to 0.47 between 1978 and 2010).

Rural poverty was the result of radical ideological and political choices: forced collectivizations in the countryside, massive transfer of resources from the countryside to the cities to fund the primitive accumulation of industrial capital (1950 to 1990), in conjunction with a ban on migration from rural areas to the city by the *hukou* system (residence permit providing access to social benefits and rights). The 1978 reforms transferring management responsibility of farms to families have resulted in a continuous increase in agricultural incomes since the 1980s. Thus a comparison of changes in rural and urban incomes shows that average rural income rose from 39% of the average urban income in 1978 to 54% in 1985, before collapsing back to 36% at the end of 1990s due to the boom in the urban economy.

Over the past 60 years, with its political and demographic realities and its very prominent integration into the global economy, China has followed four successive types of anti-poverty policies. Zhang Lei (2007) identifies:

– The general policy of poverty reduction in the context of centralized planning (1949-1977). The objective was to reduce poverty-generating mechanisms of pre-revolutionary rural China. The main measures included regulation of tenancy and sharecropping rates, land reforms allocating land to farm laborers, then successive waves of collectivization (group workshops, cooperatives), and state control of marketing and of inputs;

– Institutional policy reforms for large-scale reduction of poverty (1978-1985). Policies promoting economic growth were harnessed to fight against poverty. Priority was accorded to institutional reforms: the definition of a legal framework for family farming, decollectivization of the commerce in agricultural inputs, marketing of agricultural production, and an end to the practice of mandatory cereal cultivation;

– "Pro-poor" policies during the economic boom (1986-2000). Poverty reduction became a priority political issue with growth alone no longer being seen as a sufficient condition to achieve it. Public action was targeted at the poorest areas of the country (districts, provinces) in the form of a systemic approach combining public investment and social policies (in particular, vocational training for young people, management of migration, etc.);

– Reduction of poverty on the road to a welfare society (2000-2005). It was based on a targeting of poor populations and the expansion of social policies in addition to existing territorial policies in order to take into account the new social context characterized by the overall reduction in the number of poor, the growth of urban poverty, the end of the effectiveness of *hukou* and extreme mobility of the workforce (nearly 300 million Chinese live outside their official place of residence).

In addition to redistributive policies, which concerns urban, rural and agricultural populations, policy changes have also resulted in greater attention being accorded by public actors to organizations representing family farmers. These organizations have strengthened their structures, expanded their alliances in academia and NGO circles and worked towards appropriate political representation in governance bodies (Chap. 9). This growing organization of the rural world sometimes finds political and administrative arguments in the decentralization and deconcentration of the State and in the strengthening of forms of territorial governance. These changes have been facilitated by a stronger democracy and the desire of some donors to encourage the professionalization of the rural world. These changes have been observed in Latin America, Southeast Asia and West Africa.

Another significant change in this period pertains to the changing political discourse on agriculture and its different roles. Limited for a long time to the role of provider of products, sometimes of labor and land in countries undergoing a process of industrialization, family farming is increasingly being recognized as a key player in the dynamics of rural and territorial development. It plays a role in providing employment and in the diversification of activities through pluriactivity, migration and its better integration with value chains. It also plays a role in the management of resources. This latter function could be further developed if farmers were remunerated for it since they are best placed to fight deforestation, sequester carbon and conserve biodiversity using suitable mechanisms. As providers of environmental services, they would receive additional income, which would reduce the economic risks family farms are susceptible to and improve their sustainability (Chap. 6).

This "return" of the State and public action in the social and environmental sphere and in matters of territorial development imply that it has the financial and political resources required for its intervention strategy. However, there are very marked differences across countries in this regard. While public authorities in many Latin American and Asian countries have managed to engage in this manner, the situation in sub-Saharan African countries has proven to be much more difficult. Increased fragility of nation states due to their recent creation (half a century) as compared to countries in other regions, greater political instability, a greater impact of the structural adjustment years, and a much narrower tax base have long led to an increased dependence on international cooperation (multilateral, bilateral and with NGOs) not only for means but also of the policy content. This has sometimes acquired an autonomy of action which has confined States to a rhetoric discourse, as illustrated by the case of sustainable development policies (Bonnal 2010; Bosc et al. 2010). However, the slow process of regional cooperation and strengthening of coordination at the continental level is opening up new perspectives, as shown by the agricultural policies of regional economic communities or by Nepad's agricultural program.[12]

---

[12] CAADP (Comprehensive Africa Agriculture Development Program), implemented by NEPAD since 2003, has long been the subject of influences of and competition between funding entities, but it is increasingly becoming a means for States in the region to regain control over their policies (NEPAD 2013). There is however a marked mismatch between public investments and the needs and commitments of States, a situation which is causing concern to producer organizations (Roppa 2013).

10.2.2 Supporting Family Farming and Rural and Territorial Development

The differences in productivity observed between family farming in the North and in the South are not only the result of national trajectories of structural change driven by markets and demand, but also of affirmative actions by individual countries. Investment and support policies for the agricultural sector, including price support and market protection over time, have been the main ingredients of the modernization of agriculture and rural development in countries of the North (Chang 2009). These policies have often relied on sectoral compromises which have underpinned the major stages of change (Chap. 2) and which can be facilitated today by improving the institutional and policy environment.

Agricultural modernization policies implemented by countries of the South have taken two directions. Some, like the European and North American structural policies,[13] have aimed at progressive support to all farms, most notably through the extension of new technical packages and sometimes through the implementation of major infrastructure projects. This is the case of Green Revolution policies in many Asian countries (South and Southeast Asia) which were at the receiving end of massive international aid programs. But these approaches have also run up against limitations. Other policies have instead adopted a more segmented approach to the extent that they have not sought to promote a new agricultural model and transform the agricultural sector as a whole, but rather to respond in a differentiated manner depending on the production environment, territories and types of farms, or even by creating new types of farming from scratch[14] (Bélières et al. 2013).

Thus, in Latin America and to a lesser extent in other regions, family farming is now often the subject of specific government programs. This is the especially the case in Brazil, whose government formalized and institutionalized the duality of its agriculture by differentiating family farming, on the one hand, and entrepreneurial and family-business farming, on the other, with their interests therefore coming under the ambit of two separate ministries. The Brazilian government has created innovative public-policy instruments dedicated to family farming in the technical domain (credit, insurance, advisory services) and the commercial domain (reserved quotas in public markets, intermediation for the supply of food to poor consumers through specific market mechanisms). These instruments complement minimum price policies which are the oldest in the world.

In some Latin American situations, though much more rarely in Asia and Africa, the difference between family and entrepreneurial farming's impacts leads to a questioning of the productivist model. This contestation, driven by family farming actors, proposes alternative models which emphasize the preservation of the

---

[13] These two regions implemented very proactive policies in the second half of the twentieth century, but they benefited from an overall environment that offered various options to exit agriculture, which has facilitated the sector's modernization.

[14] This is often the case with government farms or agro-industrial enterprises.

environment and social development in rural areas. For example, agroecology is the subject of social mobilization and intense media coverage, even at the international level, and this preoccupation is sometimes incorporated into mechanisms of research on agricultural support and advice.

All these policies for the modernization of agriculture require significant means that many countries simply do not have, given the lack of a sufficiently developed tax base, and which exceed any possibility of support from international aid. The most difficult situation is that of low-income, agriculture-based countries in sub-Saharan Africa and Asia, which is precisely where agricultural modernization is most needed. Indeed, if public policies to support the development of other sectors (for example, industry) are necessary for their structural change, an increase in farm incomes – which concern the majority of the workforce – is and will remain the main lever for poverty alleviation in the coming two to three decades. Increased agricultural incomes will drive rural demand for goods and services, on which depends the diversification of activities.

The topic of an increase in farm incomes raises the crucial question of priorities of actions and their sequencing. Indeed, it is unrealistic to expect farmers whose incomes barely cover their basic needs, including those of food, to make any investments on their own. And yet, the list of types of support necessary is long and the financial resources of most governments are unable to cover them all: information, training, advisory services, loans at very low interest rates, improved supply and marketing conditions and costs, support for equipment, etc. Choices will perforce have to be made given the limited public funding available.

Three priorities can be highlighted: improving conditions under which family farm undertake their activities, improving the environment and the functioning of markets, and, finally, boosting local economies.

An improvement in operational conditions of the farms more particularly pertains to the securing of rights (access to land and water), the provision of public goods (information, training), support for organizations which can help bring about economies of scale and increase the producers' negotiating capacity, and investment support mechanisms (HLPE 2013). In the technical field, while agricultural advice is important, it has often reached its limits and has been called into question by many farming communities. As shown in some countries, when public policies take into account the usable value of endogenous knowledge and of local experiments conducted by family farmers, whether linked or not with producer organizations or external institutional actors (NGOs, universities, research centers, etc.), the results are dramatically convincing (Chap. 14).

Improving the environment and the functioning of markets depends not only on the implementation of information and regulatory systems, but also those for market protection. These methods have figured very prominently across the world in the history of the modernization of agriculture. The issue here is to reduce the risk of farmers by implementing price-stabilizing mechanisms (Box 10.5) and offering them suitable insurance schemes.

> **Box 10.5. Guaranteeing minimum prices for cereals to stimulate investment in family farming in developing countries.**
>
> Franck Galtier
>
> In order to increase the efficiency of family farming – and its incomes – in developing countries, investments are necessary. But farmers often face very volatile prices that make these investments very risky. Producers are therefore understandably reluctant to invest and banks do not want to lend to them. To overcome these obstacles, it becomes necessary to protect producers against the risk of sharp price declines.
>
> Some international organizations recommend the use of private price-risk hedging instruments (futures, put options). But these tools are not designed for staple products. They are expensive, difficult to use and suited only for large-volume operators. Last but not least, they do not provide an effective cover for farmers located far from existing futures markets.
>
> Another option would be to protect farmers through cash transfers when prices fall too far, modeled on the system of deficiency payments in the United States. But such counter-cyclical transfers are difficult to implement in developing countries because of the lack of databases on the production and incomes of farmers.
>
> A final, more realistic, option is to guarantee minimum prices to producers. Depending on the context, this guarantee can be implemented through public stocks (whose procurement price can guarantee a minimum price) or through trade policies (most notably the use of variable levies on imports or exports). In the past, such policies guaranteeing minimum prices to producers have played a decisive role in North America, Western Europe and Asia in stimulating investment in cereal production (Demeke *et al.*, 2012) and remain crucial, even in the present, in a number of countries, such as Brazil. Because they are expensive, these actions should be implemented primarily for basic food commodities, particularly cereals (Galtier, 2012).

Finally, boosting local economies through suitable investments in rural towns and small cities helps strengthen links between cities and the countryside by offering infrastructure and services necessary for economic and social development (social services, training, information, support to SMEs). These investments facilitate the production, marketing and local processing of products (Parrot et al. 2008), but also the gradual diversification of activities and, ultimately, growth in employment. These approaches help better connect and link different economic activities – including agriculture – in a perspective of overall territorial development (Losch and Magrin 2013).

10.3 Considerable Challenges and Uncertain Outlook

All of family farming in the countries of the South does not therefore face the same challenges. In many countries, particularly in Latin America, East and Southeast Asia, the rural situation has improved significantly over the past two decades. And this has taken place across domains: economic, under the effect of the growth of public investment, industrialization and the expansion of the domestic market; political, due to reforms of governance systems and improved redistribution of the fruits of growth; and social, thanks to improvements in services and increasing organization of the rural world. In a number of countries, the role of family farming is now acknowledged – though it is not always subject to specific policies – and this recognition has provided effective levers for action. Strengthening the integration of producers into market supply chains gives them a key role in supplying domestic markets, themselves growing because of the reduction in poverty.

Nevertheless, two sub-continents have more uncertain outlooks: South Asia and sub-Saharan Africa, mainly because of the extent of rural poverty and the continued demographic growth of their rural populations. Sub-Saharan Africa, in particular, is home to many handicaps: its lack of diversification and weak industry, the lack of budgetary resources necessary for public intervention that could make a difference, and a continued dependence on external funding entities. Without real economic alternatives, the working population is still mainly absorbed in an agriculture that remains almost exclusively family oriented. International migration has shown its limitations and does not constitute a viable long-term option given the magnitude of the requirements. Thus, the risk remains of an acceleration of the exodus to the cities, which, in the absence of sufficient employment opportunities, will merely have the effect of relocating poverty. In such a context, family farming remains – and must remain – a sector that States must concentrate their efforts on.

Chapter 11
Energy Challenges: Threats or Opportunities?

Marie-Hélène Dabat, Denis Gautier, Laurent Gazull, and François Pinta

Intertwined with agricultural issues and closely linked to the challenges of poverty discussed in Chap. 10, the production of and access to energy are major issues for the growth of the South and for the attainment of the Millennium Development Goals. They are among the foremost prerequisites for economic and human development (Sachs 2005). In the agricultural and agrifood domain, energy remains a key factor for improving the production, harvesting, storage, processing and marketing of agricultural products (FAO 2000). Access to energy in rural areas is, in general, synonymous with improving the conditions of life and the diversification of income generating activities.

Yet the vast majority of family farmers in the world is located in areas with limited access to electricity, where the supply of fossil fuels (diesel, gasoline) is irregular and often of poor quality, and where gas distribution is virtually nonexistent. Rural areas in the countries of the South and in some emerging countries (most notably India) are characterized by low energy consumption, especially of electricity and fossil fuels. For example, the primary energy consumption of a Cameroonian or Cambodian citizen is 11 times lower than that of a French citizen and 20 times less than that of a U.S. citizen (IEA 2012). The main energy sources available to the rural populations are traditional biomass (wood, agricultural residues, animal waste), animal traction (oxen and donkeys for tillage and transport) and human labor.

M.-H. Dabat (✉) • L. Gazull
ES, Cirad, Montpellier, France
e-mail: marie-helene.dabat@cirad.fr; laurent.gazull@cirad.fr

D. Gautier
ES, Cirad, Ouagadougou, Burkina Faso
e-mail: denis.gautier@cirad.fr

F. Pinta
Persyst, Cirad, Ouagadougou, Burkina Faso
e-mail: francois.pinta@cirad.fr

But since the early 2000s, the global energy landscape is undergoing a change, with significant implications for agriculture. The global economy is indeed faced by three major challenges: coping with increasing energy demand, mainly due to emerging countries; dealing with the predicted depletion of conventional oil, easy to extract and use; and reducing greenhouse gas emissions by at least half as part of the fight against climate change.

In this context, and in response to these challenges, bioenergy can play a major role. It can, in fact, partially or completely replace fossil hydrocarbons; it can be carbon neutral when produced in a sustainable manner; it can be produced almost everywhere on the planet and thus help countries become energy self-sufficient; and it can be used in many forms, to produce, for example, electricity, heat or liquid fuels.

Many governments and private operators are already engaged in the search for land which could be favorable for bioenergy production. Nations are drafting and implementing new public policies conducive to this production. This has sparked numerous debates and concerns about land expropriation, especially in the South and in emerging countries. Energy demand translates into a massive demand for new agricultural products which can represent an opportunity for small farmers. Pressure on arable land and natural resources also creates new situations of competitions in which small farmers find themselves at a disadvantage as compared to urban actors or local and international industrial operators. This configuration of the existing demand very logically leads to the issue of the integration of family farming into these new markets and of the organization of bioenergy sectors.

To these global challenges must be added others, more specific to the countries of the South and their agricultural sectors: of improving access to energy in rural areas, and to electricity in particular, and of supplying of wood energy in the context of high rates of urbanization and population growth. Indeed, policies for the privatization of electricity implemented in the 1980s have failed in virtually all countries of the South. Private companies have no interest in rural electrification and connection rates have risen little in 20 years. Major development agencies are calling today for mixed solutions (centralized and decentralized), the development of local renewable resources and stronger local institutional anchoring. In this new context, bioenergy is seen as an alternative solution to reduce production costs in regions which are difficult to hook up to national grids and to help create new local value chains. Moreover, the opening of urban and foreign markets to many agricultural products has increased the demand for rural energy for production, processing and transport.

Much like policies for rural electrification, energy transition policies initiated in the 1970s in the developing countries and some emerging ones – which aimed to substitute wood energy by oil or gas – have not produced the desired results. In sub-Saharan Africa, as in India or Southeast Asia, and to a lesser extent in some rural areas of Latin America, over 80 % of domestic energy needs (mainly cooking energy) of the rural population are met by biomass and in particular by wood energy. The latter is mainly harvested by rural people, primarily from natural forests and savannas. Another major challenge for many countries is to ensure the sustainability of wood-energy supply chains in which many farmers are involved.

Within this overall energy context, family farming is ultimately confronted by four major challenges: participating in the global biofuel market; maintaining a dominant position in supplying wood energy to towns and cities while improving the sustainability of its supply chain; participating in the production destined for domestic markets and improving access to energy in rural areas in developing and emerging countries; and, finally, making better use of energy in their production systems and their sectors in order to improve agrifood production.

The first three of these challenges pertain to the ability of family farming to insert itself into energy supply systems (biofuel or wood energy) for new markets (urban, industrial, local communities, etc.) or for itself, without fundamentally changing the modes of agricultural production. The fourth, on the other hand, challenges the ability of family production model to intensify under the impetus of increased energy consumption, entailing the adoption of new technologies and recourse to new knowledge.

11.1 Participating in the Biofuel Market at the Global Scale

The market for liquid biofuels is characterized by a mass demand, by prices indexed to those of petroleum products and by significant economies of scale. In 2010, biofuels accounted for 3 % of the consumption of fuels for road transport. This share has been steadily increasing since 2000 and could rise to 8 % in 2035 (IEA 2012). Land allocated for the production of biofuels in 2006 accounted for about 1 % of arable land, and, according to forecasts by the International Energy Agency (IEA), this share could reach between 2.5 % and 4.3 % by 2030 depending on consumption scenarios (IEA 2006). If we wanted biofuels to reach 10 % of total oil consumption, it would be necessary to plant an additional 200 million hectares of wheat (which is currently the area sown) or 85 million hectares of maize, or 33 million hectares of sugarcane (i.e., 1.5 times the current area) (FAO 2008). The European Union has estimated that to achieve its target of 10 % of biofuels in the transport sector by 2020, four to seven million additional hectares will be necessary, approximately the surface area of Ireland (Bowyer and Kretschmer 2011). The International Institute for Applied Systems Analysis (IIASA) and the FAO estimate that between 250 and 800 million hectares of land is available once forests, protected areas and land needed to meet the increased demand for food crops and livestock are excluded. Most of this available land is in the tropics in Latin America and Africa. There is therefore significant potential for the expansion of cultivated lands for the production of biofuels and investors are therefore currently attracting much interest (Deininger and Byerlee 2011).

These investments are directed mainly towards large-scale agro-industrial projects, with Brazilian sugarcane and soybean cultivation models being often cited as examples. These models are based on a concentration of production land around processing factories (about 20,000 ha within a radius of 30 km), heavily mechanized cultivation techniques, a large workforce, investments in equipment and integration into several markets (electricity for ethanol, animal feed for biodiesel) in order to optimize the entire value chain. Small farmers are being gradually excluded from this market or have to conform to the requirements of agro-industries. For jatropha, for example, an oilseed plant producing vegetable oil which can be converted into biodiesel, a global census of plantation projects shows that average surface areas per project are of the order of 40,000 ha, whereas no agronomic tests have ever been conducted on this plant at this scale.

However, these projects often encounter local resistance, implementation difficulties and low profitability. Indeed, most of the land theoretically available is already being used for traditional purposes or is subject to customary ownership rights that investors are ignoring. This land, though theoretically available, is usually inaccessible and is of low fertility. Investment in the necessary road and other infrastructure will burden operating budgets. In fact, in Africa in particular, these difficulties can prove insurmountable, and many of these large agro-industrial projects never see the light of day.

However, this focus on agro-industrial projects should not obscure the fact that family farming is already involved in the production of biofuels. For example, over 40 % of Indonesian palm oil comes from small individual farmers and this proportion is similar for Cameroon. In Colombia, it has now reached 20 % due to a conscious State policy. These sectors are based on alliances between processing industries (Chap. 8), which generally have their own plantations and oil presses, small farmers under contract and the State (or its local representatives), responsible for facilitating dialogue between the various actors and for ensuring access to land. This sort of contractualization provides an outlet for farmers who perceive it as a way to diversify their incomes through an assured market. The Indonesian and Malaysian experiences show that small farmers benefit from the know-how, techniques and seeds used in industrial plantations. They also show that small farmers tend to empower themselves in this manner and become independent suppliers, no longer reliant on industrial groups (Feintrenie and Rafflegeau 2012).

Such contractual models for integrating family farmers in large production chains have also been developed in Brazil in the case of biodiesel from castor, soybean or palm oil (Box 11.1). But the difficulties in implementing these programs show how hard it is to establish relationships between industry and small planters and how industrial strategies may be out of tune with those of farmers (Favareto 2011). The failure of the cooperative experience of mini-mills for processing castor oil in Floriano, in the region of Piauí, involving 4,000 family farmers with each cultivating a few hectares of castor, also shows the difficulties that small farmers can encounter in their attempts to master this sector's processing tools and mechanisms.

> **Box 11.1. Integrating family farming into a national biofuel production chain. The Brazilian example.**
>
> Marie-Hélène Dabat, Denis Gautier, Laurent Gazull, François Pinta
>
> The National Program for Biodiesel Production (NPBP), launched in 2004 by the Brazilian government, aims to meet the growing domestic demand for fuel and ensure the country's energy security while promoting employment and incomes in rural areas. Within this framework, the Ministry of Agrarian Development set up an action program in 2005 to promote production of oilseeds (castor, soybean, jatropha, sunflower, palm oil) by family farmers. It relies on tax incentives offered to processing companies (including to Petrobras, the national oil company) which buy products grown by family farms. These tax rebates are conditional on a commitment on the part of industrial processors to purchase a minimum production from family farms, and the establishment of contracts between industry and farmers to fix prices and terms of delivery and of technical assistance.
>
> The program had a difficult start due mainly to problems of deliveries to factories and very low price levels. But in 2013, it appeared to have become a success, both in terms of production – Brazil has now decreased its diesel imports by more than 40 % – as well as in terms of participation of family farming. Since 2010, over 100,000 family farms have become part of the program and produce 95 % of Brazilian biodiesel.
>
> However, the program still generates controversial opinions between farmer organizations that share the goal of inclusion but do not share the "mining" approach nor the role of subcontractor to industry that is assigned to small-scale agriculture; and other farmer representatives who perceive the program as an opportunity to diversify and a positive factor for food security.

Since the purely agro-industrial model excludes small-scale agriculture, the contractual model is the most common way to involve family farming in the mass production of biofuels. However, this model does bring with it some risks for farmers: allocation of land for energy crops rather than for food crops, dependence on a single buyer (case of jatropha producers in Zambia or West Africa, or sugarcane in Mozambique), unfavorable contracts (commitments for very long periods, limited market, low level of support, penalty clauses, etc.), slow return on investment, etc. To reduce the negative social and environmental impacts of biofuel production, good-practice guidelines and product certification schemes have emerged. This is the case for palm oil with the RSPO certification (Roundtable on Sustainable Palm Oil), for soybean with Proterra, for sugarcane with Bonsucro, and in general for all biofuels with RSB (Roundtable on Sustainable Biomaterials). These schemes propose social and environmental standards, some of which may be favorable to small farmers. Nevertheless, these schemes are still struggling to establish themselves in the absence of local versions of standards adapted to different contexts of production and because of lack of political and economic consensus on their use or imposition.

In the end, it is clear that very few countries are relying on family farming alone to respond to the global demand for biofuels. Some countries in West Africa have adopted policy frameworks clearly advocating family farming models, but the few measures that are implemented usually originate from ministries or departments of energy rather than those concerned by agriculture or rural development. And yet, the experience of inserting family farming into energy-oriented agricultural biomass production systems shows that public policies to support biofuels must be multi-sectoral and multi-stakeholder (public, private, PO) and must be sustainable over time (Djerma and Dabat 2013). The experience of Brazil (the world's second largest ethanol producer and fourth largest biodiesel producer) shows that the success of its various programs for biofuels is the result of public policy that stretches over 30 years and which needed real cooperation between industrial, research, agricultural and environmental domains to succeed.

11.2 Improving the Sustainability of Wood-Energy Supply Chains for Cities

Wood energy is the primary energy resource in rural areas and in many cities in Africa, Latin America and Asia. More than three billion people depend on wood for cooking food and heating their homes. Although the figure is difficult to pin down with any accuracy, more than 30 million people worldwide – rural producers, transporters, sellers – form part of the wood-energy chains for supplying cities. Almost all rural families in developing countries collect or otherwise obtain firewood or other solid biomass (Openshaw 2010). In 2030, wood energy will still account for 80 % of the energy consumption of African households. At the same time, consumption of urban households is expected to surpass those of rural ones (IEA 2006). In urban areas, more than half of the population uses firewood or charcoal, which have become commercial products. These wood fuels are currently the least expensive when compared to their main alternatives: gas, oil or electricity. Many analysts believe that they are the only credible medium-term resource for many countries in the South.

At present, the supply to cities depends on production by rural actors with distribution taken care of by urban actors. These sectors are often informal but represent a primary economic activity. At equal energy production levels, the bioenergy sectors employ 10–20 times more people than centralized oil or electrical sectors (Remedio and Domac 2003). In most sub-Saharan countries, the turnover of trade in wood energy is higher than that of electricity and hydrocarbons. Throughout West Africa in particular – and also in Madagascar, Latin America and Southeast Asia –, rural populations are actively involved in cutting wood, charcoaling and the sale of this essential product. Rural women and youth have, in particular, developed these activities to earn money throughout the year, and many families are dependent on the exploitation and marketing of wood (Hautdidier and Gautier 2005).

There is a lively debate on the impacts of the use of wood energy. The practice comes in for heavy criticism by international institutions and by most national forest administrations. Many see it as a source of forest degradation, deforestation and loss of biodiversity. It has also been noted that the cutting and extraction of firewood is often linked to clearings for the purposes of agriculture, which is the dominant driver for deforestation. But many analyses and observations disagree. They show that the use of wood energy does not lead to deforestation and that it can be managed sustainably (Openshaw 2011). In Africa and Asia, most of the wood energy comes from areas where the resource is little – or not at all – managed and poorly monitored or controlled by national forest services. For example, 80 % of the wood entering Bamako, Niamey, and Ouagadougou originates from uncontrolled areas.

In this context, the sustainability of exploitation and supply of wood energy is a major issue for rural populations, especially the poorest amongst them. The supply of fuelwood requires a priori little technology and little initial capital. These sectors are easily accessible to the poorest people and form a safety net against extreme poverty, even if their sustainability remains uncertain. Faced with the informal nature of these activities, the proliferation of actors involved, the difficulty of controlling them and the ever-present fear of massive deforestation, many governments are tempted to impose oligopolistic and fully controlled supply chains, but these tend to exclude the poor, especially the rural poor.

In West Africa, for the past 20 years, the local management of natural resources has remained the dominant paradigm, but only through the compliance with technical standards imposed by governments and with highly controlled rights for access to and usage of the concerned spaces. At the beginning of the 1990s, the majority of West African countries undertook forest reforms that gave local communities a much bigger role in forest development than previously (Ribot 2004). These new forest policies are part of ongoing decentralization processes that have succeeded to varying extents but at least recognize the importance of a partnership between local communities and forestry departments. This principle is embraced by most funding entities, especially since it is rooted in a neoliberal economic perspective through which community involvement can reduce the transaction costs of forest management and since it comes adorned with a pro-poor discourse (Gautier et al. 2012). However, new forest models adapted to knowledge and practices of local actors, especially those of family farms, need to be discovered (Wardell 2003).

On a global scale, the recourse to dedicated industrial or family plantations seems inevitable given the quantities necessary and the environmental constraints inherent to the intensification of the exploitation of natural forests. However, these new energy crops will have to compete for the best land with food crops, which will inevitably hamper their development. Many countries have developed smallholding plantations, including Madagascar, Benin, Ethiopia, Brazil and the Democratic Republic of Congo, but results have been mixed. Planting of woody species is an act that marks land appropriation and one that is not allowed to the entire rural population. Furthermore, the plantation represents a set-aside for many farmers. Although technically accessible and economically viable over the long term, the plantation solution is far from being a panacea and it can hardly be considered in isolation.

The increase in urban demand for wood energy is therefore an opportunity for the rural world. It provides an accessible and stable market and a source of additional income. But this opportunity can only be truly sustainable if rural actors are part of a production approach that combines good operating practices with investments in wood plantations.

11.3 Producing for Domestic Markets and Improving Access to Energy

The increases in hydrocarbon prices (conventional oil and gas) between 2000 and 2009 and the uncertainties pertaining to oil supplies have had direct effects on the energy environment of non-oil-producing countries. An increasing number of sub-Saharan countries (Mali, Senegal, Burkina Faso, Niger, etc.) are finding it very difficult to absorb these increases in order to ensure reasonable consumer prices and regular supplies. This is especially true for gas, which is heavily subsidized for domestic use (up to 80 % in Burkina Faso).

Consumption levels of modern forms of energy remain low in the South, especially in rural areas where access to electricity is limited and its infrastructure badly outdated. The costs of producing electricity here are higher than in urban areas and current investments are largely insufficient to achieve the Millennium Development Goals. More than 80 % of the 1.5 billion people who do not have access to electricity live in rural areas (IEA 2004), and rapid population growth in many countries of the South which have not yet reached their demographic transition point can only worsen this energy poverty. The issue is also relevant in emerging countries such as China and India where per-capita rates of energy consumption are increasing rapidly.

Thus the search for alternative energy, adapted to the conditions of production and life of these populations, and energy independence, have become an important issue for several countries. In this context, family farming faces the challenge of producing enough agricultural biomass to meet the energy needs of local populations.

At present, a large number of countries in the South and in the emerging world are involved in bioenergy production.[1] But in fact, aside from a few countries which have an advanced level of production or have implemented export-friendly public policies (Brazil, Argentina and Indonesia), most countries are concerned with the domestic markets, with exports limited only to surpluses. Such is the case of Colombia, Peru, Paraguay, Thailand and even China, whose domestic needs are increasing and whose exports are therefore declining (Maltsoglou et al. 2013).

In West Africa, the limiting of production only for the domestic market is due to the media coverage of the 2007 and 2008 international food crises, which led to the cancellation of several large bioenergy projects which were accused of undermining food security. Furthermore, private investors doubt the profitability of energy crops given the drop in oil prices since 2009, as well as the numerous technical and agronomic problems (for example, with jatropha) encountered by projects seeking a quick return on invested capital.

---

[1] The main biomass sources are: jatropha and sugarcane in Africa; palm oil, sugarcane, maize, cassava, wheat in Asia – with specializations by country such as maize in China (which has recently diversified into cassava in particular), cassava in Thailand or palm oil in Indonesia; and sugarcane, maize, palm oil, soybeans, and other oilseeds in South America.

Farms usually allot only part of their land to biomass production, which can be food-oriented (such as sunflower, soybean or palm oil) or dedicated and inedible (such as jatropha) as a monoculture or in association, etc. (Box 11.2).

Box 11.2. Reconciling food and energy production at the farm level: a major challenge for the participation of family farming in the bioenergy market.

Marie-Hélène Dabat, Denis Gautier, Laurent Gazull, François Pinta

One of the main criticisms of the introduction of biofuels in family production systems stems from the fear of competition for arable land, thus ultimately endangering food security, both globally as well as locally. However, this either/or perspective overlooks the possible synergies. In many parts of the world, family farming has long used cultivation practices that cater to different uses: production of human food, animal feed, fiber, litter, and energy. To reconcile food and energy production, synergetic production techniques are already being used.

Intercropping. Energy productions are associated with food crops on the same plot. This is the case, in particular, of castor in Brazil, intercropped with maize and beans (Favareto 2011); jatropha in Mali and Benin, with maize or millet (Jatroref projects belonging to Geres and Iram NGOs); or acacia firewood plantations in Congo, with cassava (Makala and Mampu projects).

Crops grown on field edges. Energy crops are grown on the fringes of cultivated spaces, as hedges or live fencing, or on less fertile areas normally intended for grazing. This is the case with jatropha in West Africa, for example, the Jatropha Mali Initiative, or smallholder eucalyptus plantations in Madagascar.

Secondary crops. Energy crops are planted between two main crops (cereal or leguminous) during the year. This practice is widespread for fodder production, but it can be extended to ligno-cellulosic energy crops. Short-cycle sorghum, for example, tends to grow well in this manner.

Multi-purpose plants. These plants provide food, energy and often fodder. This is the case of certain sugarcane varieties cultivated for their sugar content but also for their ligno-cellulosic biomass. Sweet sorghum is currently undergoing a genetic selection for production of grains for human consumption, sugar for ethanol production and fodder (Sweetfuel and Biomass for the Future projects).

The web sites of these projects are: <http://jatroref.org>, <http://makala.cirad.fr/>, <http://www.eco-carbone.com>, <http://www.sweetfuel-project.eu/>, <http://www.biomassforthefuture.org/>.

Among the various existing models of production, two take family farming into account (Von Maltitz and Stafford 2011).

The first model concerns small farms with less than 10 ha of land producing biomass as a cash crop, for selling to larger farms which thus supplement their production, to marketing intermediaries or, most often, to processing units. These farmers are bound by annual (or longer duration) contracts to industrial operators (committed quantities, delivery times, prices, etc.). For the most part, the family[2] supplies the labor and it is all manual.[3] Industrial operators can provide support to farmers in the form of credit for inputs, access to equipment and technical supervision. This form of production requires close coordination between the various actors of the supply chain who are usually spread over a vast territory. Experiments

---

[2] Salaried employees are sometimes hired for labor-intensive tasks, such as soil preparation and harvesting.

[3] Sometimes independent operators or service providers make available mechanized services such as plowing, harvesting and transportation to the family farms.

have been conducted using this model for, for example, sugarcane, where small producers supply the bigger plantations in South Africa, Tanzania and Kenya. Some have been conducted on jatropha, where family-farm production takes place in close connection with the oilseed processing sector in West Africa (Geres-Iram 2012), or oil palm plantations, for which small-scale plantations are set up in mixed agroforestry systems and whose produce the family farmers sell to intermediary networks (Chaps. 5 and 7). From the point of view of industrial actors, contracts with family farmers offer several advantages: access to greater quantities of raw material, ability to set up operations in the host country, a commercial advantage when a certification system is in place, etc. In return, the family farmers benefit from guaranteed markets, can negotiate stable prices and have access to technical advice and financial support.

A second model, very localized, concerns family farms oriented towards supplying electricity generation projects, often organized into collectives (cooperatives in particular). In general, electricity related installations (generators, power plants, etc.) are appropriately sized to supply electricity to the village or to multifunctional entities that provide diversified services (milling, pumping, welding, etc.). This decentralized community model is often the result of an initiative of the government or of NGOs. It forms part of a perspective of a project of rural development, partially based on commercial incentives for producers (Von Maltitz and Stafford 2011). We can cite examples of projects using jatropha oil in Mali and Burkina Faso, or ethanol in Ethiopia, Brazil and Tanzania. However, these projects struggle to attract private funding since returns on investment remain generally low and uncertain. Indeed, the technical and financial support available to producers is not as good as in the previous model. Dependence on project funding is unsustainable and not sufficient effort is expended into transitioning to private ownership or stronger collective forms.

However, compared to the centralized model, the rural electrification model allows for greater technological flexibility (gasifiers, solid-biomass power plants, etc.) with an opportunity for private enterprise and, for more multifunctional agricultural production (rice husk, cassava waste, dedicated plants, etc.). Processing equipment designed for small-scale production, suitable for family farming, has been developed, mainly by the Indians and the Chinese who export it widely, especially in Africa. Economies of scale, normally decisive in the energy sector, are less important in this type of scheme and can be offset by the necessity of transporting biomass or the finished product over long distances. This ensures greater profitability and a positive ecological balance. Family farming is thus more competitive in neighborhood energy-producing systems for local development purposes. Furthermore, these systems contribute to rural employment, whereas some examples, such as in Mozambique (Fig 2011), show that the large-scale model often ends up creating very few jobs.

Studies have also highlighted the impact of short energy circuits (proximity between farmers and energy users) on the potential demand for biofuels (Litvine et al. 2013). A survey in rural Burkina Faso has shown that engine owners express a greater demand for jatropha vegetable oil when it is produced locally as a substitute for diesel fuel normally used. They are aware of and sensitive to the local creation and use of revenues and their spillover effects.

Family farmers are subjected to risks when they become part of short circuit models built around local cooperatives managing supply of energy to villagers or operation of multifunctional distributed platforms with the support of UNDP (United Nations Development Program). Magrin and N'Dieye (2007)) point out that the problem is similar to that of the management of boreholes, irrigation schemes or energy equipment (solar, wind) installed in the 1970s and 1980s during the first oil shock and periods of drought in the Sahel. For every success story, there are many failures, often related to organizational problems. Recovery of costs, amortization and maintenance are the weaknesses of the dynamics involved. These dynamics hardly survive once the initial impetus of the externally funded project has run its course.

Despite the growth in demand for biomass for energy purposes, models of energy production which are based on family farming are not very widespread globally. These types of agricultural production models encounter many technical, financial, organizational and market-related difficulties and therefore current production models are far from being stabilized. Farmers are thus exposed to many risks: competition with food crops, dependence on industrial entities from non-agricultural backgrounds, subordination to a volatile market – with prices dependent on the price of oil and gas. And yet, the benefits are also potentially many for small farmers and local populations: income, access to energy services, deriving value from by-products (oil meal, fertilizer, soap, glycerin, etc.), anti-erosion action, land protection, local development, social balance, land management and planning, etc.

Family farms can also increase their flexibility and reduce their vulnerability by diversifying into these energy crops provided that these do not affect the existing family rationale-based production systems: income opportunities permitting access to food markets, openings to new markets (energy), possibility of choosing between different markets (food and energy) for the same product depending on the prices being offered, maintaining the link between the family (user of energy) and the farm (self-sufficiency in energy biomass).

National policies express the willingness to support improved access to energy for populations and advocate energy independence, but it is clear that in African countries, especially in West Africa, they have had little impact in the biofuels sector. Even though the Brazilian program mentioned above, which was aimed at greater inclusion of family agriculture in the biodiesel sector, failed to achieve all its objectives, it at least had the merit of generating debate on the relationship between family farming and energy production and of opening up interesting avenues. Most often, however, biofuel policies do not pay heed to the family nature of biomass production. They focus instead mainly on the technical and economic dimensions, without considering production systems and the farmers' multidimensional objectives: choice of the plant and production level, production and usage incentives (subsidies, tax exemptions, subsidized interest rates, etc.), blending with conventional fuels, concessions for plantations and construction of production units, etc. The insertion of family farming into production systems of agricultural biomass for energy purposes remains fragile and requires support from public action.

Very often too, these policies are out of step with other sectoral policies (environmental and social), although we do note in several countries a recent trend in public policies towards greater sustainability (environmental as well as social): United States, Brazil, China, and India. Given the already mentioned failure of policies of the privatization of electricity services, development agencies are now engaged in the development of local renewable resources. They have also begun advocating public action in support of decentralized projects such as the construction of road infrastructure and the creation of networks for the distribution of inputs and for facilitating access to markets for family farmers. Such public goods can also be of benefit to other family or farm activities and more generally to other local activities and rural development. In the absence of government action to facilitate the development of these facilities and services, the risk is that the private sector will limit itself only to profitable investments, as we have seen in other sectors (mining, dams, etc.).

11.4 Better Energy Use to Improve Agrifood Production

Together with problems of supply, improving availability and increasing the use of sustainable energy are the key issues of the energy problem. An exploration of these aspects reveals possibilities for the growth and development of agricultural production, for improvements in the quality of agrifood production and for extraction of value from agricultural waste. But family farms and agricultural and agrifood enterprises in the South and in some emerging countries like India are characterized by low energy consumption, especially of so-called modern energy, most notably fossil fuels. This is due to low mechanization, the virtual absence of climate-controlled livestock housing and pressurized irrigation infrastructure, and to limited investments in crop storage and processing facilities.

11.4.1 Uneven Distribution of Energy Use, but Needs Everywhere

A significant challenge for the poorest family farmers is to produce larger quantities of food products in order to go from a status of deficit to one of surplus. This will help improve their incomes and will contribute, through their market integration, in a more effective way to the food security of populations in areas where they are located and in the cities. Conventional solutions designed to increase production require greater energy consumption.

At the production level, the adoption of animal traction and the motorization of certain stages of the technical crop itinerary (soil preparation, sowing) when combined with the use of chemical fertilizers, especially nitrogen fertilizers,[4] imply that these are systems with higher energy content. However, impacts need to be carefully considered. Thus, the use of animal traction and motorization – which policies tend to encourage everywhere – have indeed led to the expansion of cultivated areas and increases in production. But it has not led to agricultural intensification, when defined as the increase in production per unit of inputs (labor, agricultural surface area, fertilizers, seeds, capital, etc.). In contrast, the use of motorized pumps increases yields in irrigated areas, especially if it is combined with the use of inputs. But the perspectives are not obviously the same across continents: only 5 % of arable land in Africa is irrigated, against 40 % in India, for example (Pingali et al. 1988).

At the global level, access to and use of energy thus help explain substantially the differentials in agricultural productivity (Chap. 2). The proportion of mechanical energy (i.e., other than provided by man and animal) used in agriculture is 50 % on average in the world, but it is only 10 % in sub-Saharan Africa (Clarke and Bishop 2002). Between 1980 and 2003, the number of tractors in use per 1,000 ha of arable land even decreased from 2 to 1.3 in sub-Saharan Africa, whereas in Asia and the Pacific, this level went up from 7.8 to 14.9 (Mrema et al. 2008).

The mechanization of processes upstream and downstream of production – threshing, transport and processing of products (milling, husking, grating, grinding, pressing, etc.) – allows farmers and their families to handle any potential increases in production and therefore market surpluses in local, regional or international markets. This results in additional income for farmers and owners of agricultural equipment (Havard and Sidé 2013).

Finally, an increasing share of agricultural products is transformed by small businesses and farmer groups. Many sectors are part of this trend such as those of shea, cashew, mango, cassava, locust bean, and this for different post-harvest processes such as the drying, hulling, cooking, pasteurization, steaming or smoking of products (Box 11.3). These income generating activities turn the agrifood processing sector into a major contributor to the food security of populations (Alpha et al. 2013). They can be key to leading populations out of poverty (Jacquet et al. 2012). But here too, many studies have shown that difficulties in accessing energy are a major factor limiting efficient, quick and effective post-harvest processing operations, and hold back improvements in the sanitary and food qualities of products (Madhlopa and Ngwalo 2007; Rivier et al. 2009) and that the countries of the South continue to lag far behind in this area.

---

[4] The energy content of nitrogen fertilizers (called embodied energy), due to a very energy-intensive manufacturing process, is particularly high.

Box 11.3. Energy for food conservation and processing: the case of drying mangoes in Burkina Faso.

Marie-Hélène Dabat, Denis Gautier, Laurent Gazull, François Pinta

Mangoes are mainly grown by family farmers in Burkina Faso, although there do exist some processing units and some operators who own orchards near cities. Dried mangoes are one of the specialties of the Burkinabe artisanal food sector and are exported to foreign markets.

However, the rising cost of fossil fuels (dryers typically use liquefied petroleum gas) and the lack of a strategy for technological adaptation have led dry mango producer groups to lose competitiveness and market share. The share of Burkinabe mango drying units in the European export market has fallen significantly since 2007, mainly due to poor quality (browning of mango) and very high prices.

Rivier *et al.* (2013) show that drying of the mango is decisive in obtaining the best possible quality and this process determines the final product's price. For food with high water content, increasing the speed of drying by installing a ventilation system in the dryers helps limit color degradation, ensures homogeneity of drying and increases thermal efficiency.

However, the amounts of energy required are very high. The challenge is to provide access to cheap energy to improve the quality of products, not only for mangoes but also for other sectors such as of shea butter or dried fish.

The use of motors and engines (on fixed or mobile platforms) has been increasing in the South for the past 30 years in the context of labor reduction programs, especially those targeted at women, and the growing demand from urban markets. This is the case in Asia and also in Africa for operations such as pumping water, phytosanitary applications, harvesting, threshing and processing of agricultural products. This equipment, owned by individuals or groups, is mainly used to provide services by family production units (Havard and Sidé 2013). It offers interesting perspectives, but access to available and cheap energy is a pre-requisite to their use and their contribution to development.

One way to address these energy challenges of the South is therefore to increase energy use in small units by overcoming situations of insufficient or irregular availability and high costs. It is a matter of finding ways of a logical intensification that is suitable to the production conditions of the beneficiaries, in particular by avoiding high-tech and poorly mastered solutions which may tend to increase the dependence of family units. This concerns agriculture as well as the main post-harvest operations: drying and steaming of cereals, suitable packaging for the storage (even for the short term) of sugar or starch plants, etc.

11.4.2 Agricultural Byproducts as an Energy Resource: Experiments to Confirm Feasibility

Given the soaring prices of fossil fuels (starting in the 2000s) and faced with the risk of disruption of energy supplies, the countries of the South are exploring several types of energy. The use of agricultural and agrifood products and byproducts is one possible energy alternative for achieving greater independence for these countries, as well as for improving the quality of products and increasing the overall energy efficiency of agricultural and agrifood processes.

Decentralized Rural Electrification (DRE) from biomass illustrates the potential of agricultural byproducts to address energy challenges. In particular, it shows that specific solutions are achievable because they are adaptable and low cost. In 2013, 95 % of the rural population in sub-Saharan Africa still had no access to electricity. Not only electricity is rare there but it is also among the most expensive in the world. Given the costs and difficulties of extending the national electrical grids, stakeholders of the sector have all agreed to promote decentralized energy obtained from renewable solutions. Amongst these solutions, biomass-fueled energy appears to be technically robust while generating specific economic activities (value chain for the supply of local biomass). This specificity reinforces biomass technologies' reputation as promoters of local development. They are often amongst the cheapest renewable energy technologies available.

Several technological DRE solutions are possible depending on available renewable resources (crop residues, wood, dedicated plantations), the amount of energy required and the structure of the habitat (geographic dispersion): solutions based on conventional diesel generators and liquid biofuels, solutions based on syngas/diesel dual-fuel generators, and steam-based solutions.

The first of these types of solutions are those based on conventional diesel generators and liquid biofuels, such as Multifunctional Platforms in Mali and Burkina Faso. Implemented thanks to funding from UNDP from 1993 onwards, these platforms consist of diesel engines connected to modular components forming a self-contained system capable of providing various services: mechanical energy for the processing of agricultural products and electricity for lighting. The engine can be run directly on jatropha oil, although this is not yet commonly the case in the field.

Solutions based on syngas/diesel dual-fuel generators also display promise. Syngas is a gas produced in a gasifier, a reactor invented originally in the nineteenth century, from solid matter such as wood and agricultural residues (rice husk, straw). According to Ader, the Rural Electrification Agency of Madagascar, three power plants, each combining a gasifier with a generator, are used to supply three villages with electricity generated from rice husk. When manure, feces, plant waste and water are available, it may be worthwhile producing biogas using fermenters. This gas can then be used in a generator in the same way as syngas.

Finally, steam-based solutions couple a biomass boiler to a turbine generator or a steam engine. In 2012, CIRAD and its Malagasy and Brazilian partners launched the first biomass thermoelectric power plant in Madagascar. This type of power plant can operate either on wood (eucalyptus plantation) or rice husks.[5]

Even though research studies have shown ways to improve the energy efficiency of equipment, reduce its costs and scale its size down to small rural structures, these technologies are available and operational today. However, experience shows that

---

[5] <http://bioenergelec.org/>.

the success of these biomass DRE solutions depends primarily on the ability of rural populations to organize themselves to manage power generation facilities and coordinate to ensure sustainable supply of raw material.

More generally, ligno-cellulosic waste represents 20–40 % (dry matter) of agricultural production. The main resources in West Africa are shells of peanuts, cashew, shea and balanites, and maize cobs and rice husk. We should not, however, overestimate the availability of this waste since some of it is already destined for other uses by family farming, such as for animal feed or fertilizer, but it is a significant source of usable biomass energy, which may have a decisive impact in meeting the energy challenges of the rural worlds of the South.

11.4.3 What Public Policies for Increased Energy Use on Farms?

The challenge of energy access for family farmers, understood in the sense of "accessibility in sufficient quantity and at reasonable cost," requires an unambiguous environment and clear policies from the government for rural areas to support agricultural intensification. This applies as much at the level of the regulatory framework as of taxation and technical assistance. Access to energy for family farms is an important factor for food security for rural and urban populations and deserves the implementation of strong and structured public policies.

To promote mechanization and an increased use of mechanical energy by family farms, governments should no longer focus on the acquisition and financing of tractors or animal traction equipment, as many West African countries currently do. Havard and Sidé (2013) show that public policy must respond in the form of the establishment of a favorable socio-economic environment by taking the perception and motivation of farmers into account; by improving infrastructure, rural facilities and land management; by facilitating access to financing for mechanization; and, finally, by a reflection and action on the substitution of oil in the mechanization of agriculture.

In order not to compromise the positive impact of mechanization on agricultural production in a context of rising oil prices, some initiatives have be encouraged by governments (Gifford 1985). These initiatives pertain to the development of alternative fuels – and also to a judicious combination of human, animal and mechanical energy – and to more efficient fuel consumption. Energy savings of 25–50 % are possible by using equipment with a capacity suitable for agricultural operations to be performed, by offering appropriate and efficient programs for the maintenance and fine-tuning of the equipment, by improving working methods and fuel-storage conditions and by replacing equipment once its useful life has ended (Havard and Sidé 2013).

Finally, to fully take advantage of the opportunities of the local production and use of energy – especially from the agricultural byproducts mentioned above –

collective action is essential. It is, in particular, a solution to individual economic and financial inability to acquire equipment for intensification or valorisation of products and to cover associated costs. The bringing together of farmers in professional organizations (groups, associations or cooperatives) can play a fundamental role in the relationships between energy, family farming and increases in food production.

11.5 A Plebiscite for Public Action to Facilitate an Energy Transition

The complexity of analyzing interactions between energy and family farming and associated challenges stems from the specific links between the specific links between the domestic sphere and the productive sphere. This form of agriculture can be seen at the same time as an energy producer at various geographic scales, as a user of various forms of energy (on the farm and in the household), or as a self-sufficient energy producer for self-consumption. For this type of agriculture, energy can be, at the same time, a final product, an intermediate product, a production factor and an item of consumption.

Family farmers have become part of energy systems in many parts of the world. Family farming has advantages which allow it to implement an agricultural production model that has a low environmental impact, is employment-rich and better integrated into the territories where it is practiced. But faced with several, sometimes competing, alternatives – for example, land allocated for biofuel production may already be in use for the collection of fuelwood –, the governments must step in and arbitrate. Studies on opportunity costs and socio-economic effects of implementable choices, particularly in terms of employment and income in rural areas, could help these policy decisions.

Bioenergy is one of many options to boost agricultural growth and hence to reduce poverty, provided that family farming is involved (Thurlow 2010). But for many observers and policy makers, the inclusion of family agriculture faces the obstacle of low agricultural productivity in the South: low level of human capital, inadequate access to credit, and unsupportive legal and regulatory systems (Pingali et al. 1988). Thus, countries are tempted to support the industrial model, which is specialized and large-scale, to produce larger quantities. Furthermore, this model is easier to control and monitor since a smaller number of actors is involved. Dealing with energy markets – characterized by mass demand and economies of scale – means accepting high risks. Countries thus tend either to orient family farming towards more industrial forms of bioenergy agriculture, which would lead them to lose their specificity, or limit bioenergy activities to only family-business or entrepreneurial forms of agricultural production.

A proactive public action is essential to support sustainable bioenergy development for it to be, at the same time, competitive with conventional energy, inclusive of family farming and mindful of food security. But such action must form part of the wider debate on the rational use of natural resources across sectors. This debate is the subject of Chap. 13. It follows the discussion of another major challenge in the next chapter, that of the control of sanitary risks.

Chapter 12
Health Challenges: Increasing Global Impacts

Sophie Molia, Pascal Bonnet, and Alain Ratnadass

Health issues are finding a special and increasingly critical place in agricultural debates and discourse, alongside those pertaining to economic and social risks and the uncertainty of energy transitions. They are directly related to the nature and type of methods used to deal with animal and plant health in agricultural production systems, with family farming drawing particular attention. These issues encourage research at various scales, ranging from that of animal and crop species all the way to relevant international governance system.

Health challenges in agriculture refer to the different types of threats that can affect agricultural production (animal diseases and plant pests) or the health of farmers, breeders and consumers (zoonoses,[1] microbial contamination, mycotoxins, exposure to pesticides or drug residues). Some health crises have had economic and social consequences that have been etched in the history of mankind; others create global headlines today. Examples include the potato blight, *Phytophthora infestans*, which led to a deadly famine in Ireland in the nineteenth century and resulted in an exodus of a large proportion of that country's population, famines caused by destructive locust attacks (migratory, desert and red locusts) on crops in Africa and Madagascar, and, more recently, the global crisis caused by the influenza virus (avian and human flu).

Agriculture and livestock rearing in every continent have, since the 1950s, been impacted by an unprecedented proliferation and spread of pests and diseases.

---

[1] Zoonose: infection or infestation that can be transmitted naturally from animals to human beings and vice versa.

S. Molia (✉) • P. Bonnet
ES, Cirad, Montpellier, France
e-mail: sophie.molia@cirad.fr; pascal.bonnet@cirad.fr

A. Ratnadass
Persyst, Cirad, Montpellier, France
e-mail: alain.ratnadass@cirad.fr

Several interacting factors have been responsible for this situation. To begin with, globalization increases the movement and exchange of people and goods which, in turn, favors the spread of pathogens, pests and vectors. Climate change alters their spatial distribution and can affect the resistance of hosts (animals and plants) to pests and pathogens. Food demand, especially for animal proteins (resulting from urbanization and increased incomes), and non-food demand for bioenergy are growing rapidly. This encourages intensification (development of monoculture, specialized breeding techniques with increased livestock densities) and affects the resilience of production systems and sectors (Chap. 6). Finally, increased pressure leads to landscape fragmentation before uniformity is introduced (Chap. 7). Such changes in spatial organization give rise to contrasting health situations. Certain conditions can reduce the spread of natural enemies of pests and bring some plant pests closer to their alternative wild host plants. New interfaces between agricultural land and natural areas which result from spatial organization can lead to contacts between livestock and wild animals which serve as a reservoir for new pathogens.

Farmers, actors of upstream and downstream sectors and consumers are increasingly aware of the importance of food-related health and environmental issues and evolving social demands. The emergence of an urban middle class in the countries of the South has led to changes in habits and diets. Health scandals that have taken place over the past 20 years (mad cow disease, chlordecone, melamine-contaminated milk, etc.) have made people more demanding in terms of food safety. With forms of production being viewed through the prism of these issues, the strengths and weaknesses of family farming in being able to identify and control these new risks are being reexamined.

12.1 Multi-level Health Risks

While the various forms of agriculture are all susceptible to health threats, family farming, in particular, is affected at different levels and thus becomes more vulnerable, not only because of its structure and functioning (family labor, home consumption, mutual help, low investment capacity, etc.) but also because of its large extent, widespread distribution and the diversity of functions it performs for society.

12.1.1 Impact on Production Means and the Assets of Farmers

Pests and diseases affect the economy of agricultural households through a combined effect on different types of assets via mechanisms in dynamic interaction.

They initially result in decreased productivity of herds and crops due to their impact on technical assets: annual harvest losses, reduced meat or milk yield, loss of breeding stock due to mortality, decreased nutrient transfer from livestock to agricultural fields resulting in lower harvests and increased costs of chemical fertilizers. For instance, the respiratory infectious disease CBPP (contagious bovine pleuropneumonia), caused by *Mycoplasma mycoides* subsp. *mycoides* (variety SC: *small colony*) has wreaked havoc, since the 1990s, in the mixed crop-livestock farming highlands of Wellega region, in western Ethiopia. Family farming systems here are based on a mix of agroforestry farming (coffee), cereals (maize) and livestock rearing. They rely heavily on draft power to till arable land or home gardens and on nutrient transfers (manure) from herds to fields. "Smoked" fields (*Kae'e* "close to the house") are actually fertilized by a rotation system known as *dëlla*, a mobile enclosure for guarding livestock at night; the enclosure is moved to a new place once in roughly 3 days. The exchange of draft cattle for collective labor is a key element of practices that combine technical efficiency and community and social cohesion. This, however, led to chronic carrying, acting as a source of re-infection (Lesnoff et al. 2004), since CBPP-affected animals were not treated effectively with antibiotics through private veterinary care (failure of the State and private healthcare services in ensuring quality animal health care), and since vaccination had yet to become widespread or the established standard (Bonnet and Lesnoff 2009). In such a situation, CBPP has significantly affected the efficiency of draft cattle, leading to higher mortality and thus considerably impacting family capital and production systems.

The prevention and control of health risks entail added costs (termed as direct health costs) since family farmers must pay for remedial or preventive treatments and spend time treating their fields and herds, activities that could interfere with the normal schedule of agricultural operations. Some types of impact also have delayed effects because of changes in herd composition and demographic parameters as a result of lower reproductive performance (increased rate of abortion, lower fertility rates) (Lesnoff et al. 2002). Finally, if the product quality changes (agricultural and livestock products, or trade of live animals), demand and supply mechanisms can lead to a revision of selling prices, or even a reconsideration of whether a product can even be sold (post-harvest losses).

These combined effects, known as indirect health costs, limit the quantity and quality of food available for family farmers and affect the income from the sale of agricultural products. They aggravate household food insecurity and deplete savings (financial capital) amassed from commercial activities and accumulated assets (small livestock for cash or large livestock for capital). Some farms may thus resort to overexploitation of the natural capital (firewood, slash-and-burn to make way for additional crops, etc.) to compensate for losses from livestock and crops, resulting in erosion of natural capital (Chaps. 6 and 13).

Finally, health risks can directly or indirectly influence the working capacity of farmers, family members or farm employees. A decrease in available labor (productivity) and inability to work (availability) not only affect work within the farm but also outside it (additional employed work) as well as community work (social capital is thus affected) in regions where field labor is shared. The effects mentioned can be explained, on the one hand, by the reduced availability of food for family farmers who rely primarily on home consumption for their food needs, and, on the other, by lower farm incomes and their reduced purchasing power of food items in their exchanges (terms of trade: livestock against grain). In the worst cases, young children are severely malnourished, leading to nutritional deficiencies as well as cerebral and physical developmental disabilities.

We can also mention the direct impact of certain zoonotic diseases on the health of producers. Work can be disrupted due to short-term morbidity or even mortality, or long-term disability (for example, loss of vision from Rift Valley fever). Other short- or long-term effects arise from consuming food contaminated by microbes (salmonella, Q fever, *Escherichia coli*, etc.), mycotoxins or pesticides and drug residues, or contamination through inhalation during phytosanitary field applications. Mycotoxin-related health impacts can be even more severe for farmers and their families when they consume products too degraded to sell. Finally, we must draw attention to the resurgence of vector-borne human diseases (malaria and arboviruses) resulting from the resistance of vectors to insecticides used in large-scale phytosanitary treatments in peri-urban areas, mainly on vegetable crops (Chap. 7) or irrigated crops (paddy fields) or in vector control (N'Guessan et al. 2007; Yadouleton et al. 2011).

12.1.2 Impact on the Intensification of Agricultural and Livestock Systems

Health threats restrict the "conventional" intensification of agricultural systems. They impede the adoption of genetic varieties that are more productively efficient, but which often are more susceptible to health risks.

Some diseases endemic to a region or perceived as a risk lead agropastoralists to adopt diversification strategies to manage this risk (multiple animal and crop species). Some of these strategies are conducive to the conservation of hardy breeds and indigenous varieties (see Sect. 12.2.2 later in this chapter). They, however, also tend to limit complementary investments in breeds and varieties that are more productive and respond better to market demands. Consequently, pests and diseases reinforce a feeling of confinement for some family farmers in poorer areas without access to technological innovations by limiting their capacity for rational intensification.

For example, animal trypanosomosis, a disease transmitted by the tsetse flies and a zoonosis (a disease that affects both man and animals), limits the development of livestock rearing in sub-Saharan Africa in areas where it is present (de La Roque et al. 2001). In the absence of affordable treatment or trapping techniques to reduce vector pressure, farmers are constrained to rear breeds that are naturally resistant to the disease, but are less productive (for example, Ndama cattle or Djallonké sheep).

It is estimated that the area currently affected by tsetse flies could support an additional 120 million heads of cattle (Touré and Mortelmans 1990). The perceived risk of parasitic diseases transmitted by ticks (East Coast Fever (ECF), caused by *Theileria parva*) is another example of a disease that limits dairy intensification in areas with otherwise favorable markets (Kenya, Uganda). Yet another example is of food crops such as sorghum in West Africa, where the susceptibility of improved compact-panicle varieties to bugs and grain mold prevents intensification, especially for productive hybrids (Ratnadass et al. 2008).

However, these limitations to conventional intensification, albeit severe in the short-term, encourage us to pay greater attention to ecological intensification. The model based on high performing varieties and breeds, in addition to its attendant health risks, also relies on irrigation or resources that are non-renewable or hard to obtain, such as fertilizers and synthetic pesticides, in contrast to the model based on hardy varieties, less productive but significantly less susceptible to biotic stresses (see Sect. 12.2.2 later in this chapter).

12.1.3 A Difficult Integration with International Markets for Health Reasons

Agricultural products with organoleptic or nutritional characteristics degraded by diseases or pests have low commercial value. In addition, their access to international markets is decreased because international trade in agricultural products is regulated by sanitary and phytosanitary agreements (commonly referred to as the SPS Agreement) of the World Trade Organization (WTO). The presence of certain pests or diseases or of mycotoxins or pesticide residues can, in fact, lead to a restriction on or suspension of trade between countries (embargo). The control measures put in place at the national level in the event of a transboundary disease cause monetary losses that are both short-term (lower added-value of value chains thus contributing less to GDP) and long-term (systemic effects associated with a loss of markets and disruption of value chains).

Family farming models are particularly vulnerable. Health measures implemented and justified by international standards established by OIE (World Organization for Animal Health), CPM (Commission on Phytosanitary Measures) or the Codex Alimentarius Committee on behalf of the WTO, directly affect family farming systems that depend on such international trade. This is, for example, the case of farmers who rear sheep and goats in the Horn of Africa region, and supply the livestock markets in the Arabian Peninsula during the annual Haj pilgrimage to Mecca. Any embargo based on health issues deprives pastoral farmers of the region (Somalia, Ethiopia, Djibouti, Sudan, Yemen, Saudi Arabia) of access to these markets and pushes them into a deep food security crisis, as was evidenced on several occasions since 1998 during outbreaks of Rift Valley fever in this region (Bonnet et al. 2001; Pratt et al. 2005; Gerbier et al. 2006; Chevalier et al. 2009b).

Some countries that depend on international beef markets to sell meat produced essentially by family livestock farmers (in Botswana) or ranching enterprises (in Namibia) are affected by outbreaks of foot-and-mouth disease or CBPP. Since the total domestic production of beef cannot be consumed within these countries with small populations, the surplus is exported, and these countries have put the proper infrastructure and services (market places, animal identification) in place to meet international standards. They are among the only countries in sub-Saharan Africa that export to Europe (Faye et al. 2011). They can be seriously impacted by measures stemming from health policies linked to epidemics in these regions, despite the fact that they have established zoning techniques and systems to exchange health information (Bonnet et al. 2010).

The countries of the North sometimes close their borders to imported fruits when these are infested or infected by quarantine pests or pathogens, resulting in devastating consequences on family farmers who grow them in the countries of the South. An example is the Haitian mango which is grown mainly by family farmers. The production of mangoes collapsed in 2007 when the United States, the major importing country, rejected consignments of Haitian mango containing the *Anastrepha obliqua* fruit fly larvae, leaving family farms to deal with an issue that is both technical and institutional in nature. It is a matter of ensuring product quality (including food safety), using the help of research efforts to secure an approval or certification of production processes and their risk levels, and being well represented in forums that implement science-based health standards and regulations (some of which could be considered as overquality). As far as fruit cultivation is concerned, post-harvest losses due to flies have a lesser effect on producer incomes than very or overly strict import barriers (such as the requirement to implement a detection and control program in production areas) and the expenses that exporters have to bear for hot water treatment of fruit.

Finally, international measures and standards can also affect family farming systems which operate only in local and national markets (direct sales in the neighborhood and through city-supply chains), and which are located either in areas that may or may not be affected by epidemics. Thus, during outbreaks of avian influenza caused by the H5N1 virus in Vietnam, a precautionary ban was imposed on trade in ducks raised traditionally, impacting all producers, including those whose ducks were not affected by the disease (Box 12.1). Standardized risk management affects everyone involved in a given production system in the absence of the ability to target protection actions to a specific region or on a particular farming system. Fragmented fields and scattered houses are often typical of the spatial organization of family farming, a situation that can favor the spread of pathogens. Programs to control transboundary diseases in such a context are designed to work by limiting the movement of livestock, products or people. In sub-Saharan Africa, enterprise farming is often concentrated in certain regions and benefits from a potentially higher level of biosecurity than that enjoyed by family farming because of its superior capability to invest in health protection and to concentrate its efforts in smaller areas.

Box 12.1. An example of an environment where different health issues come face to face: the association of duck breeding and rice cultivation.

Julien Cappelle, Pascal Bonnet, Alain Ratnadass

The use of ducks in paddy fields is an age-old Chinese practice that was adopted in Japan by Furuno (2001), from where it spread to other countries. Ducks consume specific plants, pests and intermediate hosts that are part of parasites' life cycles. This provides phytosanitary and animal health protection by managing weeds, pests like brown leafhopper or diseases like sheath blight of rice (Ratnadass *et al.,* 2013a). Ducks can, however, only be introduced in paddy fields where pesticides are not used, as these chemicals cause high mortality in domestic duck populations (Chapter 15).

In addition, the association between paddy fields and ducks introduces the risk of avian influenza (H5N1) outbreaks, since paddies are also frequented by wild birds that come to feed there and who can easily transmit strains of the influenza virus (Gilbert *et al.*, 2007; Paul *et al.*, 2011). These virus strains are categorized according to their pathogenicity in chickens, and are grouped as low pathogenic avian influenza (LPAI) and high pathogenic avian influenza (HPAI), which is capable of causing outbreaks with high mortality. Wild birds are carriers of the LPAI viruses and such meeting grounds can facilitate the transmission of these viruses to domestic animals. And following a circulation of the virus in domestic populations, particularly in industrial or partially industrial sectors that support higher animal densities and are linked to family farmers, these LPAI viruses can become more pathogenic and develop into HPAI viruses, resulting in devastating epidemics among domestic poultry, particularly in family farms. Similarly, an interface between wild and domestic populations facilitates the return transmission of these HPAI viruses like H5N1 to wild birds. Some species are then capable of transmitting these viruses over long distances during migrations. These interface areas then allow a stepwise long-distance propagation of viruses like H5N1.

12.2 Family Farming in the Countries of the South: A Particular Susceptibility to Health Risks

It must be remembered that the context in which the majority of family farms in the countries of the South, and in some emerging countries, develop is defined by an overall health situation that is less favorable than that in the countries of the North: intense pest and pathogen pressure, especially in humid tropical regions; the presence of a greater number of diseases (trypanosomosis, certain tick-borne diseases in ruminants, or the millet head miner are only found in non-temperate zones); and lower training and support by Government phytosanitary and veterinary services, especially subsequent to reforms and structural adjustments (Chap. 10).

However, depending on the type of health risk and the environmental and institutional context, family farming can exhibit a greater susceptibility, or conversely a greater resistance, to health threats than other forms of agriculture.

12.2.1 A Vulnerability That, At Times, Is Greater...

Compared to standards of international agencies, the technological level of family farmers in the countries of the South is considered low. They have limited access to recent innovations to help prevent and tackle health threats or damage to produce. Moreover, their individual and organizational capacities for initiating and implementing international-level prevention and control programs are limited. For example, techniques for cultivation, drying and storage of crops grown in family farms may be more conducive to mycotoxin contamination for a wide range of agricultural commodities: cereals, nuts, coffee, cocoa, peanuts, spices, etc. These mycotoxins – secondary metabolites produced by fungi (especially of the genera *Aspergillus* and *Fusarium*) – are toxic (mainly carcinogenic) to human beings as well as to monogastric livestock. This is especially true for cereal grains which need to be dried down to a moisture level of about 12 % immediately after harvesting. This is not possible with existing drying facilities in family farms under humid conditions. Furthermore, grains or seeds affected by mold are not necessarily discarded after sorting, but are only removed from marketable lots (used for home consumption in the form of edible oil or "butter" in the case of peanuts,) or from lots destined for human consumption (used as poultry feed in the case of maize or sorghum, Ratnadass et al. 1999). These practices increase the risk of human or animal poisoning.

Family farmers often do not vaccinate their livestock adequately due to a lack of knowledge (lack of training, misrepresentation), of collective organization, paucity of pooled funds to purchase vaccines and non-availability of products that are properly stored by authorized distribution networks. Since the State has assigned itself the sole responsibility for the distribution of certain vaccines, its failures in fulfilling this responsibility deprives farmers of access to such technology. Some animal diseases that can be easily controlled with vaccination can thus end up causing devastating outbreaks in the poorest family farms. For example, Newcastle disease is commonly responsible for a 50–90 % loss of poultry in backyard flocks or on traditional farms, while large poultry farms that use vaccination are barely affected, if at all (Miguel et al. 2013).

Some plant diseases, particularly those that cause a systemic invasion, like sorghum smut or pearl millet downy mildew, are also controlled easily by seed dressing with a fungicide. It has been generally shown that this treatment benefits the producer and has minimal impact on human health and the environment because its application is specifically targeted and localized (Sidibé et al. 2011). These diseases have an economic impact only in areas where family farmers do not use treated seeds, a situation common in Sahelian Africa. Lack of knowledge on biosafety is also responsible for the spread of numerous animal pathogens and plant pests, particularly in the context of peri-urban agriculture, large animal densities or collective herd management (at the concession or village level).

In some areas, family farming is associated with pastoral farming methods based on mobility and collective herd management. These systems are inherently more conducive to the spread of various pathogens (parasites, bacteria, viruses, rickettsiae and protozoa) and, due to their very nature, have little, if any, access to conventional veterinary healthcare systems. The use of common grazing and watering areas in the Sahel or the Horn of Africa is commonly associated with the transmission of contagious diseases. These are areas where herds intermingle

(Waret-Szkuta et al. 2011) and transhumance of livestock takes place. These diseases entail a high economic cost and include rinderpest (now eradicated), peste des petits ruminants (prevalence currently increasing), Rift Valley fever, CBPP and foot-and-mouth disease, or fevers with high zoonotic potential (brucellosis). Certain types of transmission (NTTAT – *non tsetse transmitted animal trypanosomosis*, due to *Trypanosoma evansi* and transmitted by Tabanids) are exclusive to some arid pastoral areas. Finally, seasonal transhumance movements are dynamic networked phenomena which produce numerous epidemiological interfaces with sedentary breeding systems coming in contact with these migratory herds (for example in highland escarpment areas in the Horn of Africa in Ethiopia).

Family farms are often located in environments which are not very anthropized or use natural open spaces. This can lead to an increased risk of infection by pathogens associated with these environments, especially through vector transmission. Family farms characteristically rear livestock outdoors or in open shelters, as opposed to a protected system with livestock being housed in enclosed and monitored buildings. Such a livestock system is conducive to contacts between domestic animals and wildlife that is often a reservoir for certain pathogens and favors the transmission of diseases from the latter animal population to the former. Such is the situation in southern Africa (Botswana, Zimbabwe, Namibia) where family farms are located on the periphery of protected areas, which naturally leads to an epidemiological interface between livestock and wildlife, including with the African buffalo (*Syncerus caffer*). These wild populations are considered to be a reservoir of several bovine diseases such as foot-and-mouth disease, brucellosis, tick-borne diseases (cowdriosis, babebiosis, theileriosis) or bovine tuberculosis and pose a threat to the livestock of small farmers (Caron et al. 2013; De Garine-Wichatitsky et al. 2013). Asia too is subject to the same problem: in southern China, some wetlands include both protected areas, with their large populations of wild birds, and rice growing areas where large numbers of domestic ducks are bred (Fuller et al. 2013). The risk of transmission of avian influenza strains is particularly high here, and it is by this route that some transmissions of the H5N1 viruses took place during the avian influenza crisis (see Box 12.1).

Family farmers in oases, or in areas of recession crops in the Sahel and the Maghreb, as well as nomadic pastoralists who depend on grazing areas bordering the Sahara desert, are particularly vulnerable to infestation of pests or grain-eating birds (Benfekih et al. 2011; Ratnadass and Djimadoumngar 2001). Locust invasions develop in outbreak areas where surveillance is low in times of armed conflict or security threats, and swarms that develop destroy these "green islands" first, which are these farmers' cultivation or grazing areas. In addition, conflicts force families and herds to migrate temporarily, thus increasing the risk of reduced surveillance.

The proximity between fields and homes in some forms of family farming limits the options of spraying harmful plant diseases. This is the case in the French West Indies, where banana plantations in family or traditional farming are close to homes and contiguous with fields of other crops. Aerial spraying to treat cercospora leaf spot becomes impractical because of the risks it will pose to people living near sprayed areas, and the potential adverse effects it may have on crops other than banana on diversified farms (Bonin et al. 2006).

12.2.2 ... But Sometimes an Improved Resistance Too

Family farming generally incorporates cultivation methods based on varieties or breeds which are hardier and are more socially and economically efficient because of their low input use, and on systems that are more diverse and flexible. It also often manages agricultural biodiversity better and makes better use of localized opportunities that ecosystems offer (Chap. 6). Multispecies systems resembling natural ecosystems are for the most part more robust and resilient (Ratnadass et al. 2012a).

Family farmers often select their own varieties based on multiple agronomic and economic performance parameters as well as social criteria (Chaps. 13 and 17). They grow plants resistant to biotic and abiotic hazards, including pests and diseases. Mass selection has, voluntarily or involuntarily, led to the accumulation of genes resistant to pathogens and pests in local ecotypes of cultivated plants. A balance is consequently established between traditional cereal varieties (obtained mainly through mass selection) and pests that attack them in agroecosystems. This is particularly the case of photoperiod-sensitive, loose-panicle varieties of sorghum (of the Guinea race) traditionally grown in West Africa (especially in Mali), which are generally not attacked by insects. These varieties are suitable for tillering, mainly in response to the formation of dead hearts, allowing the plant to compensate for attacks by shoot flies and Lepidoptera stem borers. Furthermore, their photoperiod sensitivity results in a uniform and synchronized flowering which reduces midge damage and in a ripening of grains in dry periods which reduces infestation by panicle-feeding bugs and infection by grain mold (Ratnadass et al. 2003).

Levels of infestation by the stem borer *Eldana saccharina* in the Sezela and Felixton regions in South Africa are two to three times lower in smallholder sugarcane plantations than on commercial farms. This is explained by crop diversity (sugarcane, gardening activities, banana, maize, etc.) and by the presence of very diverse natural vegetation (shrubs, hedges, grass strips) which promotes the abundance and diversity of pest predator species such as ants and spiders (Draper and Conlong 2000; Goebel and Way 2009). In contrast, some agronomic factors in large farms – such as nitrogen overdosage in fields, stretching of the normal harvest period of sugarcane to obtain greater yield – aggravate pest infestation.

In the context of livestock breeding in family farms, a small herd size is combined with the use of breeds that are adapted to the local environment, are equipped with genetic characteristics to resist heat, pests and diseases, especially those transmitted by ticks, and able to use locally available food resources in a more energy-efficient manner. These factors limit the impact of communicable diseases and production related diseases. In addition, livestock breeding in family farms is frequently characterized by the association of different animal species, which constitutes a type of economic diversification and risk management (cattle, small ruminants, poultry, pigs, etc.). While this species multiplicity makes them more susceptible to non-specific pathogens, it enables farmers to preserve part of their heritage livestock asset in the event an epidemic completely wipes out all individuals of a particular species.

In general, the organic links between family and production spheres characteristic of family farms, promote closeness between the farmer and the tasks involved in running the farm, system self-regulation, rapid decision-making and the use of agroecological technologies favorable to a health balance. Thus, for example, the relatively small storage units associated with family farming – such as used to store grains or tubers – allows the quasi-entirety of stock to be easily checked on a daily basis when a part is removed for use or consumption. This process limits the development of pest infestation since the most infested part can be removed first. Moreover, contaminated portions are not necessarily "lost" as they can be used to feed livestock, provided due care is taken, as described earlier (Ratnadass et al. 1999).

Similarly, some agroecological strategies like push-pull (Cook et al. 2007), developed mainly for vegetable crops but potentially applicable to livestock rearing too, are more suited to labor-intensive cultivation methods and the small-scale of family farming. Thus, in Réunion, where vegetable flies (*Diptera, Tephritidae*) are considered major pests of horticultural crops, the limitations of chemical control methods that were used for many years have become apparent: loss of effectiveness, high cost, and risk to the environment and human health. An agroecological method to manage infestation by these flies has now been implemented and successfully adopted by farmers (Deguine et al. 2012). It is based on including maize plants in the vegetable agroecosystem to attract the Cucurbit flies. Maize acts as a trap crop by effectively attracting flies; the efficacy of this trap cropping effect can be enhanced by applying a food attractant mixed with biological insecticide.

Some risk management techniques practiced at a large scale in Asia were developed through the observation of family livestock farmers. These farmers realized that different animal species (pigs and cattle) attracted vectors differently. This allowed the combating of a severe infection in a productive – and thus economically valuable – species through intermingling with another species used as a lure.

The different situations presented clearly demonstrate that family farming is a rich source of traditional local knowledge, and a place to test processes of innovation based on a real ecologization of practices (Temple et al. 2013).

12.3 Health Risk Management Strategies and Measures

12.3.1 The Role of Governments

Governments have a major responsibility in managing health risks, especially those having a significant impact on the public. This is particularly relevant where health risks pose a threat to human health (zoonoses, food contaminated with microbes, mycotoxins, pesticide and drug residues), food security or markets (scourges like locusts or rodents, invasive or quarantine pests, animal diseases). The State intervenes at various risk management steps: definition and assessment, surveillance and prevention, control, establishment of standards and communication.

The plant and animal health situation of a country is diagnosed by expert committees that provide informed points of view – normally independent of public or private influence – on the risks faced by populations. The government then uses these opinions to take decisions to improve public health situations. Expert committees can also propose recommendations on legislation related to international documents (Codex Alimentarius, Terrestrial Animal Health Code of the OIE, etc.) and directly or indirectly contribute to developing national standards, estimate difficulties in applying them and analyze dysfunctions.

The State may also act by imposing standards for phytosanitary use, by monitoring the rigorousness of their approval process and by helping expose hidden public costs (diseases, pollution) associated with excessive use of synthetic inputs or by ensuring proper compliance with international laws. These actions involve public investment in analytical laboratories and in adequate surveillance systems.

The implementation of national programs for monitoring the most important health risks, however, should involve family farmers more comprehensively as they represent the vast majority of farmers, and therefore a potentially solid base for the establishment and functioning of health information systems.

Governments often subsidize awareness and training interventions related to major health risks to ensure the effectiveness of surveillance systems. They also support family farmers by organizing and financing, partially or fully, massive vaccination campaigns against animal diseases that, if left unchecked, entail high economic or social impacts. For instance, the annual national vaccination campaign against CBPP in Mali is launched by the country's president himself. The government may also act indirectly by imposing quality standards or, even by ensuring price stability for inputs like pesticides.

Control measures for animal and plant health adopted by governments aim to limit or eliminate health hazards present in their countries, regardless of the type of farming affected. However, these measures are often established within frameworks conducive to the protection of economic interests of businesses capable of influencing appropriate political lobbies. The effect of such lobbying is higher when public institutions are weak. Thus, during the avian influenza crisis in Thailand in 2004–2005, the government decided not to use vaccination, contrary to the case in Vietnam or China, but to resort only to culling infected livestock to ensure a quicker reestablishment of access to international markets for the industrial poultry-meat exporting sector.

European phytosanitary legislation applies in the French West Indies. Even though pest and pathogen pressures in the humid tropics are very different from those on the European continent, it is this legislation that strictly determines the chemicals than can legally be used. Consequently, only 4 chemicals are authorized to be used to control cercospora leaf spot in the French West Indies, as against 88 chemicals that can be used in neighboring Dominican Republic, where production conditions are, in fact, drier. These regulatory measures penalize producers, including family farmers, in the French West Indies. While this has led to an economic disadvantage in terms of regional competition, it has also helped reduce the exposure of farmers and the public to the harmful effects of certain chemicals.

Furthermore, some risks, such as emerging diseases, now tend to be managed according to international regulatory policies mobilizing farmers beyond the scope of their local issues. They result in the imposition of authoritarian measures that sometimes have dramatic consequences for small farmers whose living conditions are very insecure. An example is the culling of the entire population of domestic pigs on Hispaniola (the island on which Haiti and Dominican Republic are located) to prevent African swine fever, introduced to the island in 1978, from spreading to other American countries.

12.3.2 An Often Limited Compliance

A family farmer may actually neither be informed of health risks nor of the measures to prevent and control them. It may also be the case that he may himself not feel concerned with collective control programs against these risks and this for various reasons (Box 12.2). For instance, agriculture may represent only a small percentage of his overall income. This was the case, for example, in the French West Indies, where the control program against the tick *Amblyomma variegatum*, responsible for two serious cattle diseases (heartwater and dermatophilosis), failed to elicit the involvement of several cattle owners who did not consider the sale of these animals as being their main professional occupation. Another reason may be that the cost-benefit ratio assumed by the government may prove to be inapplicable to family farms due to their socio-economic status, and thus regarded as unsatisfactory (minimal direct impact of the threat or it does not concern the purpose of production). For example, in southern Africa, especially in Zimbabwe, the cost-benefit analyses which justify the control of foot-and-mouth disease, a disease that results in export restrictions of meat to the lucrative European market, showed that only entrepreneurial cattle breeders – who represent a mere 2 % of the total cattle breeder population – are involved in this market. The vast majority of livestock farmers, mainly belonging to poor rural classes, would derive only about 16 % of the benefits from the control of the disease (Perry et al. 2003).

> **Box 12.2. Veterinary compliance on family farms.**
>
> Muriel Figuié, Aurélie Binot, Marie-Isabelle Peyre, François Roger
>
> Many experts lament the fact that farmers, particularly the non-entrepreneurial ones, apply preventive measures (biosecurity, animal vaccination) or control measures (limitation of exchanges, destruction of contaminated crops or culling of infected animals) recommended by health administrative authorities in a limited or inconsistent manner (Conan *et al.*, 2012). This limited compliance is often attributed to a lack of awareness on the part of producers or insufficient information supplied to them and to psychological or cultural factors. However, more in-depth analysis indicates that the reasons are far more complex.
>
> On the one hand, the implementation of measures to control and manage health threats can be hampered by physical constraints that affect family farms more than entrepreneurial ones. These constraints include isolation, dependence on government phytosanitary and veterinary services that are few in number and dispose of limited resources, high costs and impracticability of available control methods (for example, the requirement of a permanent cold chain for certain vaccines).
>
> On the other hand, a limited compliance with health requirements can also stem from the desire of family farmers to protect their most immediate benefits, such as, for example, a refusal to adopt measures that involve destruction or extensive culling. It may also result from reactions of defiance towards various authorities (national, international) who farmers suspect are trying to use opportunities like those provided by health issues to interfere with their systems and introduce changes (sociopolitical, economic, or even bioethical), leading to a standardization of their practices, with the aim of strengthening the authorities' control over their activities (through mechanisms of epidemiological surveillance, of traceability, etc.).

The consequences of this limited compliance with the prevention/control of risks can sometimes be dramatic. They can obstruct the control of health crises, a fact that was highlighted by the circumvention of the ban on the marketing of traditional poultry during the avian influenza crises in Vietnam and Nigeria. This circumvention contributed significantly to the spread and persistence of the disease in these countries, as well as in neighboring countries, highlighting the need for better financial compensation from the government. Non-compliance with operative recommendations for phytosanitary and veterinary treatments (pesticides, herbicides, antibiotics, trypanocides, etc.) can lead to failures in treatment (poor conservation, incorrect application) or resistance to treatment (mainly due to incorrect dosages).

Reporting plant infestations may entail an economic loss for family farmers when this results in the destruction of infested/infected produce that may not have been intended for sale, nor is eligible for any compensation incentive from the government. However these measures, accompanied by various surveillance systems, curative treatment processes and strict product control procedures, may help lift certain export barriers, for example via the recognition of pest-free regions in certain countries by trading partners. Such is in fact the case for certain fruit flies. Conversely, in view of restrictions on the use of chemical inputs in organic farming sectors, especially with production destined for export, an organic producer who faces a massive pest or pathogen attack will prefer to lose his harvest rather than his certification, a luxury that subsistence farmers in the South who practice agroecology can ill afford.

12.3.3 A Management That is, Nevertheless, Active on the Part of Family Farmers

The lack of observance of health measures mandated by the national authorities does not mean that farmers do not implement their own mechanisms to fight against health threats (Box 12.3).

> **Box 12.3. Breeders managing risk within the "surveillance territories": example of avian influenza in southeast Asia.**
>
> *Aurélie Binot, Muriel Figuié, Marisa Peyre, François Roger*
>
> Even though family farmers sometimes do not wholeheartedly cooperate with administrative procedures for managing health risks, or follow operative recommendations only partially, this does not mean that they are inactive, individually or collectively, when faced with health risks. They display reactivity as well as a certain degree of autonomy in the way they estimate, evaluate and manage risks. Several studies conducted in Asia on poultry farmers (Binot *et al.*, 2012; Delabouglise *et al.*, 2012; Valeix, 2012) have shown an active exchange of information on animal health through networks that are largely independent of national epidemiological surveillance networks and which have their own governance standards and rules. The composition and functioning of these networks are closely linked to the local socio-technical context. In this "informal" surveillance (Desvaux and Figuié, 2011), livestock farmers rely on their own definitions of cases (in the epidemiological sense of the term), based on a collective representation of animal diseases founded on local experience. The composition and functioning of these networks indicate the existence of what can be called "surveillance territories." These territories are, in fact, areas of spatialized interactions within which health information is exchanged, allowing livestock farmers and others to take a comparatively coordinated decision to protect against or prevent a disease, or limit its effects. Within these territories, the actors may develop their own practices to manage animal health which may be completely independent of the expertise of public veterinaries. These practices reflect a socio-territorial belonging with its own rules of governance.

Risk management measures implemented by family farmers are based on a logic that combines health information collected through these networks, individual and societal perceptions of the risk and its consequences, and technical (know-how), economic (cost-benefit) and social (reputation) constraints. These measures include recourse to traditional and socially rooted and trusted knowledge (ethnobotanists, ethnoveterinarians, shamans, etc.), diverse and more resistant cropping or livestock systems (hardier varieties and species, diversified systems) and collective action. However, this action is dependent on the availability and accessibility of formal public or private services for medical technologies regulated by law (especially vaccines, certain antibiotics or anti-parasitic chemicals). Family farming is often capable of organizing itself effectively to gain access to such technologies and training in their use. This is how community health (paravets, community animal health workers (CAHW)) in sub-Saharan Africa has emerged. It is a system in which livestock farmers are trained and encouraged to provide primary health services to their community. They are supervised by the government which gives them access to certain drugs.

Despite all these efforts, such organizational innovations are yet to obtain official recognition (Bonnet et al. 2003). Legal restrictions in accessing certain technologies have spawned informal parallel markets for drugs, a large part of which are considered to be counterfeit (those originating from Nigeria, for example) but which use distribution networks that are more efficient than official ones. These fake drugs mainly end up with sub-Saharan family farmers. They may have contributed to a disillusionment of family farmers with modern health technologies, since most of these drugs do not demonstrate any effectiveness. In places where medicinal plant biodiversity was not lost, this has meant a return to traditional treatment methods (ethnoveterinarians).

Livestock health protection groups, trained and supervised by governments, have also been established in other countries to undertake certain control operations, especially those related to collective control.[2] These associations are now leaders in animal health control in France; their activities are comparable to those of the State. Cooperatives have also played a determining role in influencing pharmaceutical policies and laws. They have acquired the rights to use technology that was, until then, reserved for private veterinarians, either by employing the latter or by obtaining exemptions to access and use such technology. These institutional arrangements have also been developed in sub-Saharan Africa and Asia along the same principles in the form of various models depending on political acceptability and support from development funding entities.

12.4 For a Greater Integration of Family Farmers in Health Control Programs

Despite significant recent advances (Chap. 9), family farmers are still largely cut off from power centers, and are only involved as executants or unimportant actors in programs to manage health risks. However, as we have seen, in many countries, actors in the private sector, especially family farmers, already play a major role in managing health risks in a priori problematic situations. Thus, in situations where sanitary services have limited resources, where public and private health facilities are unevenly distributed in the region in terms of infrastructure and personnel, or where the State's response may be slow in coming (for example, in emergencies such as pest invasions or outbreaks of diseases), the administrative authorities could improve their effectiveness – and probably even their legitimacy – by relying on organizations deployed by family farming.

Family farmers must be better involved in collective monitoring, prevention and control programs against health risks. These actors already possess networks and associations (for example, tribal) that form part of the social capital which they can call upon to accomplish such tasks. Indeed, this social capital was one of the key factors in the success of early initiatives in pastoral areas of Chad based on the "one medicine" concept, which attempted to provide integrated human and animal health

---

[2] An example is the control of brucellosis in pastoral sheep breeding areas in France.

care services, and which was a precursor to the "one health" integrated approach (Wiese and Wyss 1998). This approach can be facilitated by the establishment of community-based organizations for agriculture, especially in the field of health protection. These could be structured as associations, cooperatives or collective action groups, depending on the country, able to address social and economic issues, in addition to health ones (features like market access and organization). A good example is Animal Health Groups (GSB) in Haiti that proved to be so effective that their scope was extended to cover phytosanitation, resulting in the groups being renamed Animal and Plant Health Groups (GSBP). Once established, these organizations provide services to their community. They offer advice, help improve the supply of inputs and facilitate the removal of barriers to agricultural trade, for instance through participatory production certifications. By coming together in federations, they also have their voices heard at the national level. They provide actors who can be part of national committees, or other such groupings, to coordinate actions regarding food safety and security, thus improving the understanding and protection of the interests of family farming.

The participation of family farmers and livestock breeders seems increasingly crucial in an environment of decreasing public funding for health programs. The successes of several control programs against African trypanosomosis in the West African sub-region have also highlighted the value of involving beneficiaries. Thus, village committees in Burkina Faso have contributed cash and labor to build, install and maintain insecticide-impregnated traps and screens to control tsetse flies. Some livestock farmers participated in the control program by subjecting all or part of their herds to trypanocide-based treatment or prevention programs (Bontoulougou et al. 2000). The experience gained from these participatory approaches was used to develop and implement the Pan-African Tsetse and Trypanosomosis Eradication Campaign (PATTEC). Similarly, the creation of committees for participatory surveillance and control of the H5N1 avian influenza virus has significantly helped control the disease in Indonesia, resulting in a decrease in human cases and deaths (Azhar et al. 2010).

Health risks must mobilize, at the same time, the State, representatives of producers and other members of civil society, and their participation must be regulated by several institutional methods (contracts, laws, regulations, conventions). NGOs thus have an important role to play with family farmers in terms of raising awareness, training and the implementation of measures to deal with health threats. This can take the form, for example, of paraveterinarian guide books and manuals they develop for animal health workers in different countries of the South which provide clear and simple instructions relevant to the local health and social context, thus helping improve the communities' health situations.

Actors and entities should be mobilized against health risks not only at the local level but also at the regional level, particularly with reference to transboundary infectious diseases. Regional control programs such as SEA-FMD (Southeast Asian control program against foot-and-mouth disease) or CAP (Caribbean *Amblyomma* Program, control program against the *Amblyomma* tick in the Caribbean) improve the effectiveness of surveillance and control initiatives through economies of scale (establishing vaccine banks or offering treatments negotiated at reduced rates), by facilitating the coordination and standardization of control techniques, and by improving communication and building up trust between health actors from different countries. Among other success stories in recent years that reflect the involvement of family farms, we can cite the eradication of rinderpest from its last outposts in Africa and of the screwworm from several Central American countries. The control of invasive pests such as certain fruit flies (Vayssières et al. 2007) is equally illustrative in the phytosanitary context. These successes show that the eradication of health threats is possible when all health actors – governments, family and other farmers, professionals of the agricultural sectors, NGOs, international institutions and funding entities – are involved in health control programs.

Chapter 13
Challenges of Managing and Using Natural Resources

Danièle Clavel, Laurène Feintrenie, Jean-Yves Jamin, Emmanuel Torquebiau, and Didier Bazile

A major challenge facing the planet's agricultural systems is to manage land without opposing agricultural practices and efforts to conserve natural resources. This is the crux of the "land sharing vs. land sparing" debate (Grau et al. 2013) which, for example, influenced the Agrimonde prospects implemented by CIRAD and INRA in 2008 (Hubert and Caron 2009). The "land sparing" approach (i.e. preserving land) is known as the Borlaug hypothesis (after the American Nobel laureate who is known as the father of the Green Revolution), which states that a per-hectare increase in agricultural productivity allows the reduction of the surface area under cultivation and thus helps prevent the clearing of new land and exploitation of virgin resources. This hypothesis appears somewhat simplistic today (Pirard and Treyer 2010). The extreme modernization of agricultural systems in developed countries and the Green Revolution in some areas of the South have ignored 'difficult' lands – limited accessibility, climatic limitations, mountain areas, low fertility, isolated regions – while over-exploiting fertile ones, leading to detrimental effects on the environment. Moreover, the land cannot be spared everywhere, in the way Borlaug advocates it, given social and human contexts, especially not in South Asia, where the rural population depending on family farming will continue to grow, or in sub-Saharan Africa where, in addition to strong demographic growth, labor productivity is likely to remain low in the coming decades (Chaps. 2 and 3).

D. Clavel (✉)
Bios, Cirad, Montpellier, France
e-mail: daniele.clavel@cirad.fr

L. Feintrenie • J.-Y. Jamin • D. Bazile
ES, Cirad, Montpellier, France
e-mail: laurene.feintrenie@cirad.fr; jean-yves.jamin@cirad.fr; didier.bazile@cirad.fr

E. Torquebiau
Persyst, Cirad, Montpellier, France
e-mail: emmanuel.torquebiau@cirad.fr

In a rapidly urbanizing world, global population growth and food demand will, in the medium term and regardless of productivity gains, require more land, more water, more energy, including for family farming which remains the predominant form of agricultural production. However, constrained by demographic pressures and the market, family farming, especially when not supported by appropriate policies, can also manage natural resources in a way that impacts the environment (Chap. 6). Rural areas are not only becoming increasingly residential, but are also being targeted by other productive sectors, including extractive activities (Chap. 7), and various forms of agriculture often end up competing for them. Therefore, the challenges of using and managing natural resources which confront family farming and other forms of agriculture must also be examined through possible alternatives to land occupation.

Furthermore, the challenges of natural resource management are part of particularly encompassing and significant processes: scarcity of fossil fuels which is encouraging an energy transition (Chap. 11), globalization of health risks (Chap. 12) and climate change and its potential implications, both in terms of mitigation and adaptation.

This chapter concludes the third part of the book by examining the challenges of the use and management of natural resources by family farming models, both at the level of their system of production and in relation to their competition with other agricultural and non-agricultural forms of resource use. Since we cannot obviously address these challenges in an exhaustive manner in this chapter, we intend exploring the distribution of the world's resources between productive and protective activities, focusing especially on the diversity of cultivated plant species and varieties, access to land and water, and challenges posed by environmental change. Given these global challenges, the practices adopted in family farming systems are described and put in perspective in relation to their alternatives and the role of public policy.

13.1 The Loss of Agricultural Biodiversity: A Global Issue

According to the Millennium Ecosystem Assessment (MEA 2005b) world report, the past centuries have witnessed a species extinction rate that is a thousand times greater than levels considered natural. This alarming trend is equally true for species and varieties used in agriculture. The modernization of agriculture that began in the 1950s and its variations in the form of the Green Revolution have been widely responsible for a loss of crop biodiversity. Initially justified by the global food needs of the post-war period and subsequently by developmentist rationales, these systems have historically been based on the creation, introduction and widespread use of hybrid commercial varieties produced by professional plant breeders. As these varieties are not genetically very different from other, the loss of agricultural biodiversity in developed countries seemed in some ways to be inherent to the intensification logic. Given the remarkable increases in yields obtained by the application of this model, it became, within a few decades, *the* way to go, thanks to the concerted support of agricultural policies, research and regulatory regimes. This rapid increase in the availability of seeds specific to a few agricultural products has gradually induced nearly two-thirds of the world's family farms to abandon the process of creation-selection of new varieties, resulting in a massive loss of agricultural biodiversity (Hainzelin 2013; Chaps. 2 and 17). This system was not designed to be used in "marginal" areas or ones less strategic in geopolitical power relations. Thus, it is natural to find a great variation in the levels of agricultural biodiversity across the world, depending on the spread of the Green Revolution and the local resources dedicated to it.

These systems were subsequently adopted widely in America, Europe and northern Asia, but varietal and seed systems of food crops in the countries of the South are still largely maintained by family farming, whose practices help conserve varietal diversity. Since these systems often result from marginalization in the dynamics of intensification, they could potentially provide solutions to the challenges of maintaining biodiversity (Chaps. 6 and 17).

Agricultural biodiversity, or agrobiodiversity, pertains to the genetic diversity present within biological and microbiological systems that contribute to agricultural production. In the process of producing seeds and renewing their livestock over centuries, farmers have bred plants and animals on their lands with a specific composition and genetic structure by maintaining a greater or lesser diversity, depending on the context, within and between plant populations. The specific character of such a breeding system is its close relation to local production, processing and consumption conditions. In West Africa, for example, farmers still maintain a high genetic diversity of food grains such as sorghum and fonio to cater to the diverse and specific needs of family farming, and because it is considered to be a hedge against environmental and climatic variations (Vigouroux et al. 2011). Because of erratic rainfall regimes in several regions, such as the Andean and Sudano-Sahelian zones, factors like yield stability and production security through the use of different varieties acquire a strategic importance.

As much as 90 % of the food crop seeds in West Africa are self-produced in order to be planted in family farms the following year. The seed management system is based on non-market exchange networks (self-production, inheritance, gift, barter, etc.) which operate at various levels, ranging from the farm to small natural areas, with varying seed qualities depending on the care taken by each producer (Coulibaly 2011). Knowledge and expertise related to seed production are based on hands-on learning and knowledge sharing within the community group. During the harvest, the head of the farm, often in discussion with his family, conducts a massal selection of seed according to his own criteria such as crop cycle time, yield, resistance to diseases and pests, soil adaptation, taste, ease of processing of the grain, preservation in granaries, etc. (Bazile and Soumaré 2004). Social relationships and the networks that underpin them play a major role in the movement of genetic resources and the sharing of related information. They help everyone in the small farming region gain access to seed diversity (Bocci and Chable 2008). These networks also promote and increase seed exchanges between families, and allow continuous experiments on local varieties under various environmental or socio-technical conditions (Pautasso et al. 2013). Any loss of crop biodiversity thus reduces the options of managing agricultural and food-security risks, and affects the social organization that often unifies family farms. The management of biodiversity ultimately pertains to social, cultural, economic, health and environmental challenges, regardless of the scale considered.

Ambiguities concerning policies for protecting intellectual rights vis-à-vis seeds began to emerge at the international level with the 1992 Rio Earth Summit. The Summit specifically highlighted links between agriculture and biodiversity by introducing the concept of agrobiodiversity. In a break from the principles of the Green Revolution, the role of biodiversity and related knowledge in helping boost the capacity of biological systems and human societies to adapt to global change, especially environmental ones, was highlighted. At the same time, it became clear that we had moved from a system based on diversified species and varieties to one that was constrained by requirements of genetic stability and homogeneity, i.e., of commercial standards that had become mandatory in agricultural systems of the countries of the North. The Green Revolution model of varietal breeding that combined the conservation of a large germplasm collection outside the area of production (ex situ) with a professional in-laboratory plant breeding and seed production system, was thus logically challenged as the sole model to adopt (Hainzelin 2013). However, this shift in thinking is not devoid of ambiguity. The generalization of the provisions of standardization is, nevertheless, deemed necessary, even for in situ conservation. Such provisions are simultaneously seen as a threat to the prerogatives of family farmers to use their own seeds from open-pollinated varieties, which by definition are non-homogeneous, and as a potential lever for family farming models.

More specifically, around the same time (1993) that the Convention on Biological Diversity (CBD 2013; Article 8j) began recognizing the role of the knowledge and expertise of indigenous human populations in maintaining biological diversity, the development of molecular genetics accelerated the implementation of policies for protecting intellectual rights on living organisms. The professional and commercial logic behind varietal breeding is based on the ex situ conservation of genetic resource banks. However, we believe today that knowledge on the condition and status of various in situ agricultural species must be developed on a priority basis, as is most notably suggested by Bioversity International (Jarvis and Hodgkin 2008). The multilateral protocol on access to genetic resources and the fair and equitable sharing of benefits arising from their utilization, embodied by the Nagoya protocol to the CBD (2010),[1] is a step in the same direction. It provides inputs to the

---

[1] The Nagoya Protocol on Access to Genetic Resources and the Fair and Equitable Sharing of Benefits Arising from their Utilization (ABS) to the Convention on Biological Diversity (CBD) is a supplementary agreement to the Convention on Biological Diversity (CBD). It provides a transparent legal framework for the effective implementation of one of the three objectives of the CBD: the fair and equitable sharing of benefits arising out of the utilization of genetic resources. The Nagoya Protocol on ABS was adopted on 29 October 2010 in Nagoya, Japan and will enter into force 90 days after the fiftieth instrument of ratification. For more information, see: http://www.cbd.int/abs/.

current international debate between a system based on commercialization and another designed to protect the environment and collective goods (Vivien 2002). This confrontation explains, to a large extent, the enhanced interest in mixed solutions, such as payment for environmental services (PES), which propose the commercialization of ecosystem services provided by biodiversity (Chap. 6, Box 6.3).

Going beyond this opposition between market regulation and a logic of conservation, the protection of crop biodiversity relies today mainly on resources and knowledge developed mainly by family farming. This therefore calls for an integration of local knowledge in the planning and execution of research programs, as suggested by the IAASTD international assessment (Caron et al. 2009; Chap. 17), and of development programs. Darwin's theory has helped us understand that mechanisms for adaptation to a constantly changing environment function through natural or artificial selection of the progeny of natural or artificial crosses or hybrids. Evolution, through a succession of adaptations, is the condition of the survival of the species, and it is exercised at all levels of the system (microorganisms, varieties, species, ecosystems, landscapes, regions, the planet itself). This adaptation is crucial in re-establishing the functional capacity and resilience of ecosystems after a disaster (Barbault 2008). It is therefore necessary to maintain the evolutionary potential at different levels of organization of biodiversity (Blandin 2009). Family farming has developed this evolutionary pool as an agricultural heritage, and its use must be made on the basis of long-term sustainability and in a varied manner, depending on systems and the pressures they are subject to. These family farming systems are, in fact, not immune to the imbalance they cause due to the pressure they exert on natural resources, as in the case of excessive slash-and-burn practices in equatorial forests, or in the case of the elimination of fallow periods in fragile agroecosystems in the dry savannas of Africa (Chap. 6).

Recent studies conducted by CIRAD and its partners in Kenya and West Africa on food grains highlight the multidimensional nature of agrobiodiversity management, by combining genetics and social anthropology (Leclerc and Coppens 2012; Vigouroux et al. 2011). It emerges that human societies have a strong effect on the genetic structuring of agrobiodiversity. The genetic study of crop varieties, now being perceived as social objects, is described in a different manner keeping in mind that genetic recombination is mainly influenced by the social structure of the human groups using genetic resources. The manner in which varieties – as social objects – are reproduced, exchanged and inherited often result in very complex combinations and genetic structures that no "professional" pattern of crosses from ex situ germplasm would have created or imitated. Plant breeding and varietal selection programs are beginning to take family farming criteria into account in the processes of adaptation, conservation and dissemination of agrobiodiversity in very diverse contexts, and by mobilizing the participatory plant breeding and improvement techniques which form part of rural seed management systems (Chap. 17).

Perspectives on the role of family farming in preserving and enriching the evolutionary potential of crop biodiversity and the knowledge associated with such management pertain to the concept of sustainability and, ultimately, to larger public policy issues designed to maintain the autonomy of family farming models.

This also raises critical questions on the sustainability of agro-industrial agriculture that favors an artificial homogenization of the environment through external inputs. Knowledge on the dynamic management of the evolutionary potential of in situ crop biodiversity has to be preserved and enhanced in the same way as the genetic resources themselves. This is an important lesson to help define agricultural policies that would allow the conservation of this evolutionary potential.

13.2 The Race for Land, a Growing Competition

Competition for land is also a major challenge for family farming. Access to land and land tenure security are key criteria that determine family farmers' choices and shape their strategies and agricultural practices (Chap. 4). But land is not merely a means of agricultural production. It is also fundamental to the identity of individuals and communities, and the social and cultural structure of territories (Chap. 7). It is also necessary for the expansion of urban and rural settlements, development of public utility infrastructure such as transportation networks or energy production. Finally, it is intimately linked to the sub-soil zone and the resources contained therein: fossil fuels, minerals, clays, metals, precious stones and rare earth elements. Thus the earth is a resource with multiple issues put forward by multiple actors with diverse interests and capabilities for action (Chaps. 5 and 7, Box 13.1).

> **Box 13.1. Central Africa, rich in untapped natural resources.**
>
> Laurène Feintrenie
>
> In Central Africa, agro-industries and extractive mining industries have become key sectors in national economic strategies, while the forestry sector and environmental policies find a decreasing place in deliberations to revive the economy (Megevand, 2013). Mercer *et al.* (2011) estimate that over 85% of the platinum reserves, 60% of the cobalt and 75% of the diamonds still lie unexploited beneath the soil of the region. This treasure remains untapped because production costs are considered too high due to factors like site inaccessibility and lack of transport infrastructure. However, the increasing value of these resources in the international market could change this situation.
>
> Until now, industrial investments in the region have been limited because of poor governance, political instability and its associated risks, and the lack of infrastructure (Megevand, 2013). These conditions have started improving significantly since the beginning of the century and more and more projects are seeing the light of day (Feintrenie, 2013). Liberalization of the last few decades now allows consumers and enterprises to seek natural resources they need from all over the world (for consumption and production). Countries need no longer depend solely on resources available within their borders; they can rely on multinational corporations to carry out explorations in untapped regions of Eastern Europe, Africa and elsewhere to expand their activities (Anseuuw *et al.*, 2012). Deininger and Byerlee (2011) estimate that the countries in the Congo Basin account for 40% of the arable land in sub-Saharan Africa that is neither cultivated nor protected and with a low population density, and 12% of such land that is considered available in the world according to these criteria. If natural or planted forests are excluded, the region still accounts for 20% of available land in sub-Saharan Africa and 9% globally. Finally, large areas of high environmental value (especially natural forests) are set aside for conservation purposes, thus reducing land area available to agriculture in general and family farming in particular.

What is the place and future for family farming systems in this scramble for land and its known or hidden riches? Will family farming that uses locally available resources and feeds rural populations be gradually replaced, as predicted at the turn of the twentieth century, by agro-industries that better meet urban population demands, both for food and manufactured goods? Although the relative importance of family farming is difficult to estimate from an economic, social and food-availability point of view because of the informal nature of their activities and the lack of statistics (Chap. 8), we are still far from a global territorial domination by agro-industrial systems (Chaps. 2, 7 and 8). However, it is clear that there has been an increase in industrial projects since 2000 that involve large-scale land acquisitions and in urbanization, which is a massive concomitant phenomenon that is expected to result in the 65 % of the world's population living in cities by 2015.

Family farmers regard the earth as their heritage and their means of agricultural production. It is the provider of the natural resources they exploit and ecosystem services they use or generate (Chap. 6). Their responses vary, however, when they are confronted with the development of agro-industrial projects. Some responses take the form of specialized family farming systems in supply basins of factories and agro-processing industries[2] working in conjunction with them:

- A first strategy is to form partnerships with industries. A plantation and an agro-processing industry can be associated to family farms via a contract, which could be based on patterns of nucleus estates and smallholders or alianzas (Chap. 5);
- Farmers can also form professional agricultural organizations (PAO), such as cooperatives or farmer groups, and the PAO can then enter into a partnership with an enterprise[3];
- Farmers may also choose to take advantage of a demand for a specific agricultural product required by an industry in the area, without entering into a contractual relationship (Chap. 7).

Conversely, family farmers may choose to oppose industrial projects and put up a collective resistance, often with the support of local or international NGOs.[4] And finally, urbanization or industrial projects can displace family farming to new agricultural frontiers, leading to problems of deforestation, as seen in the Amazonian region where family farmers are pushed further into forests while their land is taken over by agro-industries producing soy or sugarcane (Chap. 7).

Beyond the question of sharing the land and its products, we must also find solutions to issues and consequences of the massive manpower needs of industrial projects that require large land surfaces. The Atama Company project in the Republic of Congo, for example, plans to hire 27,000 people for its oil palm

---

[2] The great eastern plains on the island of Sumatra in Indonesia were thus gradually covered by independent family plantations, first of rubber and then of oil palm as an extension of agro-industrial projects in contract-based partnerships with family farmers.

[3] This model is now very common in cocoa production in Cameroon (Chap. 9).

[4] The strong opposition to Herakles Farms, a proposed oil palm plantation, in Cameroon illustrates this type of response (Feintrenie 2013).

plantation spread over 180,000 ha. Similarly, Olam-Gabon, a company that grows rubber, oil palm and log timber, plans to hire 19,000 people by 2020. The local population density in both these cases is low in the exploitation sites and an immigrant workforce will be necessary, either from within or outside the country. If such a migration of labor force is not well planned and managed, the resulting human pressure and ethnic and cultural diversity can become sources of conflicts.

More generally, the increasing pressure on land caused by agro-industrial projects, mining projects or population growth upsets equilibriums and challenges existing methods of managing natural resources. The population density in the Atama plantation, for example, will rise from less than two people per square kilometer to about eighteen in the long-term, resulting in a sharp increase in demand for food crops and animal protein, and consequently a pressure on natural resources, especially bush-meat. This will jeopardize the sustainable use of natural resources practiced by family farming systems that are already located there.

These challenges often highlight asymmetries of resources and bargaining power which put family farming at a disadvantage. Various procedures have thus been established by governments or international organizations (ILO) to regulate industrial investments that require large amounts of land (such as for mining and agro-industry) in order to ensure the sustainability of the projects as well as a due respect for the right to land. These procedures are designed to reduce the risk of land grabbing and pressure on resources at the expense of local populations. For example, the acquisition of a large area (long lease, concession, lease or ownership) for an agro-industrial or mining project generally requires environmental and social impact assessments (ESIA) and the submission of an impact management plan. It is also becoming increasingly common to get a free, prior and informed consent (FPIC) in writing from potentially affected communities. The FPIC specifies what the company has committed to doing: undertaking public works, developing social support systems, and possibly including interested people from the impacted communities in the project (direct jobs, partnerships in production, small plantations under contract with the factory, etc.). Certification processes link these legal processes by providing market incentives for their implementation. Thus, impact studies, FPIC and accompanying specifications, consequently, become prerequisites for certifications like RSPO (Roundtable on Sustainable Palm Oil, for certification on sustainable palm oil) and FSC (Forest Stewardship Certification, for timber). Certifications are official and transparent procedures based on voluntary application that reinforce controls and the supervision of impact management plans, and closely monitor commitments made by companies in their specifications. They thus help local and national authorities who do not always have the means and skills to monitor projects on a regular basis.

Much remains to be done to improve these measures. However, it has been observed that when the negotiations with local actors for establishing an agro-industrial project are conducted in a transparent and participatory manner, they often lead to productive partnerships between an enterprise and family farming systems, while reducing risks of land expropriation (Chap. 8). Without regular monitoring by the government of compliance with laws and adherence to regulations pertaining to social and environmental impacts, the implementation of impact management plans and compliance with specifications agreed upon with farms and local people will largely depend on the company's goodwill. It is therefore the demand by the end consumer targeted by the company which has the greatest influence on the industries' behavior. However, in such cases too, it is currently difficult to measure the impact of this sudden increase in standards on curbs on or, on the other hand, the accelerated growth of agro-industries at the expense of family farms.

13.3 Access to Water for Agriculture: Beyond Technologies, Policy Choices to Be Made

Agriculture accounts for, on an average, 70 % of worldwide water consumption. Many countries are thus pressing farmers to reduce their consumption in a context of increasing global demand for both water and food (Brauman et al. 2013). About 18 % of all agricultural land makes use of irrigation, and produces over 40 % of the global food supply. Although irrigated agriculture's per-hectare productivity is higher and more regular than that of rainfed agriculture, it also implies from greater requirements. Irrigation is essentially based on the intensification model of the Green Revolution which is input- and energy-intensive. Reducing the use of inputs to protect the environment runs counter to decades of effort to promote agricultural intensification (Heong et al. 2013). A reduction in water consumption and better management of inputs in order to preserve the environment form an intrinsic part of the challenges of managing natural resources in agriculture. For family farming, especially irrigated family farming, the agronomic, economic and environmental challenges that need to be addressed are immense.

Irrigation infrastructure necessitates variable – but always large – investments (3,000–20,000 Euros per hectare), a cost that will need to be recouped over the long term (20–50 years). Since small family farmers normally cannot afford such investments, they must be subsidized by the State or by external funding entities (who can then greatly influence public agricultural policies) or by public-private partnerships (PPP) which, under the guise of investments, are likely to monopolize resources (Box 13.2). Access to water also significantly impacts land prices and sometimes even its legal status, with major social impacts. The rationale behind the management and inheritance of land and water rights is the result of a progressive social elaboration. In many cases, the introduction of irrigation, its supervision by the State or its modernization through public-private partnerships challenge pre-existing social structures.

Box 13.2. When a rush hides another: land, water and capital in Office du Niger (Mali).

Jean-Yves Jamin, Thomas Hertzog and Amandine Adamczewski

Following the withdrawal of major international donors from irrigation projects in Office du Niger in Mali, the Malian government approached domestic and foreign investors with the idea of developing a "modern" agribusiness model alongside existing family farming systems. The last few years have thus witnessed a veritable "land rush" with all kinds of alliances developing. Each investor used contacts with one or other government ministries to attempt to corner the land with the best conditions; the most influential among them interacted directly with the President's office. Farmers in the area quickly reacted to what they described as land grabbing. They organized themselves and mobilized civil society* on the one hand and, on the other, used their networks to get their own land concessions and develop them (Adamczewski et al. 2013a; Chap. 7, Box 7.5).

However, behind this land rush lies a scramble for water. Water is a basic necessity for irrigation. Yet, in the Sahel, while there is an abundance of water in major rivers like the Niger, their flows are erratic, both during the year and from 1 year to the next: the flow of the Niger river can thus vary from more than 5,000 m3/s during floods to less than 50 m3/s in the dry season, or even as low as 10 m3/s once every 10 years or so. Following the construction of the Sélingué dam, this minimum flow has risen to about 80 m3/s and flood levels have been lowered (which has reduced recession agriculture surface area). Since the dam is primarily meant for electricity generation, a multi-usage issue is added to water sharing issues between farmers, as it is for most such major rivers.

In much the way they have done for land, investors have forged alliances at different government levels to ensure their access to water. The Sino-Malian sugar company signed an agreement with the Office du Niger bureau to obtain the "water required" for its needs. A South African company obtained a guaranteed water flow of 20 m3/s. Libya obtained rights to unlimited water supply from Mali to meet the demand of its project in the region (Hertzog et al. 2012). This jeopardizes small farmers' access to water in the dry season, thus limiting their ability to grow a second rice crop and to undertake vegetable gardening during the off-season, which are important aspects for their economic survival given their small plot surface areas.

A more equitable sharing of water is not impossible, but it would require all stakeholders in the area to come to a consensus, and to balance power and information asymmetries. This would also entail, to be equitable on a larger scale, the taking into account of the needs of downstream regions as well as future upstream development projects, in Mali and in other countries of the Niger river basin.

*The "Kolongo Appeal" is an example: http://pubs.iied.org/pdfs/G03055.pdf (retrieved 13 June 2014).

The high costs of operations and maintenance associated with irrigation schemes compel family farmers to be more dependent on external actors for inputs, credit and marketing channels (Jamin et al. 2011). The flip side of this dependence on external sectors, on the other hand, is that farmers with irrigated lands enjoy a higher standard of living compared to those with rainfed systems, when supply and marketing conditions are favorable. Irrigation's positive impact on income is undeniable, despite the fact that poverty still needs to be alleviated in such irrigated areas, especially in sub-Saharan Africa (Bélières et al. 2011).

In most agricultural contexts in the countries of the South, it is only the large agro-industrial farms that manage to obtain individual access to surface water, either by pumping it or from a dam. For the majority of small family farmers, a collective management of irrigation systems is necessary to distribute water, and share its costs, among beneficiaries. This entails major constraints related to collective organization in accessing the resource, conveying it and distributing it among the many users. Such constraints often result – and have historically done so – in the involvement of governments in the form of strict control measures.

Irrigation is a determinant strategic lever of agricultural development and its challenges pertain to several dimensions, including a significant political one. Moreover, governments have been compelled to abandon, either on their own or under pressure from donors and international institutions, a significant part of their control in managing collective public areas (Jamin et al. 2005). They were constrained to transfer all or part of the infrastructure and water management functions to water users' associations (WUA). Such associative structures, often created as required and based on an imposed model, are re-appropriated and adapted by farmers and various stakeholders of the irrigated sectors. WUAs are thus very diverse, as seen in North-West Africa (Riaux 2011) or in post-collectivist Asian societies like in Vietnam (Jourdain et al. 2011).

Going beyond the context of agriculture, the sharing of the water common resource between competing uses requires coordination with the rest of society, a coordination that is far from easy. Demand for water not only comes from the agricultural sector, but also from constantly growing urban and industrial sectors. In addition to its direct consumption, water is used in several other ways, for example for energy needs, most notably through the development of the hydropower and biofuels sectors. Its essential use for domestic purposes, the key role it plays in the functioning of agroecosystems and its strategic importance in every sector of activity accords this resource the status of a public good and justifies government control over its access and utilization (Perret et al. 2006). However, the massive expansion of irrigation in recent decades, especially the individual pumping of waters from rivers and groundwater tables, has taken place without any effective regulation nor any consideration for a balance between growing demand for and limited availability of water resources (Jamin et al. 2011). The need for such regulations has led to proposals for dialogue platforms, where different actors could agree on a common vision for the resource they share (Farolfi et al. 2010).

The necessity to limit water withdrawal has been translated into strong incentives for adopting water saving techniques such as drip irrigation. This technology is often seen as complex, expensive and designed for large farms that either have a large capital or receive government grants and expert guidance from advisory firms. Ameur et al. (2013) have however shown that in Morocco, once large foreign investors made the initial investment and set up systems, large Moroccan farmers, and later small family farmers, were able to adopt – and even adapt – this innovation. To explain this success, Benouniche et al. (2011) and Bouzidi et al. (2011) not only highlighted the role of the Moroccan government's subsidy programs but also those of local initiatives that are often informal in nature, and of the renewal of community links that allow, for example, the recycling of used material and the exchange of information.

Groundwater exploitation represents a revolutionary method of obtaining water. Water can be drawn from aquifers for irrigation purposes, thus making it less dependent on the collective constraints and strict rules governing the use of surface irrigation networks. This development has taken place with tacit, sometimes even formal, government authorization. Governments find groundwater exploitation to be an easy way to satisfy farmers and minimize, in the short-term at least, conflicts arising over water sharing and water saving. Such individual irrigation systems account for half of the irrigated areas in South Asia (Giordano and Villholth 2007), and are also well represented in the North-West Africa (Kuper et al. 2009). Nevertheless, in the absence of regulation, such easy access also poses dangers of the overexploitation of aquifers. As water table levels are dropping, in the near future only the richest farmers will be able to afford to drill deeper borewells. Any attempt to regulate groundwater use will have to take ecological, economic and social issues into account.

The goal of new agricultural and water policies should now be to support the local dynamics of family farming or small private companies (Jamin et al. 2011). However, the formalization of these dynamics by appropriate regulations or subsidies remains difficult because governments are not in a position to offer incentives. Besides, they tend to promote the public-private partnership model proposed by major donors and private and international investors. However, in a context where land (especially irrigated land) and water resources are already heavily exploited, there is a real danger of water grabbing, which, when added to the effects of land grabbing, could threaten the entire family farming sector. The danger is even greater in countries of the South where the governments' regulatory capabilities cannot measure up to the financial might of investors who seek to acquire land and water resources, and to the pressure from funding entities who favor public-private partnerships.

13.4 Climate-Smart Family Farming Systems?

Climate change poses several challenges to agriculture, especially to family farming, pertaining as it does to a wide range of interconnected processes which have yet to be fully assessed and understood. These challenges, in fact, encompass the ones described above. Farmers, researchers and policymakers alike are going forward without too much clarity but with a shared conviction that they are confronted by fundamental changes, and that it is only by adopting inclusive, multi-dimensional and systemic approaches that they will be able to mitigate climate change or adjust to its effects.

There is discussion now on a climate-smart agricultural model which focuses on food security by concurrently considering technical, environmental and policy issues, an approach that appears to be relevant given the multiple contexts of family farming systems. This model incorporates approaches that simultaneously respond to objectives of adaptation to climate change, mitigation of greenhouse gas emissions and achieving the key objectives of food security and well-being. These approaches are designed to implement the technical conditions of public policies and investment requirements that are necessary for sustainable agricultural development and food security in a scenario of climate change (FAO 2013b). Among the emblematic techniques – keeping in mind that no single technique can make agriculture climate-smart –, there are those, of course, which address issues of conservation and sustainable management of water and soil (responsible irrigation techniques, water harvesting, erosion control, organic enrichment, increase in soil biodiversity, cover crops, tree component of fields and landscapes, etc.).

Ongoing research on the intensification of irrigated rice cultivation in family farming contexts have shown, for example, that it is possible, by employing intermittent irrigation techniques, to reduce water consumption (adaptation) and to cut down methane emissions from the anaerobic decomposition of organic matter (mitigation), while boosting yields and improving quality (food security) (Joulian et al. 1997; Bouman et al. 2007). Potential inherent qualities of family farming, like the judicious use of natural resources and the selection of an optimum blend between available resources, which includes the climate, can easily enable it to become climate-smart. Such a transformation, however, can be all the more successful if technical and organizational innovations are made part of the necessary transitions.

An agricultural system that takes into account its carbon balance – especially carbon sequestration in aboveground and belowground parts of plants – mimics local natural ecosystems (Chap. 6). For example, cultivating cereal in monoculture would make little sense in a wooded savanna; it would be better to develop "perennial agriculture" (Perfecto et al. 2009) based on woody plants, cover plants or perennial grasses. Similarly, roots and tubers grown under shade are more suited to hot and humid climates. Decentralized generation of energy from available biomass or crop residue also provides a compelling example. It is a strategy of mitigating climate change which can also facilitate adaptation in the case of reliance on locally available biomass resources, and can contribute to food security by reducing expenditure on fossil fuels (Tatsidjodoung et al. 2012). Climate-smart agriculture is, finally, multifaceted: it cannot be reduced to a monoculture or specialization but rather takes the form of an integration between different agricultural activities, such as in farms that associate crops and livestock, or fish farming in rice fields.

The landscape dimension is particularly favorable to this integration and is a key criterion of arguments in favor of climate-smart agriculture (FAO 2013b). A landscape-level analysis helps in developing a rational integration of various components of a farm or a territory, and allows to better benefit from the potential of different agricultural systems and their adaptation to climate change. This scale allows the management of production systems and natural resources to cover an area that is large enough to produce ecosystem services, but small enough to be managed by users of the land that produces these services (FAO 2013b). One of the most remarkable examples in recent times of climate-smart agriculture at the landscape level is the "greening of the Sahel," especially in Niger (Reij et al. 2009). Development research activities, the implementation of decentralization and the transfer of the ownership of trees from the government to farmers have all helped revive the practice of assisted natural regeneration of trees in fields in Niger. The increase within a few years in per-hectare tree density has been remarkable, and has helped change the microclimate and soil fertility (adaptation), increase standing biomass (mitigation) and improve farmers' incomes and livelihoods.

A multi-purpose agricultural landscape will, for example, consist of a patchwork of fields, pastures, natural areas, wooded parcels, and protected areas. Each of these categories could include a diversity of crops, species and technical systems. Hedges and drainage channels could be maintained to separate the component parts of this mosaic, riparian vegetation (wooded areas along watercourses) could be protected, grassy strips could be planted along the periphery of tilled plots, etc. This spatial diversity can also be introduced within a farm by staggering planting dates between neighboring plots, by alternating plots or diversifying rotations in the case of mixed cropping, by planting bee-forage plants or even installing hives, or by maintaining hedgerows between fields, etc. This strategy to harness and use a wide range of natural resources is consistent with one of family farming's common goals, that of diversifying one's technical options in order to reduce risks (Chap. 6).

These diverse landscapes, mainly created by family farms, which actually protect natural resources, provide ecosystem services just as natural ecosystems do. In the context of family farming in developing countries, these ecosystem services pertain to the supply of drinking water, crop pollination, soil erosion control, development of local genetic resources and, of course, adaptation to climate change, all attributes that improve the resilience of these multipurpose landscapes. However, feedback on PES (Chap. 6; Box 13.3) reveals difficulties in application and negative effects or, at least, unexpected ones. This feedback raises questions regarding public policies that were intended to make agriculture climate-responsible, and suggests changes to payment mechanisms, which can sometimes be replaced by rewarding mechanisms, in-kind benefits (for example, access to land) or collective support (for example, construction of infrastructures). As highlighted in Box 13.3, new public policy instruments are currently being proposed and new avenues are being explored.

Box 13.3. The new REDD+ mechanism to fight against deforestation: a benefit for ecological intensification of family farming?

Alain Karsenty

In the perspective of including the REDD+ (Reducing Emissions from Deforestation and forest Degradation) mechanism in the post-Kyoto climate agreements, PES are assigned the key role of implementing the concept of "avoided deforestation" and are developing rapidly within the framework of REDD+ projects. In fact, in addition to the policies and measures that governments will have to establish to counteract deforestation factors, it will also be necessary to offer direct incentives to local stakeholders to encourage the maintenance of forest cover, regardless of whether it takes the form of conservation or sustainable use. REDD+ could provide the necessary funding to establish national PES programs aimed at reducing deforestation and promoting reforestation.

PES used in the fight against deforestation usually consist of compensation paid to land users to persuade them to comply with a given land-use plan and to pay them for planting trees. The amounts paid to "service providers" in the framework of a PES are determined through a more or less balanced negotiation and, in principle, should cover at least the basic cost of foregoing an activity (the opportunity cost) due to restrictions or changes in land use. Simply paying the opportunity cost to very poor farmers, however, raises ethical questions and justifies the exploration of different bases for compensation. In addition, compensating a shortfall in income resulting from giving up certain subsistence activities can free up work time but this is not helpful in the long term as it does not promote sustainable agricultural innovations.

An intensification of agricultural by itself is not enough to prevent deforestation, as was shown by the relative failure of "alternative to slash-and-burn" (ASB) programs of the 1990s: with the extra income obtained from intensification programs, farmers grew cash crops by clearing forests. This prompted the adoption of a two-pronged strategy of combining greater investment in technology with PES offers of direct incentives for preserving forests. Such payment would then help manage already cleared areas and encourage planting there of permanent crops using new and sustainable agricultural techniques. However, such grants would be meaningful only if they form part of a system that offers viable alternative agricultural technologies, rural credit programs and practices to ensure land tenure security by recording and mapping local rights. Such a system would need to be strengthened by an integrated support and agricultural training program to assist farmers and reduce the risk of failures.

Possible lines of action could include the devising of PES mechanisms for rural inhabitants who maintain these landscapes or compensating those who manage such multipurpose landscapes combining agricultural production with environmental quality through the granting of a landscape label. Thus, people working on these landscapes would be encouraged to conserve their originality and diversity (Torquebiau et al. 2013). Landscape labeling is a particular form of PES related to an act of production that adds value to all the "products" of a landscape (agricultural and artisanal products, of course, but also services such as ecotourism or accommodation).

13.5 Supporting Family Farming to Protect Natural Resources

Diversity in family farming is based on the need for local sustainability and the close relationship between agriculture and its immediate environment. The adaptation of crops and livestock to natural or manmade changes is largely based on a "proper" management of agricultural biodiversity in a broad sense, in the framework of a specific land and a specific climatic environment. The knowledge of local farmers allows them to utilize their land optimally by extending their production to fields considered unsuitable for intensive agriculture.

Despite being constrained by limited means and scarce resources, most family farming systems and their organizations at various levels have been able to preserve these resources and the knowledge associated with their operations in their diversity. However, this balance has sometimes been upset and is increasingly being threatened by changes at a global level, chiefly among them those pertaining to climate change and land insecurity. By attempting to maximize and standardize production factors, the market imperative is disrupting systemic links between agriculture and its environment's resources, upsetting equilibriums, and compelling farmers to consider new kinds of adaptations.

In a context of a global crisis, policy formulations and research efforts do not sufficiently take into account family farming's practices and the knowledge it has acquired; the proposed solutions tend to be more defensive and local in character. Despite the progress made through discussions in international conferences and the passing of in-principle resolutions, the solutions advanced are poorly understood and lack credibility. Proposals that are usually considered to be the most effective are built around market regulating mechanisms, and very often pertain to the individual sphere or, at best, to a plot or a production system. PES and their offshoots, considered the principal vehicles of such responses, do not effectively incorporate family farming rationalities. The performance of family farming – or indeed of any form of farming – should be assessed not solely on the basis of productivity, but rather on its ability to guarantee a sustainable future.

The concept of ecological intensification, discussed from the point of view of companion research in Chap. 18, foresees technical innovations and public policy mechanisms that are more consistent with the complexity of the problematic processes of depletion of, or competition for, natural resources. By starting with an objective of reducing external inputs to agricultural fields and actively relying on mechanisms intrinsic to production systems, ecological intensification finds itself at odds with principles of the Green Revolution which, despite having had a significant effect on incomes and poverty alleviation, has largely contributed to shaping current challenges humanity faces in managing natural resources.

Part IV
Research and the Challenges Facing Family Farming

Coordinated by Danièle Clavel, Michel Dulcire, Sophie Molia

The previous parts of this book show us that the structural diversity observed in the functioning of family farming models is historically the result of autonomous, diverse and non-marginal choices in the constitution of family agricultural heritage worldwide. The combination of multiple activities, including home consumption of part or all of harvested crops or of products obtained from natural resources, as well as employment and commercial activities, which are part of non-commercial and identity dimensions, have provided and still provide rural households of the planet with a livelihood and the means to feed themselves.

But the family character does not guarantee the durability and sustainability of family farming's responses, whether technical, economic or organizational; it also needs adequate public policy support, given that investment in family farming is often found lacking, especially in agriculture-based countries which are also among the world's poorest. Where the economic and social environment allows effective support to be provided to family farming systems, it gradually starts showing positive results not only at the local level, but also at larger scales depending on the types of partnerships mobilized.

But pressure is building and family farming – like all other forms of agriculture but with greater vulnerability – is faced with challenges, discussed in Part III, whose intensity and implications are too much for it to handle on its own. For example, faced with the dynamics of demography, food security challenges are even more difficult to overcome since family farmer populations are the first victims of poverty, which also tends to exclude them from conventional education and public health systems. They often have very little room for maneuver. Broadly speaking, the familial forms of agriculture are currently synonymous with vulnerability. The essential choices they have to make to meet the challenges they face pertain to the management of risk and a search for the "sustainabilization" of their individual and collective resources. It is clear that when conditions are not favorable to them, these vulnerabilities compel them to tap, usually unsustainably, into natural resources while undermining the reproduction of social equilibrium.

The paradox of family farming is that while it is vulnerable and can generate negative externalities, it inevitably also holds part of the solution. Therefore, a major challenge of targeted research is to support family farming models and family farmers in their diversity. The purpose of this support is not to offer standard solutions but to allow family farming to innovate, adapt, evolve and seize promising alternatives as and when they appear in order to express its full potential. This "custom" support thus has to take as many forms as there are of family farming itself, forcing research efforts to adapt, too, to novel, interdisciplinary and participatory intervention methods. It is a matter also of providing the direction to – and fueling – national and global debates on the possibilities and limitations of family farming, thus alerting policy makers and development decision-makers to potentially usable levers.

The work presented here is obviously not meant to be exhaustive. Many areas of research are not represented, for example, on programs for the reduction of fossil fuel consumption, on support for the negotiation of standards, on the management of natural resources, on support for the implementation of ecosystem services, etc. This fourth and final part of the book intends to illustrate, through some significant examples, the diversity of practices of targeted and support-oriented research which are coming to the aid of family farming and its organizations. Its aim is to show how the family component of agricultural production is addressed by research and orients its topics and methods. The structure of this part seeks to highlight a back-and-forth process of co-construction between researchers and other stakeholders. It also attempts to alternate between concrete case studies, oriented firmly towards practical action, and work which offers a theoretical perspective and backdrop. The research presented aims, on the one hand, to adapt the methods and protocols to the reality of the farming models concerned and, on the other, to bring about changes in family farming cropping and production systems and its activities, all the while keeping in view the challenges that need to be faced.

Chapters 14 and 15 are concerned with methodological aspects of the intervention on, for and with family farms, through the construction and implementation of research partnerships for innovation and through advisory approaches destined for family farms. They illustrate how such interventions endow actors with new abilities to confront and adapt to current changes and those to come, and how this methodological focus offers opportunities in return to learn more about family farming and its specific dynamics.

Chapters 16 and 17 show how the technical dimensions, such as the prevention of and the fight against health risks, varietal selection, or the management of cultivated agricultural biodiversity, must be firmly anchored in the complex reality of the functioning of cropping and livestock systems, and more broadly of social and cultural practices. These two chapters also show, through the use of these technical approaches with the farmers while taking their technical, economic and social constraints and assets into account, how targeted research leverages the specific local knowledge of family farming, and is inspired by it to innovate at all levels of risk management.

Chapter 18, based on the achievements and perspectives of ecological intensification, offers a more encompassing vision and a broader assessment of the practices of targeted research for development. Echoing the methodological dimensions and techniques developed in the previous chapters, it emphasizes the levers and considerable potential for improving production systems, while remaining clear and realistic about the difficulties of implementing the required changes. Much like the rest of this book, this chapter invites the reader to decompartmentalize his thoughts and to grasp fully the diversity of family farming models and their complexity.

Chapter 14
Co-constructing Innovation: Action Research in Partnership

Eric Vall and Eduardo Chia

Family farms today are confronted with uncertainties and continuously changing contexts. They are compelled to innovate constantly to develop and to adapt, not only by leveraging as much as possible their family's productive potential and often modest capital (see, in particular, Chaps. 6, 7 and 11), but also by actively maintaining and undertaking the multiple social and environmental functions that agriculture fulfills in these farms and for society as a whole (Part II). On the whole, however, very few of the many innovations proposed by research have been adopted by family farms. The reasons are that very often these innovations are out of step with the technical, economic and organizational needs of farmers, and also that family farms often have limited possibilities to change due to their rigidity and internal tensions. The combination of strategies of upstream and downstream actors, farm advisory systems and public policy (Chaps. 4 and 12) is also less than conducive to the adoption of these innovations.

Given this situation, the capacity of family farms to change needs to be strengthened. To this end, research efforts must focus on devising mechanisms that take into account the totality of the situation of the farm and the diversity of objectives pursued by the family members in order to produce actionable knowledge. This is the knowledge that helps define the technical and organizational conditions which have to be met so that the actors of family farm are able to adopt the proposed innovations. This is the topic of this chapter on action research in partnership (ARP).[1]

---

[1] The authors thank Nadine Andrieu, Michel Dulcire, Olivier Mikolasek and Eric Sabourin for their contributions in the writing of this chapter.

E. Vall (✉) • E. Chia
ES, Cirad, Montpellier, France
e-mail: eric.vall@cirad.fr; eduardo.chia@cirad.fr

14.1 A Long Tradition of Research for Development

Agricultural research has long come up with technical inventions in controlled environments (laboratories and field stations) to improve the management of crops and livestock herds. But this knowledge has not taken sufficient account of the family farm, nor of the strong organic links between the family and the production unit (Chap. 3). In practice, these links translate into combinations of domestic and farm rationales in the process of allocation of family labor and its remuneration, in the choice of product distribution between final consumption, intermediate consumption, investment and accumulation. Systemic approaches, such as research and development, started seeing the light of day in the 1980s to overcome the limitations of analytical approaches (Jouve and Mercoiret 1987). By moving research to the real-world environment and by studying practices and production systems, development research helped analyze the causes of problems and formulate hypotheses on possible solutions. However, in these approaches, the innovation proposals were prescribed to family farmers using a top-down logic. Its weakness lay in not sufficiently taking into account any dimension other than the technical, in particular the organizational dimension of innovation: the co-construction of change and empowerment of actors has have never really been addressed by research and development.

At the same time, researchers who faced the same types of problems as agronomists were developing action research (Liu 1997; see also Anadon 2007; David et al. 2001; Avenier and Schmitt 2007; Verspieren and Chia 2012). By according priority to solving the problems of actors on the ground and to producing actionable knowledge, action research seemed to the agriculture sector to be an improvement over research and development (Albaladéjo and Casabianca 1997; Sébillotte 2007). It proposes mechanisms to promote the active participation of all actors in conducting research and invites them to reflect on the options selected and the results obtained. Action research is a research approach that originates when the desire for change on the part of actors on the ground meets the willingness of scientists to undertake research (Liu 1997). It has a dual purpose: successful intended change and production of scientific knowledge. Action research is conducted within an ethical framework negotiated and accepted by all. The process is governed in a way which ensures the participation of all stakeholders in decision-making and in activities. Actors share roles to define the objectives, strategy, and planning and monitoring of activities in order to manage any tensions and to evaluate the eventual results.

However, action research does not put sufficient emphasis on the need to empower actors. Yet, in an increasingly competitive world (for access to natural resources, services, markets, etc.) and with the withdrawal of the State, family farms have a serious need to build up their self-sufficiency and autonomy in order to solve their problems and to seize any opportunities that may arise. To this end, they need to strengthen collaboration amongst themselves and with other actors in their economic and social environment. This observation led us to hypothesize that a strengthened partnership (between farmers and researchers), i.e., the deliberate decision to work together to achieve a common goal by sharing tangible and intangible resources, would better address the issue of actor empowerment in action research approaches (Chia 2004; Dulcire et al. 2007; Mikolasek et al. 2009b; Vall et al. 2012). This was the basis of the idea of action research in partnership (ARP). ARP is founded on the construction of a partnership of united and responsible actors whose work aims to understand problematic situations, to identify possibilities for change, and to select those that best meet their needs and those of future generations in accordance with negotiated and agreed upon values and objectives. An ARP pursues a threefold goal: to produce actionable knowledge (Avenier and Schmitt 2007), to solve problems of family farms, and to empower the actors concerned (farmers, researchers, etc.). ARP offers an analysis and problem-solving framework which takes the organizational dimension of innovation into account, such as the adaptation of family farms to local conditions, collective resource management, governance of innovation, or even its institutionalization. An ARP uses intermediate objects to develop a common representation of the problem, to discuss possible solutions, to facilitate dialogue with actors on the ground and to help present the knowledge produced. These are formal representations (sketches, images, text, simplified output of models, demonstrations and experiments, etc.) that are sufficiently intelligible to be manipulated and modified, and which have a direct connection with the activities. They can be used at various stages of the innovative design process to fulfill different functions: formalization, translation, mediation, etc. (Jeantet 1998). In an ARP, the researcher always participates actively in the problem's formulation. Sometimes he assumes a leadership and facilitation role in the process (Dulcire 2010). The operational aspects of this approach have been summarized in a handbook for practitioners on the ground (Faure et al. 2010).

The principles of ARP emphasize the development of relationship between farmers and their families, researchers and technicians:

- analysis within family farms of the process of allocation of resources, production, marketing and accumulation taking into account the relationship between technical and organizational dimensions;
- establishment of multi-actor (farming families, researchers, technicians, NGOs, political organizations, etc.) mechanisms, whose members set out a number of rules and define common objectives to form a collective which is united (through interest) and responsible (acting with full knowledge of the facts and in compliance with jointly established rules);
- involvement of the actor collective at all stages of the co-construction of the innovation: common understanding of the problematic situation to be resolved, collective exploration of possible solutions, choice of solutions that best meet the actors' criteria, and joint adaptation of these solutions to optimize the desired effects.

This chapter's goal is to discuss the usefulness and limitations of the application of these principles to the co-construction of innovation by and for family farms. To do so, we rely on the studies of actual cases in several countries (Burkina Faso,

Cameroon, Madagascar, Chile, Brazil, Costa Rica, Ecuador, etc.). Three sets of interventions serve mainly as illustrations: Burkina Faso (Téria, Fertipartenaires, Sustainable Intensification Options and Abaco), Cameroon (design of fish farming innovations) and Brazil (Unaï) (Box 14.1). This chapter will follow the three main stages of ARP: exploration of the situation and the formalization of the partnership, co-design of the innovation itself (milestones, outcomes), and finally the evaluation of the results and the disengagement of research. For each of these stages, we will highlight the usefulness and limitations of the ARP methods in terms of the family character of the farms concerned.

Box 14.1. A few emblematic projects of experiments of action research in partnership.

Olivier Micholasek, Éric Sabourin, Éric Vall

In western Burkina Faso (Vall *et al.*, 2012), over the course of four project – Téria (2005-2007), Fertipartenaires (2008-2012), sustainable intensification options (SIO) and Abaco (since 2011) –, ARPs have sought to co-design more productive and more sustainable mixed crop-livestock systems by using the principles of ecological intensification and by improving crop-livestock integration, cultivation techniques (association, conservation agriculture), livestock management (dairy, fattening of animals, draft animals) and the collective management of natural resources (drafting of the land charter). These ARPs relied on local committees involving farmers, researchers and technicians (village coordination committees – VCC). The context was that of family farms consisting of several households with an average of about ten individuals, of mixed farming (cotton, maize, sorghum, groundnuts, cowpeas) and livestock rearing (draft cattle, breeding cattle, sheep, goats, donkeys) and of the use of mainly animal traction equipment.

In western Cameroon (Micholasek *et al.*, 2009), the project for the design of fish farming innovations (CIP) brought together between 2005 and 2010 two Common Initiative Groups (CIG) – the Fishermen and Fish Farmers of Santchou (CIG-Pepisa with over 15 fish farmers) and the Collective of Intensive Fish Farmers of Fokoué and Penka-Michel (CIG-Copifopem with 20 fish farmers) – and a group of researchers from different disciplines and institutions (from the North and the South). It was a matter of co-defining conditions under which a transition to sustainable and durable fish farming could take place based on farming of small ponds by families of fishermen-farmers.

In Brazil (Sabourin *et al.*, 2010), the Unaï project (2006-2009) underwent a transition from a traditional development research approach to one based on ARP principles applied to the co-construction of technical innovations (direct sowing) and organizational innovations (collective marketing of milk and maize). Unaï has brought together over 100 producers out of a total of 400, seven advisors from organizations and six researchers. Unaï has established three thematic interest groups (direct sowing of maize, marketing of milk, development of Cerrado fruits) involving families of interested farmers, a coordinating technician and a researcher. The context was of agricultural families benefitting from agrarian reform, only recently installed, with heterogeneous origins, poorly educated and poorly organized (difficulty in accessing credit and markets). They were establishing mixed farming (rice, maize, beans, cassava) and livestock (dairy cattle) systems on small surface areas of often degraded land.

14.2 Phase One: Exploration and Formalization of the Partnership

14.2.1 Exploration

The exploration phase is crucial in instilling a desire for change on the part of family farmers and a willingness to conduct research on the part of scientists. The first step is to examine the situation which is causing problems to the actors by conducting a diagnosis. This is undertaken in a systematic and multidisciplinary manner. The diagnosis focuses on aspects as diverse as the biophysical conditions of the family farms, their diversity and their dynamics of change, the organization of space (access to natural resources, etc.) and the socio-economic environment (actors of services and sectors), production practices, the division of tasks, allocation of resources, the management of production, etc. It aims to understand the strategies of family farmers in their generality and in the face of the problem identified, i.e., the means they employ and the objectives they have. This diagnosis is undertaken through group interviews, individual surveys (sometimes detailed household surveys), and complemented by an assessment of available knowledge. It involves the participation of family farm actors to understand their representations of their problem(s) and of their situation as well as differences in points of view of the various members of the household. In Burkina Faso, for example, the diagnosis helped identify links between the size and wealth of families and the intensification and extension strategies implemented. This helped orient the search for solutions towards the integration of cultivation and livestock rearing for the poorest households (Vall et al. 2012). The diagnosis also highlighted very quickly the point that mechanized sowing would be difficult to introduce in Bwaba areas since manual sowing there is an activity traditionally assigned to women, and where the heads of farms having investment capacity are more likely to take advantage of low labor costs than to invest in a seeder.

During the diagnosis, key actors and potential partners are also identified for the purposes of creating a work collective. This part of the process is sensitive and takes time; it involves listening patiently and therefore many exchanges. But it helps gradually build a relationship of trust with farmers, a prerequisite to enrolling them in the ARP. To the extent possible, it is necessary to uncover and take into account the potential participants' representativeness, legitimacy, skills, relationships (potential conflicts and asymmetries, power relations or alliances), and displayed or hidden motivations in order to assess the feasibility of the ARP.

The research problem of an ARP is defined based on the outcome of the diagnosis. It actively engages family farm actors and researchers and takes place in three stages:

- developing an argument to establish links between the problems and the initial concerns expressed by the family farm actors and the problems' possible causes;

- the construction of research hypotheses to explain the causes of the problem to be solved;
- the construction of development hypotheses, i.e., possible solutions which are accessible to family farm actors, along with all the elements needed to make them feasible.

The design of innovative family livestock systems in Burkina Faso illustrates how taking the family's composition into account helps in the problematization. In general, family livestock rearing projects face issues of profitability. Feeding practices tend not to be adapted to the conditions on the farm. When several livestock projects have to coexist in the same farm – with the head of the family looking after cattle fattening using financial means and the wives looking after the rearing of small ruminants with labor being their main resource –, their difficulties are obviously not the same. For fattening projects (concerning the heads of families), the logic most often applied is of using cottonseed meal purchased from the market to reduce costs. With small ruminants (concerning the women), on the other hand, it is a matter of producing fodder and taking advantage of local biomass to meet the food needs of the animals without making expenditures.

14.2.2 Consent

This is the time participants (researchers, farmers, etc.) commit to the ARP, formalize their objectives, the reasons for their choices and the means they intend to use to achieve their goals. The actors' engagement in an ARP is first marked by mutual consent, which can be made official in a written or oral contract. This makes is it possible to take everyone's views and collective work into account. But such an agreement is not sufficient to guarantee the participation of all the actors. Indeed, the internal power relations between the head of the family and its dependents or between groups of producers (indigenous versus non-indigenous, for example) often prove to be barriers to the participation of marginalized populations (adolescents, young adults, women, foreigners, etc.). To encourage participation and reduce asymmetries between actors (farmers/researchers, family head/dependent, man/woman, etc.), technicians and researchers then put in place a governance mechanism for the ARP. This mechanism is designed, on the one hand, to ensure the widest possible participation in decision making and the research process and, on the other, to establish operational rules and an ethical framework that clarifies and embodies the values and principles that the collective's actors have agreed to comply with. Such a governance mechanism usually consists of several components:

- the steering committee, composed of representatives of institutions (research, development, producers) and farmer groups, decides the strategic orientations and plans activities. Its role may extend to arbitration in case of disputes between actors;

- the scientific committee facilitates a dispassionate distancing and methodological reflection, and assists researchers in the exploitation of results. It is composed of recognized experts on the issue concerned;
- finally, local committees are responsible for the functioning of the ARP and the implementation of the program of activities validated by the steering committee. Local committees include researchers, farmers and other actors (such as village coordination committees in Burkina Faso, the common initiative groups in Cameroon, or the thematic interest groups in Brazil). The role of farmer representatives in these committees is not easy. Not only do they need to defend the often contradictory interests of groups and families but they have also to espouse the general interest. A proper representation of family farm members in ARP governance bodies remains a methodological difficulty.

All ARP activities are formalized through an agreement or a comprehensive protocol, stating clearly what each participant has committed to – the objectives, work schedule, rules of procedure, and budget and allocations – all validated collectively (Blanchard et al. 2012; Mikolasek et al. 2009b). New developments during the ARP can lead the collective to modify the activities and recast the governance mechanism: new opportunities, limitations that were overlooked by the initial diagnosis, involvement of new actors, etc. In projects implemented in Burkina Faso, in order to ensure that cultivation and livestock activities were well represented in the local committees set up – the village coordination committees –, the actors had decided to divide its presidency and the vice-presidency between a farmer and livestock breeder. The actors also ensured that non-native communities and women were well represented in the executive office of the village coordination committees. In the Unaï project in Brazil, the problematization led to the establishment in three test communities of thematic interest groups (see Box 14.1). These modes of organization helped draw greater attention to the link between the family and the production and to better manage asymmetries between communities or even those existing within families.

In these local committees, their facilitators play a vital role in enabling dialogue, establishing a climate of trust and overcoming misunderstandings between different communities (farmers/livestock breeders, researchers/producers, etc.). They allow actors originating from different worlds – therefore from different cultural backgrounds and with divergent interests – to work together. This mediation function relies on individuals able to "translate" messages between actors, helping one understand the other. Their work helps produce a common language between ARP actors, and the taking into account of a comprehensive approach to farms and their environment, including the specificities of their domestic organization. This role is often entrusted to advisors and agricultural technicians, who, in most cases, are not trained for it. But this role can also be effectively filled by farmers who are adept at using social networks, exchanging information and knowledge and are familiar with specific family situations and who enjoy the trust of the community. In Burkina Faso, it took a few years of practicing ARP before such mediator farmers were revealed. Conversely, in some cases, locals with strong personalities can block the process when things are not going in the direction they wish, as was the case in Cameroon in some local committees of the project to design fish farming innovations.

ARP actor interactions can be arranged in the form of meetings, guided field or farm visits, study tours, open days, appraisal and planning annual general meetings, etc.

14.3 Phase Two: Co-designing Innovation

14.3.1 Stages in the Co-design of an Innovation

During the co-design phase, the ARP collective (farmers, advisors, mediators, researchers) gradually builds pathways to change by successively addressing the following questions: What are the possible options to address the problem? Which options best meet the criteria and constraints of actors and the ARP's objectives? How to adapt these options to optimize the desired effects? Are the results obtained satisfactory?

The first stage is devoted to the search for options. It calls upon the researchers' expertise and the local knowledge of farmers and agricultural advisors (Vall and Diallo 2009; Vall et al. 2009). The possible options are listed and discussed during get-togethers, including occasions such as local committee general meetings or steering committee meetings. These forums are used to decide the ARP's strategic orientations, without going into details of how options will be implemented – that step will come later.

In the next stage, we try to go from the possible to the practicable, taking into account the constraints of producers and of the research collective. This entails the collective exploration of the feasibility of possible solutions based on the results of the initial diagnosis, objectives and constraints of family farm actors and the possible modifications of the environment. This exploration requires the organization of meetings, research and training workshops, study tours and interactions with other communities, and simulation exercises. This issue is important because it pertains to the solutions to implement, with actors eager to guide this process according to their own interests. The recourse to the ethical framework may become necessary to help actors stay the course, preserve the collective interest, and deal with internal – and sometimes external – power relations. The identification of what is practicable is based on the production of intermediate objects (flow diagram of a family farm, transect of an agricultural landscape, etc.) representing the dynamics at work and the effects of proposed solutions (Box 14.2). Study tours help anchor such objects in reality (typical case studies), and tools to model family farms simulate *ex ante* the effects of changes on the performance of typical farms.

Box 14.2. An example of an intermediate object linked to the family nature of the farm: the Cikeda model.

Nadine Andrieu and Aristide Semporé

The Cikeda model, developed in Burkina Faso to simulate the operation of a mixed crop-livestock farm (Andrieu *et al.*, 2012; Semporé *et al.*, 2011), is an example of an intermediate object which takes the nature of family farms into account. Cikeda consists of seven modules reflecting the interactions between cropping and livestock systems on the farm. These are the resources of the farm (family labor, farm capital, equipment), the livestock rearing system, the cropping system, the feeding of animals, the production of organic manure, the fertilization and the farm's economics. Cikeda simulates the techno-economic operation of a farm over a year and allows the analysis of the impact of innovations such as changes to family labor, land expansion, modification of crop rotations, improvements in the production of organic manure, the introduction of a fattening workshop, etc. Based on data entered by the user (the farm's structural characteristics, strategic and tactical decisions, type of year), the model calculates three main balances: the mineral balance, fodder balance and net economic impact of agropastoral activities. Over time, this model's use has increased and a growing number of projects rely on it to explore possibilities for innovation and to simulate their impacts on farm performance.

To adapt the selected solutions to the local context and to optimize the desired effects, the process then enters a third stage. It consists of the implementation of the selected solutions in real-world conditions, i.e., in family farms or in their immediate environment if it is a matter of a collective innovation (collective construction of a product, management of a resource or infrastructure, etc.). Depending on the cases, the implementation of solutions can take very different forms.

- Some activities can be focused on the production of knowledge to strengthen collective reflection, explore possible situations, and help build decision-making capacities. It can be a matter of specific studies to look more closely at a particular sticking point (on actors' strategies, functioning of systems, territorial governance, the organization of sectors, etc.). For example, in the Unaï project, the diagnosis pertaining to the search for alternatives to maize cultivation included the careful socio-anthropological monitoring of instances and scenes of dialogue or of the divergence between knowledge of farmers and that of the researchers. This work helped reconstruct the research problem and propose innovative strategies for maize cultivation (direct seeding).
- In some cases, it is a matter of experiments conducted by farmers (selected by the local committees) at their farms to test a solution and to adapt it to the local context, taking into account the farm's strengths and weaknesses. When these experiments involve family farms, researchers consult with the farmer to select the location of the experiment on the farm's fields – which may require a reallocation of fields for the household (especially if the experiment occupies a large surface area) –, to identify those responsible for monitoring the experiment (and to explain the reasons behind the experiment's procedures, such as repetition), and to reflect on what each family member gains or loses with the introduction of the innovation. Unlike an experiment in a controlled environment, the farmer participates in the design of the protocol. More importantly, the experiment takes place in a context where the unexpected can intervene, thus making it necessary to suitably adapt data analysis procedures. Given the

variability of experimental conditions between the various family farms in the sample group, multivariate analysis techniques are necessary to analyze the results. They allow us to understand the conditions under which various innovative options can be implemented depending on the family farm context.
– Other actions can change the context of ARP actors more or less irreversibly. Examples from our work include the design of new modes of territorial governance – such as the drafting of a local land charter under Fertipartenaires – and innovative cooperation mechanisms within sectors or institutions – as in the case of the Unaï project. In this type of social experiment, it may be found necessary to change the way the ARP governance mechanism is organized as and when the various phases of the resolution of the problem are completed in order to ensure appropriation and sustainability of the results, or even their institutionalization.

The ARP includes mechanisms to monitor the results (technical, economic, social, etc.) and the behavior of family farm actors to analyze their reactions to the innovative principle being tried out. The collected data pertain to both the studied process and elements of the context in order to explain the results. Interim results are presented collectively in order to benefit from the advice of as many stakeholders as possible. This helps understand how the experiments unfolded and, in particular, strengthen the common language and develop new socio-technical references.

14.3.2 Results of the Co-design of the Innovation

Is the final outcome of the cycle of the co-design of the innovation satisfactory? This assessment often takes the form of participatory self-analysis undertaken with different groups of actors who participated in the ARP and with the different types of family farm members involved (Andrieu et al. 2011). An external evaluation can also complement this self-analysis by providing an external perspective on the ARP's relevance, effectiveness, efficiency, sustainability and impact. But it is never easy to find the balance between the commitment to action, on the one hand, and the distancing necessary to analyze the processes and to translate them into actionable knowledge, on the other (Hocdé et al. 2008). Such an assessment can be conducted at various levels.

14.3.2.1 Actionable Scientific Knowledge

An ARP produces actionable knowledge through the analysis of change. The analysis of change itself and the determinants of this change confirm – once they are ascertained – or refute the initial assumptions and provide information about the conditions under which they are valid.

An ARP enables researchers to understand the strategies of family farmers by analyzing them in real time and not a posteriori, and with their effective participation. In Burkina Faso, when the local land charter in Koumbia was being drafted, the participation of village representatives and the local authorities in the local committees set up allowed researchers to observe their behavior (cooperation, alliance, competition, domination), their representations (relationship with nature), their projects (relating to land, the development of agropastoral activities, etc.), their room for maneuver and the main determinant of the strategy of occupation of space – which is land saturation.

An ARP encourages systemic analysis. It thus leads the researchers to consider the farmers' rationalities by formalizing, with family members, their overall objectives, the planned calendar of farm activities, the rules and practices applied in order to understand the logic behind their actions (Mikolasek et al. 2009b; Chia et al. 2008). The references produced on farmer practices are based on their knowledge and representations (Vall et al. 2009; Vall and Diallo 2009), i.e., on the nomenclatures used by family farmers to manage their activities. They constitute the elements of a common language which facilitates dialogue and reduces the risks of misunderstandings in analyzing situations and finding solutions.

14.3.2.2 Solving the Problems Encountered

The ARP is deemed successful when its deliberate conduct confirms expected results, describes the methods used and the activities implemented, and specifies the path taken to arrive at the result. This then means that this path will be valid in similar circumstances, which thus confers a certain genericity to it.

Problems are solved based on the proposals of actors (researchers and farmers). The ARP allows the path to the solution to be adapted and fine-tuned as and when the objectives, constraints and strategies of farmers are discovered, since these are never known in advance, especially in family farms where family members have differing perceptions and viewpoints. Thus, at the end of the first year of the Fertipartenaires project, it was observed that the need for manure or compost pits for women could not be satisfied because the requirements for providing support to the pit construction project (commitment to build two pits) favored the heads of the farms. In the following year, the criteria for selecting applications for support were changed to allow women wishing to do so to install at least one manure or compost pit.

Through the establishment of local committees, ARP helps organize frequent meetings to discuss management methods and household strategies (work organization, family management of productive capital, product distribution and home consumption, etc.) such as crop cultivation and livestock rearing methods. The difficulties in functioning that arise within family farm households often have causes at higher scales. Local committees are also forums where these issues can be thrashed out by inviting the relevant actors. This is what was done in Koumbia, Burkina Faso, during the drafting of the land charter, where a law firm and representatives of administrative services were involved in the process. The ARP can contribute towards an improved functioning of producer organizations (associations, cooperatives, inter-professional organizations, sectors) and local authorities (land management, access to natural resources), and modes of coordination among institutions (development, research and producer organizations).

It is also common for an ARP to arrive at unexpected results due to the involvement of new actors along the way, or because some constraints and resources had been overlooked during the diagnosis phase. We can then compare the results to the initial assumptions and seek to explain the differences. In Cameroon, for example, local committees have not always worked at their best because of power games played by some local actors. Because of this, experiments which were being conducted in ponds of producers (fish density, feeding, etc.) did not produce the expected actionable knowledge. Project participants then proposed to hold a fish-farmers' competition so that they could meet, share and define the sociotechnical framework for innovation. The idea was inspired by contests held during agricultural fairs which play an important role in genetic selection and management of local breeds (Labatut et al. 2001). The fish contest proved to be a more effective intermediate object than experiments to produce techno-economic (local fish production model) and social (fish-farmer's profession) meaning and references. It led to the co-construction of a regional fish farming manual.

Sometimes the ARP does not lead to innovations, because some constraints, resources and relationships were overlooked or only surfaced once the ARP was underway. The work of problematization and the formulation of new hypotheses has to be done anew. Sometimes it may even be necessary to start a new ARP cycle. Finally, if the experiment fails without any discernible reason, it becomes necessary to repeat the diagnosis.

14.3.2.3 Capacity Building of Actors and Reducing Asymmetries Between Them

The ARP helps build the capacity of actors to undertake research and process information, mobilize partners, build alliances, and test and evaluate implemented solutions. To begin with, it helps actors construct an argument on their situation and the causes of the problem. It then allows them to join forces, pool their resources and knowledge to build collectives where they can share a common understanding of the situation, the objectives, the means to implement and the values to uphold. It helps actors acquire the know-how necessary to experiment with innovative solutions and to validate the results obtained through the formulation of hypotheses, setting of objectives and planning of activities. These lessons empower the family farms, i.e., reinforce their ability to cope with similar problems in a similar situation without calling on external support (Chia et al. 2008). Finally, the empowerment of actors contributes to the sustainability of the ARP's results. In Burkina Faso, for example, of the seven village coordination committees set up by Fertipartenaires, only one still is supported by projects (Koumbia); the other six are no longer concerned by the new wave of projects. However, 1 year after the end of Fertipartenaires, of the six village coordination committees, three remain active without external support and continue to meet and conduct experiments on the production of organic manure, on minimal plowing, single-bovine traction, etc.

The success of an ARP depends also on the capacity of the research collective to manage the various dimensions that make up the initial asymmetry between actors, especially between farmers and researchers, in terms of both intangible and tangible resources. The ARP reduces disparities by guaranteeing the sharing of information, the right to be heard, participation in decision-making, equitable access to material resources, etc. If necessary, a contract or agreement can formalizes these guarantees in writing. Thus in Burkina Faso, in the Fertipartenaires framework, all partners, including the Union of Provincial Cotton Producers for the region of Tuy, drafted a budget together. This union and village coordination committees entered into agreements with local committees to strengthen their capacity to support producers, improve their self-reliance (empowerment) and initiate a reflection on the role of the advisor.

14.4 Phase Three: Reviewing the Results and Disengagement

14.4.1 Reviewing the Results

An ARP creates a dynamic of production of actionable knowledge, problem solving, capacity building of the actors (farmers and researchers) and reduction of asymmetries (farmers/researchers), shown schematically in Fig. 14.1, through the three main phases of the process. But implementing this dynamic is not easy; too often it comes up against pitfalls and difficulties.

The production of actionable knowledge and the resolution of farmers' problems are rapid and sustained during the first phase, as they proceed from the initial diagnosis and analysis of actor networks. During this phase, actors get to know each other, and they move forward cautiously and probably more carefully than in the following phases. Consequently, the results of the initial diagnosis must be reviewed and refined as the ARP progresses. The governance bodies set up need to be flexible enough to incorporate new actors, especially those who are socially less advantageous and who are not always present at the start of the research, or to separate from "relational offenders" who do not accept the rules established by the collective. During the second phase, the rate of production of actionable knowledge varies because it depends on the success of the experiments, but it is at this stage that the solutions to problems are gradually built. Finally, in the third phase (reviewing the results and disengagement), the production of knowledge is important (scientific and technical publications). This is also a phase of exploitation of results. For researchers, the time between the start of the work and the publication

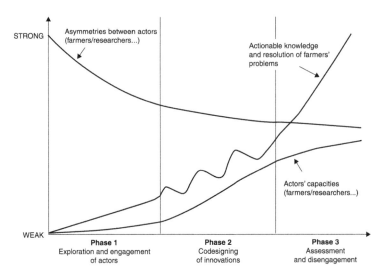

Fig. 14.1 Phases and dynamic of production of results of an action research partnership

phase is longer with ARP than in more traditional research methods. The results must also be framed so that farmers understand them, which is not easy given the cultural differences and disparate levels of education even within farm families, where one is likely to find some members who are literate in the local language, others who have had formal education (the youth) and always a significant proportion of people who have never been to school.

Actors' capacities increase rapidly during the second phase when agreements, which were entered into at the end of the first phase, have helped distribute roles. Then, if the ARP was successful, learning continues and the empowerment of actors increases during phase three.

- Farmers build up their capacity to innovate and build new production reference bases. Through ongoing dialogue between actors (researchers, farmers, technicians, etc.), exchanges of local know-how and scientific knowledge, experimentation cycles and testing of new practices, and breaks for reflection and discussion that are part of it, the ARP supports farmers in the gradual giving up of their normal agricultural practices and the adoption of new and innovative practices. But this process is often hampered by actors who do not desire change, and who occupy center stage when the research process begins. This is where a good facilitator, or a researcher's intervention, can clear the roadblocks preventing the situation from progressing.
- Researchers strengthen their capacity to produce knowledge in action. But to do so, they have to be convinced that it is possible to produce knowledge on subjects and objects they are dealing with. For many researchers, this is a difficult epistemological threshold to cross because they are used to a mode of production of knowledge based on laboratory experiments and field observations (David et al. 2001).

Finally, as far as asymmetries between actors are concerned, they decrease gradually without ever completely disappearing. Reducing asymmetries is a challenge and requires expensive and non-reproducible investments in training and prolonged technical monitoring and assistance. Nevertheless, a partial reduction of asymmetries still remains a prerequisite for a good start to and functioning of an ARP.

14.4.2 Disengagement

Every ARP has a beginning and an end! And this must be planned to avoid creating false expectations (on the side of actors) or a conversion into a system of advisors or experts (on the side of researchers). It is therefore preferable – right from the start of the ARP's contractualization phase – to clearly state the conditions subject to which the ARP is being implemented (start and end dates, if they are known, funding available, etc.) and to set realistic goals and verifiable indicators to monitor the progress of the process. It is a matter of being able to decide whether to stop or continue activities depending on the progress made.

In theory, an ARP can be terminated when the actors feel that the research objectives have been achieved and the desired change has taken place. For family farm actors, that time is when projects designed to bring about changes have been successful or when they find themselves sufficiently empowered and confident to pursue the action initiated by the ARP on their own. For researchers, that time is when they are able to validate the propositions that explain the phenomena studied and transform them into innovation proposals. But there is no guarantee that the outcomes desired by the researchers and those by the on-field actors are attained simultaneously, especially when the activities are funded by a fixed-term project, which is generally the case. It is frustrating for the ARP partners when disengagement is triggered by the funding being stopped before the desired objectives are achieved.

A disagreement or crisis between the actors can also lead to disengagement from the partnership for some of them: lack of effective interest in the project, widening asymmetries experienced by some actors, violation of the ethical framework, lack of ability to deal with the problem, etc. In such a situation, the arbitration mechanism forming part of the system of governance is invoked to help actors negotiate a disengagement, to draw lessons from the work already done, and to preserve the possibility of future collaborations.

The conclusion of an ARP is usually marked by an important event at the end of the project (workshop, conference, etc.) during which the actors present the results, draw lessons from successes and failures, sketch out possible perspectives, share the work with other actors – for the researchers, this means, in particular, the scientific community.

But in any case, the disengagement can take different forms depending on the dynamic established during the project (Box 14.3).

> **Box 14.3. Forms and changes of the disengagement.**
>
> Éric Vall and Eduardo Chia
>
> *Opportunistic re-engagement.* In Burkina Faso, after the end of the Teria project, the core of the research collective, taking advantage of calls for proposals, reformulated a series of projects based on ARP principles (Fertipartenaires, Abaco, Sustainable Intensification Options) in the same region by involving new partners, specifying the research themes and adapting the mechanisms (from local committees to innovation platforms), while remaining focused on the co-design of the agroecological transition.
>
> *Empowering disengagement.* The disengagement phase had not been part of the planning of the Unaï project in Brazil. This led to a feeling of abandonment on the part of the actors when the time finally came. But local actors (technicians and producers) took advantage of this phase to strengthen their capacities of reflection and action. The disengagement was long because of the difficulty of empowering farmers and of making thematic interest groups self-sustaining. This withdrawal phase was thus marked by intermediate steps: the training of technicians and the establishment of a technical assistance body and a cooperative for producer associations; and efforts to reduce asymmetries through information and training of farmers at the technical level for interest groups and at the methodological and strategic level for leaders of producer associations.
>
> *Planned disengagement.* In Cameroon, the disengagement of the CIP project took place gradually as foreseen in the agreement. It was used to empower local actors through the local fish contest which led to the collective development of a manual of regional fish production. But the unwillingness to follow actors and the extent of work required to rebuild a research program led to a momentary halt of the ARP.

14.5 A "Tailor Made" Approach Rather than "Off the Shelf"

The ARP is intended to be adapted to the specificities of the problems faced by family farms. It thus engages farmers actively in the analysis of the problems and the solutions proposed on issues such as the relationship between domestic (family) and productive rationales in the processes of allocation of labor or investment, and accumulation choices or of allocation of monetary resources between the production cycle and the satisfaction of family needs.

A successful ARP is able to define technical and organizational conditions that have to be met in order to unlock productive systems such as family farms and to enable processes of innovation to move forward, all the while respecting the farmers' rationalities. It allows solutions to the problems of family farms to be co-constructed by taking into account the high level of uncertainty they are subject to, by emphasizing a systemic approach (to fully understand the effects of change on the functioning of the farm) and one that is appropriately suitable (adapted to the diversity of needs and situations). The changes that family farms need to undergo must also be designed collectively by pooling the strengths and resources available not only in the producer groups but also among a family farm's members (head of the farm and dependents). The ARP helps all actors (researchers, farmers and other stakeholders) gain a shared understanding of the issues to be addressed, explore possible options for development, choose options that best meet their criteria, and adapt these options to optimize the desired effects. The ARP intervenes using multi-stakeholder mechanisms and intermediate objects. These mechanisms and devices promote the exchange of knowledge (between farmers, scientists and other stakeholders) and the development of a common language. The ARP produces hybrid knowledge (local/global) and actionable knowledge. It empowers actors and in so doing stimulates their ability to adapt to future changes.

The success of an ARP depends on the quality of the exploration, where not only the actors' problems have to be identified but also the key actors and those who are good mediators. It is these key actors and mediators who can recruit new actors, forge strategic alliances and, in particular, promote learning to build trust. Trust is critical to the ARP's success. It allows governing bodies to function smoothly, minimizes tensions and reduces asymmetries between the actors. This same logic of specific support to family farming can be applied to other forms of support, and the next chapter thus examines – while remaining vary of the application of ready-made approaches – how advisory services for family farms have evolved, and must continue to do so, through co-construction.

Chapter 15
Innovations in Extension and Advisory Services for Family Farms

Guy Faure, Michel Havard, Aurélie Toillier, Patrice Djamen Nana, and Ismail Moumouni

Extension and advisory services for agriculture are recognized by actors of agricultural development to be an important factor in improving farm performances. Their reputation is mainly based on their contribution, in a way similar to action research in partnership (Chap. 14), in weaving links within an innovation system[1] between farmers, researchers, agricultural-education actors and others from civil society. Nevertheless, extension and advisory activities are regularly called upon to defend their ability to meet the different and sometimes contradictory expectations of producers, various actors of agricultural sectors and other groups wishing to orient activities in the agricultural sector. This debate takes place in a context marked by the withdrawal of the State from providing services to farmers and by the emergence of new actors in the domain of advisory services – producer organizations, NGOs, and private actors (upstream and downstream companies). There is also recognition of the complexity of the challenges of sustainable

---

[1] An innovation system is defined as a "network of organizations, enterprises, and individuals that focuses on bringing new products, new processes, and new forms of organization into economic use, together with the institutions and policies that affect their behavior and performance" (World Bank 2006).

G. Faure (✉)
ES, CIRAD, UMR Innovation, Montpellier, France
e-mail: guy.faure@cirad.fr

M. Havard • A. Toillier
ES, CIRAD, UMR Innovation, Bobo-Dioulasso, Burkina Faso
e-mail: michel.havard@cirad.fr; aurelie.toillier@cirad.fr

P. Djamen Nana
African Conservation Tillage Network (ACT), Ouagadougou, Burkina Faso
e-mail: djamenana@yahoo.fr

I. Moumouni
Université de Parakou, Parakou, Benin
e-mail: mmismailfr@yahoo.fr

© Éditions Quæ, 2015
J.-M. Sourisseau (ed.), *Family Farming and the Worlds to Come*,
DOI 10.1007/978-94-017-9358-2_15

development in rural areas, requiring an expansion of the scope of advisory services well beyond agricultural production alone.

There exist multiple definitions of agricultural advice. In our reading, it includes, on the one hand, the actors involved in provision of advice, the means employed to deliver advice and the rules defined to achieve the objectives they have set themselves. On the other hand, it encompasses intervention methods and tools used by advisory actors, most notably the advisor, to create knowledge, skills and know-how through individual and collective learning processes. The farmer can access different types of advice, defined as much by its content (technical, economic, social, environmental, etc.) as by the way it is provided: dissemination of information and techniques, strengthening of learning processes, facilitation of interactions between actors, etc. Advice can thus be provided in many ways and from several different standpoints. It most commonly takes one of two forms: the dissemination of generic messages based on knowledge produced by research or the co-construction of advice between the advice seeker and advice provider to address a specific problem.

In this chapter, we discuss the evolution of extension and advisory services oriented towards family farming. We focus in particular, on the one hand, on institutional developments and, on the other, on advisory approaches. To this end, we rely on examples especially from Africa with experiences on management advice for family farms (MAFF) and from Latin America. We also explore the contribution of research, especially CIRAD's, in strengthening advisory mechanisms.

15.1 From the Withdrawal of the State to Privatization

The first agricultural advisory services were launched in the late nineteenth century in Europe through universities or national Ministries of Agriculture to promote new agricultural techniques, in particular the use of chemical fertilizers. After the Second World War, even though some private entities had emerged to provide technical and economic advice, especially in France and the United States (Chombard de Lauwe et al. 1969), advisory services remained predominantly the domain of the public sector. After the period of decolonization, and more so with the promotion of the Green Revolution in the 1970s, agricultural extension was largely supported by the nascent nation states through their Ministries of Agriculture or parastatal development entities. Their emphasis was on disseminating selected varieties and inputs through strongly prescriptive advice for farmers and especially for those engaged in growing cash crops. This period saw the first experiments in promoting advice more responsive to family farming's diverse needs. These experiments were generally designed to be undertaken with the support of public sector organizations. They are known under various names such as farming systems (Chambers et al. 1989), research-development (Jouve and Mercoiret 1987) or emancipatory learning in Latin America (Freire 1973).

Globally, the most significant institutional changes occurred during the 1980s in the context of structural adjustment plans to reduce public spending. They resulted in a marked decline in public extension services, judged to be not effective. The most obvious manifestation of this decline was the phasing out in many African, Asian and Latin American countries of agricultural extension projects financed by the World Bank and based on the "training and visit" approach (Anderson and Feder 2004).

This withdrawal of the State has taken various forms (Rivera 2000): a decentralization of services to the regional level, financed by regional public funds; a delegation of public services of the State to private companies, financed by national public funds; a commercialization of services by public institutions, financed jointly by the State and producers; and, finally, full privatization with companies selling their services. Promoting a private sector which is both more widespread and more effective requires rethinking the relationship between public and private advisory services and therefore rethinking the role of the State. Indeed, the transition from a public system to a private advisory system is a complex reform to undertake. On the one hand, the government may have to continue to provide services in the most deprived areas and to the poorest farmers (Anderson and Feder 2004; Kidd et al. 2000). On the other, such a transition requires a clarification of the role and functions of each actor, economic mechanisms to fund advisory activities, service providers with appropriate skills, and farmers able to clearly formulate demands and participate in the development of responses. Finally, privatization of advisory services means that the State will have to develop new functions to regulate the relationships between actors and to ensure that collective interests are properly taken into account (Rivera and Alex 2004; Labarthe 2005; Klerkx et al. 2006).

Privatization of advisory services is sometimes understood to be a means for transferring the costs of services from the government to the final beneficiaries. Some private advisory organizations with commercial relationships between clients and suppliers have indeed proven their effectiveness in the case of intensive agriculture in developed countries or of high added-value agriculture (Kidd et al. 2000). Nevertheless, it is generally recognized that the majority of family farmers, and not only those in developing countries, are unable to bear advisory costs on their own (Klerkx et al. 2009). The debate on the issue of funding advisory services thus revolves around possible funding mechanisms. In fact, it is possible to combine the provision of advice by a public or private organization with public or private funds (Birner et al. 2009), for example when the State contracts with a private provider to provide advisory services whose contents is defined by the State or when the private sector buys advisory services provided by the public sector. An operational partnership between private and public actors appears to be necessary in any case to encourage an extension and advisory system which promotes innovation in rural areas (Swanson 2006; Alarcón and Ruz 2011).

Nevertheless, the privatization of advisory services is not without risks. It can lead to negative consequences (Kidd et al. 2000; Labarthe 2005; Klerkx et al. 2006) such as limiting the dissemination of innovations that address the complexity of the production system; according priority to the most profitable advisory activities; emphasizing technology transfers over capacity building of producers; reducing exchanges between farmers who refuse to share information for which they have paid; or favoring only those farmers able to afford advisory services.

However, not all governments share the same desire or have the same resources to undertake such a process of privatization. In Africa, privatization has instead resulted in a regression of the support provided to farmers with a hesitant emergence of producer organizations in the advisory sector, agro-industries providing advisory services concerning only a few products (cotton or tobacco, for example), and a geographically uneven distribution of support provided by NGOs. In Latin America, public advisory structures have been dismantled on a large scale (Box 15.1), with some notable exceptions such as in Costa Rica with its Ministry of Agriculture or Argentina with INTA.[2] Chile is often cited as an example where privatization was planned on the basis of farmer types, with efficient farms being supported by private operators and small family farming having access to publicly funded support (Namdar-Irani and Sotomayor 2011). In Asia, private advisory services have come up rapidly, although some countries maintain significant public advisory systems (India, Pakistan, Indonesia, etc.). It should however be noted that in recent years, especially with the food riots and the return of food issues to the international stage, some countries are reinvesting in agricultural advisory activities. Examples include the Philippines, which is relying on the World Bank, or Benin and Burkina Faso, both of which recently recruited large numbers of advisors.

Box 15.1. Towards privatization of agricultural advice: implications for dairy producers in the Mantaro Valley.

Guy Faure, Kary Huamanyauri Méndez, Ivonne Salazar, Michel Dulcire

In the Mantaro Valley in the Peruvian Andes, the role of the private sector in agricultural advisory services within the dairy industry has been growing steadily following the gradual withdrawal of public institutions. The private sector consists of input suppliers, dairy processors, independent advisors and veterinarians, more present now in the valley than ever in the past. These actors disseminate information and offer individual support to all family farmers, including the smallest amongst them. But their advisory services only cover the technical aspects of production; they do not take the farm as a whole into account. There is, therefore, no co-construction of the advisory offer between advisors, producers and other dairy-sector actors to adapt advice to producer demands and requirements. The fierce competition between private providers to sell their inputs and the existence of a public advisory system, also only focused on improving production, limit the producers' technical choices and orient them towards production systems based on the use of external inputs. To regulate competition, ensure a more diverse advisory supply and ultimately improve the regional advisory system, it requires public actors to take a stand in favor of more systemic advice and improved coordination between the different actors concerned. In conclusion, this Peruvian case study shows that private advisory systems can actually develop in the context of small family farms with limited incomes, but that they remain focused on the technical aspects of a few productive activities and encourage the consumption of costly inputs.

---

[2] Instituto nacional de technologia agropecuaria.

15.2 From Simple Technology Transfers to Capacity Building

Along with the evolution of advisory systems, there has also been an evolution of advisory approaches, not only due to new extension and advisory objectives but also because of new ways of providing advice. During the period of strong State intervention, advice took the form of transfers of knowledge and techniques to farmers in a top-down manner with an objective of increasing production. Prominent examples are the World Bank-supported projects in many countries promoting a standardized approach to advisory activities, called "training and visit." This approach was first used in India, starting in 1975, and in nearly 70 developing countries thereafter. It consists of advisors disseminating simple and research-validated technical messages to target farmers, called "contact farmers," who can then share their new knowledge with their neighbors. This approach came in for substantial criticism because, besides the fact that it requires massive public funding, it focuses only on improving agricultural production and does not take the real demands of producers and their abilities to innovate into account (Anderson and Feder 2004).

Due to the lack of significant impact of knowledge and technology transfer approaches, advisory systems were called into question in different countries at different times by attempts to address issues wider than those relating solely to agricultural production (product processing, management of natural resources, multifunctionality of agriculture, etc.). New approaches favored participatory methods based on learning processes and the building up of farmer capacities. This led to the emergence in the 1970s of alternative approaches, put forward by key actors such as Freire (1973) in Latin America. They experienced a more widespread development from the 1980s with "farming systems" (Chambers et al. 1989) and "research-development" approaches (Jouve and Mercoiret 1987) by emphasizing the necessity of understanding farmer rationales and of adapting technologies to local conditions. In the 1990s, "participatory technology development" and "participatory learning and action research" methods proposed a new vision of the advisor's role (Röling and Jong 1998). During the same period, there were also initiatives to support networks of experimenter-farmers in Latin America (Hocdé and Miranda 2000), which leveraged peasant knowledge and its mode of farmer-to-farmer dissemination (*de campesino a campesino*).

It is also worth noting the strong development of the "farmer field school" approach, introduced in the 1980s in the Philippines and subsequently in Asia and Africa with FAO support. This approach was developed to address the integrated management of rice pests, and then gradually expanded to cover other topics such as crop management, labor management and improvement of the living conditions of populations. This approach consists of advisors organizing regular meetings with volunteer farmers, who acquire new knowledge and skills through field experiments and exchanges between themselves and with the advisor. The approach is based on principles of intervention emphasizing the leveraging of participants' experiences, building up of analytical capacities and autonomous decision-making (Davis 2006; Ponniah et al. 2007). However, efforts to generalize this approach to reach a wider audience run into difficulties, related especially to its financial and institutional viability (Feder et al. 2004).

Some experiments, conducted by NGOs and the research community, are focused on the need to design an advisory approach for providing holistic farm advice which takes technical and economic dimensions into account and which is intended to build producer capacities to enable them to manage their agricultural and non-agricultural activities. This approach is based on individual and collective learning processes which mobilizes decision-making tools designed for literate or illiterate farmers (Faure and Kleene 2004; Djamen et al. 2003; Dorward et al. 2007). By encouraging exchanges between producers, these experiments strengthen the decision-making capacity of farmers and hence their autonomy. Their wider use, especially in Africa, is, however, hampered by the same difficulties as those encountered by the farmer field schools.

The increasing diversity of methods for providing advice – going well beyond simple transfers of knowledge or techniques – is thus the result of changes in thinking of the international community (research, civil society, donors) on the purposes of actions aiming at supporting rural actors: promoting the dynamics of innovation, developing learning processes, building capacity for action and autonomy of producers. But these new approaches of advisory services in the South are still strongly influenced by the illusion of finding a comprehensive one-size-fits-all method, applicable regardless of the diversity of needs of family farmers, the diversity of production systems and the diversity of the countries' institutional contexts. Many voices today are calling on us to imagine advisory approaches, methods and mechanisms tailored to each particular situation (Birner et al. 2009; Faure et al. 2011).

The rationales that are orienting advisory content and approaches are often contradictory. Extension and advisory services can be oriented by producers according to their needs and demands, driven by the market and therefore by the requirements of downstream actors, defined by government entities to ensure a minimum agricultural training for producers or to focus on collective interests. Advisory services may target a small number of producers and address complex problems through a continuous process of co-construction of the service, or target a wider audience and handle simpler problems with easy-to-implement standardized methods. Finally, the advisory service may be general in nature, addressing a wide range of issues raised by the producers, or specialized, and dealing with complex problems that require specific expertise. Depending of the situation, farmers' participation may vary, especially when identifying the problem to be solved, contributing in the construction of the advice, or participating in the evaluation of the results. The advisory service may be individual or collective depending on the case, and thus promoting or not the creation of new shared knowledge through socio-professional networks of producers.

15.3 Reframing Advisory Systems for an Improved Response to the Diversity of Requirements

Experiments in renewing advisory systems have been conducted in West Africa and Latin America in an effort to better meet the diversity of producer requirements. They have shown the need to improve advisory methods and governance, and hence their integration into innovation systems.

15.3.1 Improving the Methods

Experiments in providing management advice for family farms (MAFF) have been supported by French cooperation entities, especially CIRAD, for over two decades in many countries in Francophone Africa. CIRAD has been involved, through partnerships (Faure and Kleene 2004; Havard and Djamen 2010), in developing methods, helping assess or set up projects, and in training programs to renew the skills of the various advisory actors.

MAFF is a holistic approach that aims to build farmer capacities so that they can analyze their situations and environments, and take decisions based on goals they set for their farms and families. On the one hand, the method seeks to support exchanges between advisors and farmers, allowing them to gradually change the representation of the problems they face so that they can identify solutions themselves. On the other, it offers decision-making tools based on the use of technical and economic data that each farmer has to record. Depending on the situation, individual advice can be offered, adapted to the particular circumstances of each farm, or advice can be provided in a group to foster dynamics of inter-farmer exchanges. The MAFF method always takes into account a management cycle consisting of multiple steps: analysis, planning, decision/action, monitoring and evaluation.

Research efforts have contributed significantly to the design of MAFF methods through, on the one hand, the knowledge acquired on farmer practices and strategies – thus leading to an improved understanding of their needs and rationales for action – and, on the other, the knowledge gained from research and development projects of the 1980s, which helped develop methodological reflection on participatory approaches. Thus several action-research projects were conducted in Mali, Burkina Faso, Benin and Cameroon to design and test a new method and novel tools with advisory actors (Box 15.2).

Box 15.2. Management advice for family farms in northern Cameroon.

Michel Havard, Anne Legile, Patrice Djamen Nana

In the cotton-growing region of northern Cameroon (Havard and Djamen Nana, 2010), the MAFF method was used to gradually change the attitude of farmers. They were encouraged to change from being passive recipients of advice (listening to cotton extension advisors) to becoming veritable actors themselves (analyzing their own situations and taking decisions autonomously). To this end, a program based on the "questioning technique" has enabled advisors and farmers to go, little by little, from reflection on concrete and immediate issues to topics which require looking into the future and which involve more complex concepts. Thus, the first two or three years were spent in introducing groups of literate or illiterate farmers to the basics of management and, later, to the definition and use of technical and economic indicators. MAFF participants were volunteers who felt motivated to improve their production systems. Educational material was created for advisors. Documents for producers consisted of technical sheets, grids and tables enabling them to better assess their needs and resources. For literate farmers, a farm notebook was used by the participants to plan and monitor their activities. It included information on the farm's structure, the planning, the performances, and the techno-economic monitoring of crops.

The development and implementation of the MAFF approach took place in three phases during which the roles of the various actors involved changed. For four years, research developed the MAFF approach with the support of a dozen advisors and a development project. For this purpose, research chose to rely on an action research in partnership approach (Faure *et al.*, 2010). Once the MAFF approach was developed, research shifted its focus to the training of advisors and the monitoring of field activities. The task of trying out MAFF was jointly undertaken by the Organization of Cotton Producers of Cameroon (OPCC) and the Sodecoton cotton company, which mobilized its agents to test MAFF with 450 farmers for one year. Once the test was deemed successful, MAFF was further extended to cover approximately 1,500 of the 300,000 farmers in the cotton-growing area. During these three phases, continuous feedback and exchanges between farmers, developers and researchers helped assess activities and make appropriate adjustments to the method and tools.

15.3.2 Strengthening the Governance of Mechanisms

In addition to revamping advisory methods, it became necessary to conduct research on advisory mechanisms, i.e., on the ways in which actors are involved in the provision of advice (service providers, research and development, organizations to support service providers, organizations involved in controlling and orienting advice, etc.) and the rules governing their relationships. Indeed, the design of a new advisory method – or even the improvement of an existing one – cannot be considered in isolation from the funding and governance mechanisms that are used to implement it.

Advisory services can be understood as "systems" whose functioning is strongly influenced by various interacting components: the funding mechanisms, the governance mechanisms, the quality of human resources available, and the characteristics of the advisory method used (Birner et al. 2009; Faure et al. 2011). It is these interactions, and not the characteristics of each component analyzed independently, which determine and explain the functioning of the advisory system. A change in one component affects all others.

CIRAD has intervened to support the creation or strengthening of mechanisms for providing advice to family farms, most notably in Burkina Faso, Mali and Cameroon. In these countries, two major issues have been addressed. First, the role of each actor in the implementation of advisory services was clarified. The role of producer organizations, in particular, is always a sensitive issue because such organizations can affect the orientation of the advice and the capacity to take farmer demands into consideration. Producer organizations can participate in orienting MAFF by getting involved in the planning and evaluation of activities, as in Burkina Faso with UNPCB (National Union of Burkinabe Cotton Producers) and Sofitex (a Burkinabe cotton fiber company). But producer organizations can also provide advisory services themselves, by recruiting employee advisors or by relying on farmer extension workers, as for example in Mali with service delivery centers. To ensure a better adaptation of the advice to producer requests, planning and evaluation mechanisms for advisory activities were designed at various levels: at the advisor level to better define the content of the advice and to determine the working arrangements between an advisor and the producers involved; at the local level so that all the activities of the advisors can be guided and monitored by the concerned actors (local POs, service providers, etc.); and at a more global level (national or sectoral) to ensure proper coordination between all actors of an advisory system (umbrella POs, advisory organizations, research, etc.). Second, thought was given to the funding of advisory services. Farmers are often asked to contribute in order to ensure a real commitment to MAFF activities but their contribution is mainly symbolic. Funding through a sectoral levy remains a possibility, as in the case of Burkina Faso or Cameroon, especially when agro-industrial companies occupy a dominant position in the MAFF mechanism. In most cases, however, funding is still dependent on international aid, which raises issues of institutional sustainability.

15.3.3 *Incorporating Advisory Services in the Innovation System*

Advisory mechanisms should be part of national or sectoral innovation systems, including the concerned actors (farmers, producer organizations, advisory-service providers, private companies, research, training organizations, NGOs, etc.) so that they are able to mobilize diversified skills, create new knowledge, generate learning in order to encourage technological, organizational and institutional innovations likely to improve the performance of farms or sectoral actors (Triomphe and Rajallahti 2013). In such a perspective, advice plays a special role – not to specify what has to be done, but rather to support family farmers in their projects and, above all, to facilitate interactions between the innovation system's actors.

CIRAD has intervened in Cuba in this "innovation system" perspective to transform agricultural advisory services (Box 15.3).

Box 15.3. Experience of PASEA in Cuba.

Jacques Marzin and Teodoro Lopez Betancourt

Following the implosion of the Eastern European Soviet bloc, Cuban agriculture had to endure three major disruptions on an emergency basis: first, to its Green Revolution, since its economy could no longer pay for imports of inputs; second, to the specialized production in the country and within production structures (sugarcane, tobacco, etc.), since Cuba had to quickly substitute food imports from the Eastern bloc (wheat, dairy products, potatoes, canned vegetables, etc.) and because it lost its COMECON markets (sugar, rum, tobacco, etc.); and, finally, to nationalization, since the State was unable to pay salaries and fund investments in state-owned farms, of which half were transformed into self-managed cooperatives. It is in this context that the Cuban Ministry of Agriculture asked CIRAD's help to transform the country's agricultural advisory system, until then decentralized in departments of technology transfer in the 19 agricultural research centers, divided by product type. Four principles oriented this change within the Ministry of Agriculture and the Research Directorate: taking the needs of cooperatives and their members into account; developing a systemic approach to solve problems (technical and socio-economic, funding of operations and investments, research into self-sufficiency of production systems to compensate for the lack of inputs, etc.); going beyond the standard mantra of training by adopting a diagnosis-experiment-training approach at the local level, with farmer experimentation (interest groups); and, finally, linking the diversity of forms of advice (by sector, social groups, territories, etc.) within the Cuban innovation system (agricultural research centers, universities, centres for initial and continuing training of farmers, farmer organizations and NGOs). The program to support the agricultural extension system (PASEA) was used to test a methodology (Marzin *et al.*, 2013) to create a dynamic around the profession of the agricultural advisor (*extensionista*) with the recognition of his status by the Ministry of Labor, to establish a network of agricultural advice professors (*extensión agraria*) leading to the transformation of agronomist training programs and creating a Master's program for agricultural extension and sustainable development (Sablon *et al.*, 2013).

15.4 For a Revamping and Permanent Contexualization of Advisory Systems

Advisory systems are part of innovation systems. Each advisory system depends on both the productive context – type of production systems and type of sectors, which determine the advisory content and the methods to deliver it – as well as the institutional context – standards, laws, and relative influences of the State and the private sector. Each advisory system can be characterized by its governance and funding mechanisms, its methods, and the capacities of the actors providing the advice. A standard advisory mechanism model, suitable for all situations, simply does not exist. Several advisory approaches can coexist within the same territory, with advice oriented by producer demands, by market demands, and/or by standards set by public authorities and intended to reflect social demand. Currently, debate is focused on the different modalities of privatization of advisory services, instigated by the withdrawal of the State and the emergence of new actors, with a recognition of the plurality of arrangements between advisory actors and a redefinition of the role of the State.

Our exploration of advisory methods reveals that top-down technology transfer models are still very prevalent in advisory organizations, in order to promote compliance of standards set by private or public actors. However, many experiments on advisory services have revealed the increasing diversity of advisory and capacity-building requirements for family farmers. Realization is growing that to increase the effectiveness of advisory services new types of relationships need to be forged between farmers and advisors, local knowledge has to be mobilized, and problems and solutions have to be co-constructed. Nevertheless, in the South, thinking is still largely driven by a desire to promote advisory activities – even those based on participatory methods aiming at promoting learning – but without really taking into account the local context, the actors concerned or the existing initiatives. It is therefore always necessary to strive towards the development of advisory approaches, methods and mechanisms suited to each institutional context, capable of taking into account the complexity of family farms and the diversity of situations they experience. These approaches have to incorporate participatory mechanisms and the use of new decision-making tools, including information and communication technologies. Finally, it appears worth noting that the transformation of advisory services will not be complete without a necessary effort to provide basic training, as much for advisors (acquisition of new knowledge and skills) as for family farmers (literacy, vocational training).

These principles of intervention in family farming can also be extended, of course, to targeted and specialized technical advice by taking into account the specificities of the scientific disciplines concerned. In this perspective, the next chapter uses research work on animal health and plant protection at different governance levels to examine the possibilities for adapting mechanisms for the surveillance, control and response to sanitary risks to the contexts of family farming.

Chapter 16
Support for the Prevention of Health Risks

Sophie Molia, Pascal Bonnet, and Alain Ratnadass

Research on health risks and their prevention focuses primarily on animals, plants and their pathogens. Yet, as discussed in the earlier chapters in this part of the book, production systems and activities developed by family farms call for the reorientation of this research to their specific needs and the expertise they have gained. Similarly, it is important for research outcomes to have social applicability. Even if not entirely dictated by familial forms of production systems, the work we present here is adapted to the specifics of family farming systems.

In this sense, and with reference to the lessons of Chap. 12, this chapter explores the research knowledge and perspectives that, specifically or generally, relate to family farms in four major areas: the understanding of health risks; the surveillance of health risks; the formulation of effective and efficient prevention and control strategies; policy support and the changes in scales and methods to recognize risks and manage them in an integrated manner.

16.1 A Better Understanding of Health Risks

Chapter 12 shows how the diversity of health risks and particular characteristics of family farms lead to a specific exposure to such risks and to a differentiated response to their management at various scales. Risk prevention requires an enhanced understanding of biological determinants and systemic and institutional factors that perpetuate such risks or even induce them. To this end, production of

S. Molia (✉) • P. Bonnet
ES, Cirad, Montpellier, France
e-mail: sophie.molia@cirad.fr; pascal.bonnet@cirad.fr

A. Ratnadass
Persyst, Cirad, Montpellier, France
e-mail: alain.ratnadass@cirad.fr

knowledge and models serve to reduce risks through prevention, or limit their impact through mitigation by using different kinds of intervention.

16.1.1 Understanding the Diversity of Causative Agents

It is imperative to understand the diversity of pests, pathogens and vectors prevalent in the countries and territories of the South, as well as the manner in which they are propagated or transmitted, in order to formulate effective action plans. The two following examples illustrate the need for such an understanding.

In the context of animal health, phylogenetic studies of the diversity of the Newcastle disease virus are essential to understand its evolution. This virus is among the most severe and widespread avian diseases known to family poultry farmers. Different virus strains can be found depending on the geographical location (Miguel et al. 2013). These strains evolve over time due to vaccination pressure. By characterizing existing strains of the virus and determining the extent of protection provided by existing vaccines, control programs can be better adapted to the epidemiology of different regions and production systems. The Gripavi research program, conducted in six African and Asian countries, has thus not only helped identify the existence and genetic diversity of new genotypes of the Newcastle disease virus but has also suggested modifications to existing nomenclature (Servan de Almeida et al. 2009).

In the context of plant health, the identification of an indirect impact of cultivating the genetically modified "Bt" cotton on Hemiptera populations led to fears of a gradual return to excessive use of chemical pesticides in Burkina Faso. Such a scenario would lead to severe consequences on the labor required and the finances of concerned family farms. Bt cotton, containing the Cry1Ac and Cry2Ab genes, was introduced in Burkina Faso in 2008 to combat both bollworms (*Helicoverpa armigera*, *Earias* spp.) and defoliators (*Haritalodes (=Syllepte) derogata*, *Cosmophila flava*, etc.). In 2012, Bt cotton was being cultivated on 500,000 ha and accounted for 60 % of the total area under cotton in Burkina Faso. Monitoring of insect pest populations in fields with and without Bt cotton over a period of 4 years (2009–2012) established the effectiveness of the transgenic varieties against the target pest group. The introduction of Bt cultivars helped reduce the use of insecticides by 66 % in comparison to conventional agriculture. While this led to a reduction of the target pest population and its associated impact on phytosanitary programs, there was also an increase in the Hemiptera population that was not targeted by Bt toxins and its adverse effects. Faced with this growing pest problem, Bt cotton farmers increased the number of insecticide sprays from two to three, or even four, during the crop cycle. Studies indicate several likely causes for the development of Hemiptera pests: a reduction in the number of insecticide sprays in transgenic fields, the loss of varietal resistance due to leaf hairiness, and the reduced effectiveness of chemical control programs (product quality, resistance build-up in insects). It has thus become desirable to implement integrated pest management

(IPM) programs in association with Bt cotton cultivation (Hofs et al. 2013; Brévault et al. 2013; Tabashnik et al. 2013).

16.1.2 Qualifying and Quantifying the Direct and Indirect Impact of Health Risks

Several pests and diseases affect crops and livestock, and it is necessary, especially in the countries of the South, to prioritize health problems in order to allocate resources and direct control programs to tackle the most serious threats to family farming systems.

This prioritization is difficult for several reasons. Existing diagnostic capabilities of laboratories are often inadequate or unevenly deployed over the territory. Furthermore, the scarcity of data on the frequency and on the assessment of the health, biological and economic impact of diseases and pests, remains a major constraint in prioritizing these risks, especially in relation to family farming. Consequently, without the help of an objective viewpoint based on field data, it is difficult to identify methods to prevent and control health risks that need to be implemented on priority, or even to evaluate the effectiveness of actions already undertaken. Similarly, it becomes easier to mobilize and engage stakeholders of the health sector (producers, cooperatives, government agencies and donors) when it is possible to translate losses caused by a health problem into figures, or to demonstrate the cost-effectiveness of particular programs. The following examples illustrate the importance of qualifying and quantifying direct and indirect impacts of health risks, regardless of whether they are related to animal diseases, zoonoses or microbial food safety risks.

Contagious bovine pleuropneumonia (CBPP) is a good example of a cattle disease that has consistently affected family livestock rearing, albeit with varying impacts depending on various livestock systems (pastoral, agropastoral, or intensive). Studies conducted in Ethiopia have helped quantify the severe losses suffered by pastoral and agropastoral family livestock systems as a result of the stopping of the large-scale pan-African bivalent vaccination programs. These programs had, in fact, helped eradicate rinderpest and control CBPP (Lesnoff et al. 2003; Bonnet and Lesnoff 2009). The level of positive seroprevalence of family herds ranged between 2 % and 8.5 % in agropastoral regions. The clinical disease was observed in 39 % of the seropositive animals, leading to a functional incapacity of the animal to contribute to production (loss of weight, of milk production, of capacity to work, etc.) and even resulted in a high mortality rate (mortality rate in 13 % of the clinical cases). Modeling has led to a better understanding of within-herd (Lesnoff et al. 2004) or between-herd transmission of the disease. The financial impact of vaccinating dairy cattle against another cattle disease, the lumpy skin disease or LSD (Gari et al. 2011), was calculated for the Oromia region of Ethiopia. Such a vaccination ensured breeders a net profit of 1 USD per head for local zebu breeds and 19 USD per head for the Holstein breed and its crosses, with the latter being more susceptible to the disease.

As far as microbial food safety risks are concerned, there are many types of diarrheal diseases associated with the consumption of contaminated products: foodborne diseases like rotavirus, shigellosis, *Escherichia coli* pathogens, cholera due to *Vibrio cholerae*, typhoid fever caused by *Salmonella enterica typhi* or *paratyphi A, B*, or *C*. These disorders result in about 2.5–3 million deaths per year globally. A similar assessment can be conducted for mycotoxins, toxic secondary metabolites produced by molds on a wide range of agricultural produce (cereals, nuts, coffee, cocoa, peanuts, etc.) on the field and post-harvest (Duris et al. 2010). They can be found in the more processed food products, and their various toxic effects (carcinogenic, neurotoxic, hepatotoxic etc.) can lead to acute, chronic or even life-threatening poisoning in animals and human beings. Studies lead us to stress the importance of actions that need to be taken directly by family farmers who, besides being the main producers, are also the chief consumers of their own fruits, vegetables and cereals.

A quantification of crop losses is also a prerequisite to prioritize public investment in research. Crop losses due to pathogens, pests and weeds account for between 20 % and 40 % of the achievable yield in agroecosystems (Oerke and Denne 2004). Studies carried out on family farming systems pertain to the measurement of pre- or post-harvest losses as criteria for comparing the performance of farming and processing systems. Other research studies focus on the indirect losses, more difficult to quantify, especially if we consider the use of resources in agricultural production (water, labor, know-how, energy, inputs, credit, soils) or environmental impacts of "wasted" inputs (including pollution and greenhouse gas emissions). Research is currently on-going on perennial species grown on family farms. This implies taking into account the "time" dimension in calculating production losses. In the case of coffee, for example, it is possible to predict production losses for the year ($n + 2$) in relation to events (pest attacks) that occurred in the year n. In the case of mango, qualitative losses caused by attacks by fruit flies can be estimated by building on existing research on peach trees (Grechi et al. 2010).

16.2 Producing Surveillance and Evaluation Tools

The body of knowledge accumulated on health risks (their nature, how they affect crop and livestock production, the possible methods for managing them) underpins other, more applied, research studies focused on improving tools to monitor and control health risks.

16.2.1 *Developing Suitable Diagnostic and Screening Tools*

Identifying and quantifying plant and animal diseases on the field and under laboratory conditions are the essential tasks of plant and animal health practitioners and researchers which are undertaken in collaboration with family farmers and breeders. In fact, it is important to reconcile the different ways in which scientists and farmers perceive and understand diseases and their impact. Furthermore, clinical manifestations evident to livestock breeders may not be epidemiologically very serious as far as control or prevention is concerned. The onus, therefore, shifts to the research community to develop tools for on-site screening (on the field) and mass screening using more sophisticated laboratory techniques, and also to help train breeders. This role of research is especially important as certain diseases are present only in the countries of the South, where they mainly impact family farmers and breeders, whose financial resources are often limited. These diseases are very often neglected by large agricultural, chemical and pharmaceutical companies who tend to target only large markets that ensure substantial turnover and high profitability.

It is therefore critical to create diagnostic tools adapted to the conditions prevalent in the countries of the South. These tools must be effective, easy to use and affordable to health services irrespective of whether they concern laboratory equipment or reagents (for example, ELISA kits or polymerase chain reaction (PCR) analysis primers), screening techniques (targeted sampling, enrichment techniques) or identification manuals (for example for insect pests). Thus, laboratory reagents must be based on pathogen strains present in the country, and their packaging should reflect the number of analyses typically carried out by regional laboratories. Field diagnostic methods must be applicable under condition of high temperatures or the absence of a cold chain or electricity, situations that are common in the South.

If we consider the example of CBPP – a disease that is still difficult to diagnose through clinical symptoms since differential diagnosis is difficult without examining internal lung lesions –, we can easily recognize the disease at the slaughterhouse and during autopsies (pathognomonic lesions in the lung). In the course of research conducted by CIRAD and its partners (National Veterinary Institute (NVI), Veterinary Faculty of Debre Zeit, University of Addis Ababa, International Livestock Research Institute (ILRI)) in Ethiopia, lesions in the lungs of dead animals were shown to family farmers. Screening of live animals has been improved with the use of targeted thoracic punctures for collecting lymph originating from the exudation from the infected lung. This can then be tested using slide agglutination, a very simple technique adapted to preliminary fieldwork. Veterinarians and researchers were trained in this technique using multimedia tools that were disseminated internationally (Bonnet et al. 2005). However, for epidemiological studies, "on farm" screening remains difficult and unsatisfactory without additional recourse to a laboratory. Serological tests have therefore been developed for screening live animals (ELISA and complement fixation test, or CFT) (Morein et al. 1999).

In the context of plant pathology, the development of general molecular analysis tools for fungal flora has helped explain some of the interactions between strains that produce mycotoxin molecules, which are very dangerous to humans and animals. These methods first enabled the analysis of almost all fungi strains on a food medium (coffee beans in particular) quickly and in a single step. The methods were used in combination with detailed analysis techniques for toxins (immuno-affinity/high performance liquid chromatography, or HPLC), and have led to an understanding of certain interactions between these toxigenic molds and non-toxic natural flora. The use of bio-competition between these strains with little or no genotoxic effect against toxigenic strains is a potential strategy for preventing the presence of toxins in coffee or grains produced by family farmers. These competition phenomena are yet to be fully explained, because the reduction in the production of toxins could be the result of a growth or toxin production inhibition or of the consumption of the toxin by a different strain after it is produced (Abrunhosa et al. 2002).

Tools to trap pests like fruit flies (Duyck et al. 2004) or coffee berry borers (Dufour and Frérot 2008) were also developed. Often, these are detection tools (for example, for the fruit fly *Bactrocera invadens*, Goergen et al. 2011) and tools to monitor the population dynamics of different fly species. They are usually used in conjunction with other methods and prior to a calibrated or targeted control measure, rather than as unique control tools – except for the coffee berry borer. In the context of fruit flies, research has also contributed to the development of models to predict invasions (De Meyer et al. 2010).

16.2.2 Supporting the Improvement of Surveillance Systems

Surveillance systems are a key component of programs to prevent health risks. They fulfill three functions in family farming by allowing:

- the estimation of the presence and intensity of one or more plant pests or animal or plant pathogens in a given territory and the verification of the effectiveness of control methods;
- the detection of the introduction of a new pest or pathogen, the objective being to detect the occurrence as early as possible in order to contain the threat more easily and limit its spread;
- the demonstration, using objective data, of the absence of a pest or a particular disease, thereby promoting trade in agricultural products.

As far as microbial food safety risks are concerned, the simple collection of statistical information on the number of deaths or infections due to individual or collective food toxic infections (FTI and CFTI respectively) is essential for countries in the South and their farmers. Combined with limited laboratory investigations which are, however, adequate to identify the causes of these FTIs, these measures can help justify more ambitious government policies on health and food safety in urban or rural areas. Most often, however, there is a lack of expertise at the national level and research is based on articles or sentinel sites that are supported by international initiatives to estimate the impact of microbial food safety risks on community health (WHO 2013a).

There are various kinds of surveillance systems. Passive surveillance is based on analyzing information provided by farmers through agents of health services. Its usefulness depends on the efficacy of the information chain, the representativeness of those seeking health care and the confidence placed in these services. Active surveillance, on the other hand, is based on a plan to monitor sampled observation units of suitable scale (villages, areas, herds, plots, family), with all the methodological difficulties attendant to the participation of local actors, once they are selected. Mixed systems attempt to integrate different surveillance components for increased effectiveness.

Research supports these surveillance systems in the countries of the South in several ways. Risk-based surveillance methods create a model based on information gathered on biological, agricultural, livestock and social determinants of the outbreak of a disease or invasion by pests to predict future occurrences. Areas or family farms with the highest risk of being affected by a disease or pest are identified and surveillance there is intensified. Targeted health surveillance becomes all the more important when resources are limited. Furthermore, linking these models to georeferenced data (maps, satellite images, GPS-based information) helps create risk prediction maps that provide appropriate information on the risk level to the various actors in the agricultural sector (from the farmer to government agencies).

In the case of Rift Valley Fever in Senegal, longitudinal studies conducted by Chevalier et al. (2009a) highlighted the endemicity of the disease in animals of the Ferlo region and identified conditions favorable for transmission between animal hosts and mosquitoes. This knowledge was then used to build hydrological models and mosquito population models in order to predict outbreaks (Soti et al. 2013). These prediction models are based on rainfall data and should help alert rural populations more quickly when conditions conducive to an epidemic are detected in the future.

Another form of support of targeted research is the development of systems for the collection and exchange of health information based on electronic transfer and ICTs (Information and Communication Technologies). The mobile telephony revolution in the South has helped reduce the inaccessibility of highly isolated populations. Mobile telephones are now available in even the remotest villages and have become a basic means to communicate with family farmers and breeders and to disseminate information. Early warning systems have been developed to inform farmers of impending adverse events (locust invasion, for example) or to deliver technical advice when agricultural and livestock experts are unable to travel. Conversely, other systems have been developed to send back information from family farms to information collection centers of health agencies. Some systems are developed in partnership with human health services (Baron et al. 2013). Family farmers and breeders – or even health workers based in the territory concerned – can send SMSes to a system for uploading this information to a centralized health services database.

An integral aspect of this area of research is the evaluation of the functioning of health surveillance systems. Such systems are mainly based on networks of health actors (at the central and local levels), diagnostic laboratories and farmers and breeders mainly from family farms. Research has developed quasi-quantitative methods to assess the effectiveness of surveillance networks. For example, work carried out in Mali to evaluate the surveillance system for avian flu has shown that the overall performance of the Epivet surveillance network was satisfactory, although some of its components, such as data management, could be improved (Molia 2012). A more specific evaluation of the role of family livestock breeders, who constitute the front line of the surveillance program since they routinely work with poultry, has shown that the reporting rate of avian flu indicators to health authorities was only 17 %. This consequently suggests that a lot still needs to be done for collecting and transmitting health information.

16.3 Proposing Tools and Strategies to Control Health Threats

In the context of the plant and animal health care needs of family farming systems, research has to test and disseminate technologies (screening, treatment, prevention) and intervention principles that are often seen as innovations in a given social and technical system. In continuation of the approaches of action research in partnership (Chap. 14), and in addition to any proposed technical solution, it is incumbent upon research to examine the "innovation system" that accompanies the adoption of new control measures. This can lead to an organizational innovation that is as important as a new health technology, insofar as it conditions social acceptance and transmission via existing socio-professional networks, for example, farmer-breeder organizations, community health workers, agricultural extension agents, and the devolution of health responsibilities to private veterinarians. This can also lead to a redefinition of the technology's specifications (functional analysis) so that some of its features correspond to its intended field use (e.g., heat tolerance), and that its marketing or distribution methods are in accordance with practices of family farmers – dosages suited to the size of local breeds, vaccination dispensers suitable for small herds, packaging of treated seeds suitable for small planting areas, etc.

16.3.1 Developing Suitable Vaccines, Treatments and Intervention Strategies

Public agencies often have to undertake research to fill the void left by large pharmaceutical and chemical companies who find little business sense in dealing with certain diseases and pests due to high intervention costs or low creditworthiness of target farmers. This is the case of threats faced exclusively by family farming systems, despite their total production representing a considerable percentage of the global production. Consequently, public research in partnership can ensure a better consideration of their needs and a broader surveillance of microbial strains while designing more effective packaging and better thermolability features. This has led to the detection in Malian family poultry farms of some "new" virus genotypes responsible for Newcastle disease. These genotypes were previously unknown as the focus of screening was earlier limited to large industrial livestock farms. This new knowledge has been instrumental in guiding the production of vaccines (Gil et al. 2009).

The use of vaccines that are not heat-tolerant often proves to be complex in family farms in the countries of the South – an example of a heat-tolerant vaccine was the one used to eradicate rinderpest. Other vaccines require not only a cold chain but also improved quality control during administration. Furthermore, some vaccinations need to be administered repeatedly during the animal's lifetime to prevent a loss of herd immunity that protects populations. It is thus imperative to understand the expectations of breeders when proposing interventions and techniques (Kairu-Wanyoike et al. 2013). Studies on CBPP in Ethiopia and Namibia have shown that, in a context of significant institutional change, disease control could be improved in family farms by adopting a mixed intervention approach involving improved vaccines delivered by government agencies (Thiaucourt et al. 2004) and remedies provided by the private sector that use molecules other than tetracyclines, thus reducing the risk of chronic carriage (Lesnoff et al. 2005; Huebschle et al. 2006).

It is also essential to unite actors and resources to boost research on neglected diseases that large pharmaceutical companies dismiss as business propositions of little consequence. Research is collaborating with international initiatives (GalVMed) to fight against neglected diseases afflicting animal health or against certain neglected zoonoses (WHO 2013b). Researchers use their expertise in vaccinology of CBPP or of peste des petits ruminants (PPR) (Minet et al. 2007), or in the fight against trypanosomiasis (tsetse fly trapping, biological control).

It is similar in the case of plants. The Caribbean banana, for example, is considered a minor crop because the quantities of products sold to protect it do not ensure a rapid return on investment nor high enough profits for their manufacturers (Temple et al. 2010). Moreover there has been a reduction in the range of available chemical molecules (Chap. 12) resulting in repeated application of the same chemicals, building up a resistance in the target species, as in the case of fungal infections like the leaf spot disease. Research is helping the sector, including family farmers, by conducting experiments to test different chemical or biological products in a system of specific approvals corresponding to this "minor use," as well as non-chemical control methods (de Lapeyre de Bellaire et al. 2009; Chillet et al. 2013).

In the case of plants, the failure of seed treatments (Chap. 12) may be due to the selection of strains or pest populations resistant to the chemicals used, to the variability or change in the pest spectrum at a specific level. Studies are consequently being conducted to find alternatives for treating upland rice seeds with imidacloprid to reduce its environmental impacts and to forestall the emergence of resistance to this molecule in soil inhabiting insects (Ratnadass et al. 2012b). This is particularly important in situations where the harmful entomofauna spectrum (the part of the fauna consisting of insects) can vary from site to site or from year to year (Ratnadass et al. 2013b).

Research helps countries and regions to develop, implement and evaluate health systems with regard to microbial food safety risks. Food security guidelines vary from country to country. The richest countries have regulatory systems in place (for example, regulation 178/2002 in Europe) which makes it mandatory for food manufacturers to provide safe food to the public. However, this system cannot be implemented in most countries in the South, despite increasing societal demand, mainly in urban areas, for food safety. Rules governing internal trade in these countries are generally not strict and are poorly enforced, especially in rural areas. Three broad types of situations can be identified: situations in which the health risk is not controlled; those where the health risk is managed by the consumers (overheating food items, for example, boiling milk); and, finally, situations where health risks are managed upstream in the food chain by applied and controlled regulations.

Research evaluates the impact of control measures as well as the economic and technical value added (increase in efficiency) by the involvement of farmers in surveillance systems. For example, during the avian flu crisis of 2006 in the Nile Delta in Egypt, where urban and rural family livestock farming coexist (Peyre et al. 2009). The quantitative study of socio-professional networks associated with the poultry trade in Mali (social network analysis (SNA) methods) identified their major risk areas, which facilitated targeted market surveillance (Molia et al. 2012).

New control methods against nematodes and leaf streak diseases had to be developed in association with family farming in the French Antilles following the 1993 ban on the use of chlordecone (Box 16.1), a persistent organochlorine pesticide used to fight banana weevil (Lesueur-Jannoyer et al. 2012). A technical solution involving the use of fallows and crop rotations, along with the use of micropropagated banana plantlets, helped control these health risks at lower costs for family banana plantations. Moreover, problems caused by the severity of soil pollution necessitate innovations that break with past practices and rationales. This revamping of tools to manage health risks can only be designed in partnership with farmers. Family farms supply fruits, vegetables and meat to local markets. They are particularly interested in the diversification of old and polluted banana plots and have knowledge that needs to be better mobilized.

16 Support for the Prevention of Health Risks   277

Box 16.1. Chlordecone in the West Indies.

Magalie Lesueur-Jannoyer

Organochlorine pesticides, including chlordecone, were used in banana plantations in the West Indies until 1993 to control the banana weevil. This chemical is now included in the list of persistent organic pollutants (POPs) of the Stockholm Convention and its effects on human health are documented: prostate cancer, motor and cognitive development disorders in children, etc. (Multigner *et al.*, 2010; Dallaire *et al.*, 2012).

Government agencies have taken steps to limit exposure of people to these chemicals via drinking water. It is also important to mobilize family farmers and breeders because former banana plantations which were treated with chlordecone now supply other fruits, vegetables, roots and tubers, meat and eggs to local markets. The products marketed by these small farms and family gardens must comply with defined standards, or maximum residue limits (MRLs), to ensure zero risk to consumer health. The MRL set for chlordecone is 20 µg/kg. CIRAD, in collaboration with INRA, professionals and Government agencies, has provided farmers with methods and tools to help choose the production depending on the level of pollution in each plot, the response of crops to the contamination of crops and the applicable MRLs. The transfer route of the molecule between the soil and the marketed part of the plant is divided into three categories:

- so-called "susceptible" produce: roots and tubers, where the risk of exceeding the MRL is high in soils with pollution levels exceeding 0.1 mg/kg of soil;

- partially susceptible produce: Cucurbitaceae (cucumber, pumpkin, zucchini, etc.) and lettuce, where the risk of exceeding the MRL has been proven to exist in soils with pollution levels exceeding 1 mg/kg of soil;

- non-susceptible produce: fruits (bananas, citrus fruits, mangoes, papayas, etc.), Solanaceae (tomatoes, peppers, chilies, eggplants) chayotes, okras, where the observed contaminations remains, regardless of the level of soil pollution, well below the MRL.

In the case of animals, the mechanisms involved are diverse and complex. Animals tend to accumulate pollutants (for example, accumulation in poultry, even where soil contamination is low), storing them mainly in certain tissues (liver in particular), while getting rid of some of them. Care must be taken, in such polluted areas, to reduce contamination by balancing livestock diet with a component sourced from elsewhere, or by creating confinement areas away from the polluted soil.

Consumers could adopt a few simple practices to reduce the risk of exposure. These include washing the produce to remove all soil contamination, or properly peeling vegetables listed in the susceptible and intermediate categories, thus reducing exposure by half, since pollutants are mainly concentrated in the skin through contact with the polluted soil. Substitution of various products can also reduce exposure: root (yam, sweet potato, etc.) can be replaced by "less susceptible" starchy products, such as plantain and breadfruit, that can be grown on contaminated soil with little health risks.

Tools have been developed to produce healthy food and limit consumer exposure. They now need to be widely disseminated to professionals. While health objectives may have been achieved through these measures, research still needs to be conducted to address environmental issues associated with this pollution: reducing the exposure of ecosystems, especially aquatic ones (Lesueur-Jannoyer *et al.*, 2012; Cabidoche and Lesueur-Jannoyer, 2012).

16.3.2 Proposing Alternative Pest Control Measures with Low Environmental Impact

Insect and mites are among the main causes of vegetable crop losses in sub-Saharan Africa. In order to protect their crops, farmers are currently employing chemical treatments ever more frequently, and have also increased the dosage of chemicals that are often not licensed. Such practices could result in the presence of residues in harvests and the environment.

Research has led to the testing and validation of physical control methods based on the use of nets in order to reduce insecticide use in cabbage cultivation in sub-Saharan Africa. Cabbage is a crop subject to heavy insecticide use in this region. Polyester nets of the type used against mosquitoes were found to be very effective in protecting cabbage crops against various Lepidoptera (such as diamondback moth, *Plutella xylostella*, or even the more dangerous *Hellula undalis*). This was used not only in the nursery stage with permanent cover, but also, and especially so, in the post-transplanting stage (Martin et al. 2006). Other stronger polyethylene nets with a finer mesh have helped protect against another species of aphid (*Brevicoryne brassicae*) and delay infestations by other aphid species (*Myzus persicae* and *Lipaphis erysimi*) (Martin et al. 2013).

While not effective against certain pests (for example, the *Spodoptera littoralis* caterpillars), nets have reduced by 70–100 % the number of insecticide applications by small farmers as compared to their normal practices. Following successful trials and demonstrations of the technique in Benin (Licciardi et al. 2008) and Kenya (Muleke et al. 2013), the anti-insect nets designed to protect nurseries and cabbage crops are now in the dissemination phase. Research projects are also underway to test the effectiveness of physical supports impregnated with biocides to control smaller pests that pass through the nets. Preliminary studies conducted in Benin showed a rate of return of 138 % for anti-insect nets used for vegetable crops (1 Euro of investment for a return of 1.38 Euros) as compared to conventional protection. To this must be added other positive externalities (lower health care costs due to reduction in pesticide use, lower levels of soil pollution, etc.).

Cost/benefit economic evaluations compare conventional methods (often based on synthetic pesticides) to alternative methods of protection against pests. A case in point here is the agroecological control of Cucurbitaceae flies in Réunion (Chap. 12). The common point of satisfaction for all the farmers is based in particular on reducing the cost of protection against flies using the GAMOUR technique, which is between 1.2 and 4.2 times more economical for the farmer as compared to conventional protection with the help of curative insecticides (Deguine et al. 2013).

The use of antibiotics to stabilize the population of certain pathogens in some productions systems like continental or coastal aquaculture (shrimp) – a source of employment for some family farm economies such as in Madagascar – has long been a source of problems as it leads to a buildup of resistance (Sarter et al. 2007). Alternative solutions based on the use of essential oils have been proposed, and have helped reduce the use of antibiotics (Randrianarivelo et al. 2010). The antimicrobial properties of these essential oils are derived from the local biodiversity (*Cinnamosma fragrans*) and are used to target food pathogens (*Salmonella, Staphylococcus, Vibrio, Escherichia coli*). They can be a partial substitution solution. Their effects are similar to those of conventional antibiotics: reducing vibrios and increasing larval survival rate. The delivery mechanisms and chemical variability of these oils is currently being studied.

Research is also helping develop curative control methods against plant pests, especially quarantine pests. Post-harvest treatment of fruits is sometimes necessary for the purpose of exports and to avoid economic losses. CIRAD has developed techniques based on soaking fruits (especially mangoes) in hot water (Self et al. 2012) because of the restrictions placed on the use of methods such as fumigation and irradiation (Ducamp Collin et al. 2007). Such alternative methods help integrate family farms into international markets, while reducing economic and environmental costs.

In addition to technical solutions based on improving the efficiency of conventional crop protection methods, mainly those relying on pesticides, or on the substitution of these pesticides, research is focusing on developing new cultivation systems that do not depend on agrochemicals. As far as health aspects are concerned, it is a matter of finding environmentally intensive production systems. The more comprehensive and systemic aspects of these new research directions are covered in Chap. 18.

New cultivation strategies such as push-pull (stimulo-deterrence techniques; Cook et al. 2007) are increasingly being used. Such techniques repel pests from the crop by using repellent plants (push) and attracting them to the border of the cultivated plot by using trap plants (pull). In addition, the pest's natural enemies can be attracted to the plot with the help of companion crops as part of the pest control process. These principles have been successfully implemented to control stem borers in cereals (especially maize) by ICIPE (International Center of Insect Physiology and Ecology) and its partners in East Africa in maize farms (Khan et al. 2010). It must be noted that this strategy is particularly suitable for small non-mechanized plots, a reason for its success in small family farms, especially in small cultivation and livestock farms. Its success is also due to the fact that the border plants used for the purpose of "attraction" (such as elephant grass *Pennisetum purpureum* and Sudan grass *Sorghum sudanense*) and the companion crops (such as molasses grass *Melinis minutiflora* and Spanish clover *Desmodium uncinatum*) are an excellent source of fodder. Similar techniques have been developed for sugarcane (Nibouche et al. 2012.) and vegetable crops (Deguine et al. 2012.) and their scope of application is immense.

The value of the participatory approaches has also been highlighted in Africa in the case of the integrated protection of cotton (Prudent et al. 2007.) and fruit tree orchards (Van Mele and Vayssières 2007; Sinzogan et al. 2008). Indeed, small producers have to constantly adapt to ever changing environmental constraints and societal demands. Research can effectively assist in the continuous improvement of their farming system (co-design). Tools (I-Phy indicators, Boullenger et al. 2008; Le Bellec et al. 2013) have been developed for citrus-fruit cultivation in French overseas *départements* and have become effective tools for dialogue and decision making to meet the challenges of reducing the impact of pesticides, especially herbicides.

16.4 Supporting Public Policies for an Integrated Management of Health Risks

Although research was once strictly compartmentalized – by institution, discipline, commodity, geographical area, etc. –, it now attempts to integrate innovative approaches which, despite originating from its various disciplines, are combined into interdisciplinary approaches implemented via networks. Research on health risks is conducted at an international level since its objectives are also international; health risks know no borders and, for certain major crises, must be co-managed at the regional or even global level.

16.4.1 Establishing Regional Networks and International Collaborations

Several regional networks working on animal health (research, surveillance, control) have emerged based on partnerships among various actors.[1] The pooling of strategies, experts and tools across several countries allows the formulation of a more effective public policy and improved prevention and control programs. Thus, the eradication of rinderpest, the first animal disease to be eliminated from the face of the earth, was only made possible due to the implementation of international control programs (PARC, Pan African Rinderpest Eradication Campaign, and PACE, Pan African program for the Control of Epizootics). CIRAD, in close collaboration with FAO and its regional commissions, in particular CLCPRO (Commission for controlling the Desert Locust in the Western Region), has contributed through its research projects on preventive control. Its contribution has mainly helped strengthen the use of decision-making tools to better orient surveys and improve the accuracy of projections. To this end, CIRAD has developed a model to use existing data on nine concerned West and North-West African countries (FAO-Ramses database) to better predict the risk of locust infestation (Piou et al. 2013). It has also made a survey of the desert locust's biotopes, an invaluable tool for locust survey professionals (Duranton et al. 2012). Finally, it has established a monitoring system of national preventive control systems for the western region of Africa (interactive and georeferenced database accessible via the Internet to locust centers of the ten countries in the region and to CLCPRO; CIRAD 2012).

However, the challenge is also to bring national and international health regulations into line. Regulations are no doubt necessary if countries wish to export agricultural products originating from family farming systems and there exist a

---

[1] CaribVET in the Caribbean, Resolab in West Africa, Remesa in the Mediterranean, Risk-OI in the Indian Ocean, SEA-FMD and Grease in continental Southeast Asia.

number of prerequisites for gaining entry into international markets. These include the production of quality products, an alignment of national regulations, especially on health aspects, with international regulations, and a need for agencies that set international standards to better take into account the characteristics of products from family farming. The leveraging capacity of regional economic entities is often mobilized to include various countries in these negotiations.

16.4.2 Promoting Multidisciplinary and Interdisciplinary Approaches to Research

An improved integration of different disciplines such as epidemiology, laboratory sciences (virology, bacteriology, etc.), entomology, genetics, ecology and social sciences (geography, economics, sociology) is necessary in order to control pests and diseases more effectively.

Consequently, the last decade saw the emergence of integrated approaches to analysis with the support of international organizations. An example is the "one health" concept, which encourages medical doctors and veterinarians to work together. Indeed, 60 % of infectious diseases identified in the last 50 years are zoonoses, of which 72 % originate from wildlife. The bird flu crisis of 2005 and 2006 that particularly stirred up worldwide public opinion was largely responsible for key organizations like the World Health Organization (WHO), the World Organization for Animal Health (OIE) and FAO to work together to facilitate the integration of human and animal health services at the national level. This concept is based on the need for, and benefits of, collaboration in the field of health. It has begun to spread and has led to a more effective management of specific zoonotic diseases, such as rabies, which usually results in over 50,000 deaths annually, mainly in Asia and Africa. It is especially important to promote the inclusion of the human and social sciences and their research in risk perception and analysis. Thus, the evaluation of health incentives offered to family farmers in the principal dairy production area of Uganda (afflicted by bovine tuberculosis) focused on the acceptance and adoption of control and surveillance measures (Byarugaba et al. 2010).

Collaboration between international organizations and various public and private actors in the health domain has resulted in new sociological and political science studies on health governance and on the role of uncertainties and asymmetries of influence in negotiations (Figuié and Peyre 2010). Since family farms are poorly represented in international forums or livestock sectors, they risk being ignored during developmental stages of health programs and policies that do not take into account their structural characteristics, social demands and socio-economic determinants, which will ultimately determine the success or failure of health interventions.

Lessons learnt from the effort to control avian flu in Thailand (Paul et al. 2013) have proved valuable. A study conducted in the district of Kon Krailat (Valeix 2012) identified social factors that explain the difference between the rationales of actors on the field and the official surveillance system for avian flu in the country. The consequences of the massive culling of poultry during outbreaks in 2004 and 2005 were especially highlighted. Mass poultry graves led to profound disturbance amongst the Thai people. With poultry farming being subsequently discouraged, many families were compelled to seek other sources of income or to even migrate to faraway urban areas in search of work. The study also revealed that the main health challenge faced by duck breeders were not infectious diseases but the use of pesticides by rice farmers which was the primary cause of death amongst domestic duck, a factor that epidemiological research on flu had failed to identify (Chap. 12). Such social and technical factors explain the reluctance of breeders to alert authorities in case of a new epidemic outbreak of avian flu. The study's conclusion included a recommendation to improve the surveillance of animal health at the district level while respecting existing realities of local economic and political life.

Ecohealth (ecosystem health) is another concept of recent origin. This concept considers that the health of human communities and that of domestic and wild animals are closely linked to their ecosystem (Caron et al. 2011). It recognizes that both cultivated and natural ecosystems provide several ecosystem services, in particular a regulation of health risks, that require further study (Chap. 6). These concepts also incorporate a more local aspect of intervention and territorial development with an understanding of related systems in which family farmers and their communities are actors interacting with other actors.

The ecohealth concept developed as a result of the momentum generated by the Earth Summit in Rio de Janeiro in 1992. It encourages the development of local communities more respectful towards their environment. The concept of ecosystem services in the context of health has developed, since 2000, in a more sectoral manner with the help of international organizations (WHO 2005). An intervention on one component of a socio-ecosystem may have predictable or unpredictable consequences on other components. An example is the increase in vector-borne human diseases, such as malaria and arboviruses, because of the resistance of the vector to insecticides following phytosanitary treatments on vegetable crops and cotton fields (Chap. 12, "Impact on Producers' Means of Production and Assets").

16.5 Towards New Research Avenues

Tools produced through research to address health issues must take into account the diversity of producers and their practices, in addition to studying the diversity of hosts and pathogens. As was shown for human health, diseases and health systems are also social constructs, highlighting anomalies and inequalities, with these being subject to societal control (Aïach 2010) and power relations. The situation is similar for animal health. Family livestock farmers rarely have control over processes at work. In a context of health risks becoming progressively more global and cross-sectoral, and where "good governance" is the guiding principle, the challenge is to get the increasingly diverse collectives to work together towards a common purpose and towards multi-level and multi-actor governance. The knowledge produced by research must allow the bringing together and restructuring of collectives (formal and informal) established to manage animal health, their practices and relationships in order to foster cooperation. The hybridization of agronomy and ecology has not only created new methods of working with plants (Chap. 18) but has, in the context of phytosanitation, also led to growing awareness of factors beyond plot and field boundaries, and, consequently, to a re-evaluation of the role of family farmers.

Thus, Avelino et al. (2012) highlighted different promising landscape effects on the coffee berry borer in Costa Rica. Researchers found that the abundance of beetle in coffee plots was positively correlated with the proportion of area under coffee within a 500 m radius around the plots. Negative correlation was observed with other land uses, namely forests, pastures and sugarcane fields. Since the coffee berry borer is strictly restricted to coffee as a host plant, contiguous coffee plots favor colonization of new berries by flying insects, particularly during the post-harvest period when berries are scarce. On the other hand, a fragmentation of the landscape with land uses other than coffee cultivation reduces the pest's survival rate. This effect was most marked with forest patches which act as barriers to the movement of beetles. And yet, researchers found a higher incidence of leaf rust in coffee plantation landscapes which were fragmented with pastures. The reason was found to be that the wind turbulence at the frontier of the two landscape uses detaches fungus spores responsible for leaf rust. This leads to a re-infection of coffee plots. Therefore, only a fragmentation of the landscape with forest plots will limit the spread of the beetle without favoring leaf rust. These results clearly illustrate the vulnerability of homogeneous coffee landscapes to pests, and therefore the importance of plant diversity at the landscape level to manage the risks of pests. This underlines the need for collective and concerted action in the context of family coffee cultivations.

Chapter 17
Agricultural Biodiversity and Rural Systems of Seed Production

Danièle Clavel, Didier Bazile, Benoît Bertrand, Olivier Sounigo, Kirsten vom Brocke, and Gilles Trouche

The domestication of plants and animals, and the selection and exchange of these domesticated populations between farmers, along with the movement of genetic resources during human migration and the adoption of various strategies to protect against sanitary risks (Chap. 16) have been occurring throughout the history of agriculture. This long process of domestication gathered momentum in the second half of the twentieth century under the impetus of agricultural modernization, or at least as was experienced in one part of the world in the 1950s (Chap. 2). The Green Revolution, which took root especially in the countries in the South, is based on the key role of modern varieties and hybrids (Bonneuil and Thomas 2012). No doubt, agricultural research has, on the whole, supported the Green Revolution, with the creation of CGIAR (Consultative Group on International Agricultural Research) reflecting the desire to use genetic progress as the spearhead of the agricultural revolution. In this way, agricultural research has overwhelmingly embraced modern plant breeding concepts; – concepts that are based on intensive cultivation (irrigation and chemical inputs) of high-yielding varieties.

Two strategies are combined to support this modern plant breeding model: first, the conservation of germplasm collections of major crops at locations other than the production locations (ex situ) and, second, plant breeding, varietal selection and a professionalized production of seeds (Chap. 13) (Louafi et al. 2013). This research

---

D. Clavel (✉) • B. Bertrand • K. vom Brocke • G. Trouche
Bios, Cirad, Montpellier, France
e-mail: daniele.clavel@cirad.fr; benoit.bertrand@cirad.fr; kirsten.vom_brocke@cirad.fr; gilles.trouche@cirad.fr

D. Bazile
ES, Cirad, Montpellier, France
e-mail: didier.bazile@cirad.fr

O. Sounigo
Bios, Cirad, Yaounde, Cameroun
e-mail: olivier.sounigo@cirad.fr

and development strategy, associated with research on genetics and breeding, has been widely adopted in the North and partially so in the South, within the framework of an intensive and standardized industrial agriculture.

A different viewpoint of mankind's relationship with nature has, however, emerged with the signing of the Convention on Biological Diversity (CBD) in 1992 (Chap. 13) which offers an alternative approach to current varietal research. The generalization of the concept of "agrobiodiversity" reflects the need to understand the diversity of cultivated plants in the context of agricultural history with its stages of agricultural domestication and selection, while revisiting the link between the cultivated space and its environment. The relationship between the loss of biodiversity, loss of cultural diversity and loss of knowledge is being increasingly documented (ISE 2013), despite it being subject to some controversy.[1] Recent "biocultural" studies reveal the growing importance of the new paradigm of "ecological and integrated management of natural resources for food and environmental preservation" in contrast to dominant approaches (Chevassus-au-Louis and Bazile 2008). In the report by the CBD Secretariat (Secretariat of the Convention on Biological Diversity 2010), the in situ maintenance of a dynamic agricultural biodiversity is described as the major component for adoption and sustainability of agricultural and food systems. Agricultural research must therefore offer synergies between agriculture and biodiversity that promote and support agricultural and rural development (Hainzelin 2013).

Research can help family farming find answers to these complex questions. Indeed, research on plant breeding now takes criteria defined by family farming into account for the adoption, conservation and dissemination of agrobiodiversity in diverse situations (Chap. 13). Using actual examples, this chapter illustrates this development in different plant breeding situations for crops predominantly cultivated by family farmers. It shows how small farmers are involved in the processes of creating and disseminating new varieties of coffee and cocoa in agroforestry systems which, in turn, fulfill functions pertaining to food, the environment, energy, medicine and culture. An economic dependence of these systems on export crops makes them vulnerable to price fluctuations in international markets. Moreover, these systems are constrained by the inability of farmers to invest in their cultivations. In such situations, any research intervention has to be contextualized and tailored to existing dynamics. Food problems assume enormous proportions in dry areas that support large populations. This is why emphasis is given to participatory breeding methods for traditional food crops which have multiple traditional uses, for example, sorghum in Africa and quinoa in The Andes. Finally, this chapter discusses how research, through companion modeling, encourages local seed production systems, and promotes their development by improving knowledge exchanges between actors in these systems, their training and nurturing of new partnerships.

---

[1] Thus Kohler (2011) developed the concept, now found frequently in the literature, of "a particularly forced analogy between cultural diversity and good environmental health," and stressed the dangers of such an analogy. The criticism and debates that followed this publication demonstrated the extent of the controversy.

17.1 An Innovative Partnership in Coffee Production in Central America

The fall in coffee prices in the late 1990s forced many small Central American producers out of the international market because their production costs were higher than in other coffee growing regions like Brazil and Southeast Asia (Kilian et al. 2006). In Nicaragua, Arabica coffee traditionally constitutes the bulk of the coffee production, 80 % of it being grown by small family farmers in agroforestry systems with less than 3.5 ha of land.

Research and collaborations have helped improve coffee varieties for over 50 years. Numerous research institutions (ORSTOM, CIRAD, CNRA/Côte d'Ivoire, Madagascar) worked between 1960 and 1990 to identify wild coffee varieties and analyze their phenotypic diversity. These activities were "academic" in nature, that is to say they were oriented towards gathering knowledge and not towards development. CIRAD and Promecafé[2] worked in partnership from the 1990s to breed new F1 hybrid varieties in Central America to expand the tiny genetic base of Central American coffee varieties. These hybrids were derived from crosses between American varieties and "wild" coffee varieties from Sudan and Ethiopia (identified by FAO and Orstom in 1967). Genetic material from Sudan and Ethiopia is endowed with added resistance to certain diseases and has important organoleptic qualities. "Standard" practices to improve varieties are based on intensification principles and help develop hybrids derived from crosses between lines with a narrow genetic base. F1 hybrids are designed for very intensive cropping systems that receive full sunlight, corresponding to an intensive and artificialized form of agriculture. In the case of Arabica, the hybrid program in Nicaragua sought to improve productivity and quality (Bertrand et al. 2005) while respecting the typical agroforestry systems of family farming. The results of 20 years of experiments in controlled environments and in producers' fields have shown a 30–60 % increase in yield for F1 hybrids, obtained from crossing between American varieties and wild coffee from Ethiopia, as against the best varieties in the American small-scale farming systems, an achievement that involved no fertilizers inputs which are, in any case, beyond the reach of small coffee growers (Bertrand et al. 2012).

The development of a vegetative (or clonal) propagation method through research then opened up the possibility of producing hybrid plants on a large scale, even though hybrid material is usually reproduced by seeds from crosses. However, the in vitro somatic embryogenesis technique that was developed had the drawbacks of being inherently complex and of needing a significant initial investment – with uncertain returns – from financially strapped family farmers. CIRAD was therefore on the lookout, since 1999, to team up with a private partner to commence large-scale multiplication of F1 hybrids. A contract was signed with

---

[2] An agreement for cooperation that bring together the governments of several coffee producing countries in the region with the purpose of increasing coffee cultivation as a socio-economic activity through a transition to agroforestry-based ecologically intensive agriculture.

Ecom Coffee in 2003 (Étienne et al. 2012) to mass produce F1 hybrid plants for the market. A laboratory for in vitro micropropagation was established in Nicaragua and began production in 2006. Along with the creation of the hybrids in the laboratory, they were also being tested extensively on farms. Field results confirmed that F1 coffee hybrids met the expectations of productivity and of organoleptic qualities. The model to produce hybrids in laboratories and nurseries became operational and reproducible in 2011 (Étienne et al. 2012).

All actors in the network, ranging from the research consortium to the users, have been involved since 2006 in the process of disseminating hybrids to the farmer. The research team not only provided advice to fine-tune adaptation of hybrids to climatic and soil conditions, but also initiated fresh research to develop new varieties and concepts (catalogs of varieties, technified nurseries, etc.) in order to address local issues.

A posteriori, one of the major obstacles encountered in transferring this technology was the inability to scale up as fast as necessary. The adoption of these plants at the individual level was supported by Sustainable Management Services (SMS, a subsidiary of Ecom), whose brief included disseminating plants as well as transferring the knowledge generated to help producers in their training. In return, the institutions could use field results and the changes and limitations observed to develop measures to fine-tune the approach. A process of hybrid awareness (*conciencia híbrida*) was initiated among farmers and the demand for hybrids rose steadily. This resulted in the "hybrid variety" becoming a scarce resource and Ecom's credibility as a supplier of high-yielding varieties began to suffer. The partnership with CIRAD thus became strategic in order to ensure a greater diversity of supply and to meet an increasingly specific and growing demand.

In order to leverage the value of these highly desired hybrids, the Diamond Coffee certification label was registered in 2012 for coffee made from these hybrids. This label describes original coffee that has been cultivated at a minimum altitude of 1,100 m above sea level, under shade and complying with stringent specifications. Under these conditions, Diamond Coffee develops organoleptic qualities which are considered to be unique. The hybrid coffee's high productivity and rarity seem to offer a new market opportunity to some producers, as also the possibility of creating a cluster to enjoy particular services and benefits (funding, specific business opportunities, etc.) provided by the trading company. Unfortunately, this "elitist" approach of the Ecom Group de facto cut off access to innovation for producers who were located outside the favorable areas. However, these farmers were all potential suppliers and customers of the trading company. As a result, further research was conducted to produce high yielding line-varieties that could be adapted and adopted by all types of farmers. The *Marsellesa* line-variety was thus introduced in 2013. This variety is resistant to many diseases and has a productivity midway between those of hybrid and traditional American varieties. It produces a coffee with a high sensory quality. An added advantage for producers, when compared to hybrids, is that it can be reproduced through seeds. Once they acquire the initial plants, producers have the right to reproduce the variety for their own needs. Ecom-SMS supports this activity by offering a premium to producers who cultivate *Marsellesa* coffee.

It is still too early to judge the impact of this public/private partnership on coffee-based agroforestry conservation systems or on the economic benefits for small farmers who did, or did not, adopt these hybrids or the new *Marsellesa* line. However, an initial sociological study conducted in 2012 and 2013 showed that innovation and collective learning processes that marked the release of new coffee varieties to small Nicaraguan family farmers led to innovative practices and adaptation strategies. The study clearly showed that small farmers who adopted the new varieties enhanced their production potential (Alami et al. 2013). The role of intermediation played by researchers has led to a broader vision and a transfer of the research and development model to Mexico, where the project is currently experiencing a particularly good growth. These small farmers were able to convey their requirements to the trading company and the public research support community, which ultimately resulted in a wider dissemination of genetic progress.

17.2 Management of Cocoa Varieties in Cameroonian Agroforestry Systems

Three million small farmers cultivate about seven million hectares of cocoa (*Theobroma cacao* L.), with 85 % of this production coming from family farms of a few hectares. Global demand, particularly in Europe, is continuously rising and the area under cultivation is expanding rapidly. Although the plant is native to the Amazon regions in the northern parts of South America, Europe gets the bulk of its supply from West Africa which, with two million cocoa farmers, accounts for over two-thirds of the world's total production.

Cocoa farmers in Africa have little capital at their disposal and the needs of the family are met from forest produce during the initial years of cultivation. It is an economic system based on "forest income" in which forest resources help offset the lack of income until the plantation can start producing cocoa. The family farmers' know-how and experience is what amounts to capital in this system. Cocoa is grown mainly "under cover," i.e., in the shade of trees within the complex agroforestry systems. Production is based on the association between cocoa and various perennial fruit and forest species. Fruit trees provide a valuable dietary supplement and an additional income, all the more useful to family farmers whose main income is from cocoa, which materializes only in the last quarter of the year. Forest species are useful in various ways (shade, improved soil fertility, pharmacopoeia, firewood, timber). Cocoa-based agroforestry systems are considered a good alternative land use to cope with climate change because of their high levels of species diversity, of soil cover they maintain throughout the year and the carbon they help store below and above the ground (Somarriba et al. 2013).

Low prices of cocoa in the 1990s led to a drastic decline in funds for cocoa breeding programs. The first project of the Common Fund for Commodities (CFC), associated with the International Cocoa Organization (ICCO) and the International Plant Genetic Resources Institute (IPGRI, later renamed Bioversity International), brought together ten countries: Papua New Guinea, Nigeria, Ghana, Côte d'Ivoire, Cameroon, Brazil, Venezuela, Ecuador, Trinidad and Tobago, and Malaysia. When this project got underway, most of its constituent partners and the major cocoa producing countries, namely Cameroon and Nigeria, were facing a virtual standstill of their cocoa breeding programs. Amongst these countries, only Brazil and Malaysia have extensive industrial-type plantations. The cocoa crop there came under great pest pressure which affected yields and product quality. The spread of pests and diseases was of particular concern in new areas that were rapidly being brought under cocoa cultivation (Eskes 2011). The second phase of the CFC/ICCO/Biodiversity project (Eskes 2011) continued and intensified breeding activities and developed participatory approaches such as farm surveys and the direct participation of farmers in the selection of trees on their farms. CIRAD, in association with the Cameroonian Institute of Agricultural Research for Development (IRAD), supported various participatory projects on breeding and crop production.

Several research projects conducted in Cameroon since 2003 had an on-farm research component to evaluate the cultural and sanitary aspects of these systems and suggest improvements. The agronomic and economic performance of these systems exhibited a wide variation depending on regions and cultivation techniques (Jagoret et al. 2009). A major source of variation was discovered to be the density of shade trees used which, when it is too high, greatly reduces the cocoa yield (Jagoret 2011). The effect of shade trees on damage caused by phloem-feeding insects – the mirids (*Sahlbergella singularis* and *Distantiella theobroma*) – was studied. These insects damage plants and pods, often even leading to the death of trees. Yields can drop by as much as by 40 % in West Africa. Mirids are generally more numerous in plots that are less shaded. Open, illuminated areas that are a result of canopy gaps and the presence of certain species, like the kola, encourage the presence of these insects (Babin et al. 2010). These multidisciplinary studies have highlighted the importance of proper shade management to control these major cocoa pests. A similar study is currently underway on black pod rot, a major disease caused by the fungus *Phytophthora megakarya*.

The continuous rise in cocoa prices (+66 % since 2007, according to ICCO), partly due to the negative impact of diseases and the climate on crops, has led to an increase in farm gate procurement prices. The rapidly increasing demand for cocoa has led to a vast expansion of new cocoa plantations in recent years. At the same time, access to improved varieties has been very limited. State entities, the National Cocoa Development Company (SODECAO) and the Cocoa and Coffee Seed Project (CCSP) that are in charge of managing seed production fields and disseminating these varieties manage to meet only 20–30 % of the national demand (Asare et al. 2010). Since 2006, CIRAD, in partnership with IRAD, has therefore initiated various participatory programs to breed, evaluate and produce plants to help farmers obtain the varieties best suited to their needs.

Plant breeding activities are mainly carried out in a network of 80 coparative progeny trials. These trials are spread out over three *départements* of the Center Region of Cameroon and correspond to three different environments. About 150 progenies of various origins were studied. They were either progenies of released varieties obtained from seed production fields, created through manual crossing at the research station, or created from open-pollinated pods harvested from tree species jointly selected by producers and breeders (the selection is based on productivity and tolerance to black pod rot) or, finally, originating from open-pollinated pods harvested from trees selected randomly from farmers' plots. This latter progeny type corresponds to planting material generally used by small farmers who do not have access to commercial varieties.

Another factor behind the low production of agroforestry-based cocoa farms is their advanced age of the trees. Research has been conducted since 2007 on a regeneration technique based on coppicing cocoa plants, followed by grafting shoot rejects using grafts taken from selected trees. These regenerated plots are used for clonal trials in different environments to compare clones bred at research stations with those bred on farms. Genetic resources are conserved and improved planting material disseminated using a similar method in small seed production fields managed by farmers. Pods (or seedlings from such pods) are sold by farmers or farmers organizations (FO), which may stock these resources to establish new plots. Clones and progenies thus acquired enable family farmers to regain access to planting material in an independent manner, a freedom that was forfeited when they chose to use hybrids produced at a research station (Ruf 2011). This operation began in 2008 and resulted in the establishment of seed production fields in five *départements* of the Center Region of Cameroon.

The impact of projects implemented since 1998 has been significant as much in terms of planting material quality as of the partnerships forged and the training of farmers and technicians (Efombagn et al. 2011). Participatory programs resulted in an increase in multidisciplinary activities involving plant pathologists and entomologists. A hundred small cocoa farmers were involved in these programs. Not only did they gain access to improved varieties and technical training on cocoa (nursery, plantation, plot management, grafting, vegetative propagation), but also to the development of other components and aspects of the agroforestry system (oil palm, fruit trees, plantains, quality of shade, control of predators). These activities were developed through new partnerships with local NGOs, IRAD researchers, the World Agroforestry Centre (ICRAF) and the International Institute of Tropical Agriculture (IITA). These partnerships help mobilize various actors and provide access to different funding sources.

17.3 Biodiversity and Participatory Breeding of Sorghum in Africa and Central America

Local varieties of pearl millet and sorghum developed and implemented by family farmers in the Sahel possess a high degree of plasticity toward environmental changes over time and space due mainly to their photoperiodism (Sissoko et al. 2008). These two cereals constitute the basic diet of the rural populations and are frequently grown in association with legumes such as cowpeas and groundnuts. They can either be grown monovarietal or as a varietal mixture of the same species.

Participatory breeding is a method that was principally developed in Central America and West Africa for the purpose of making available varieties that matched farmers' expectations and thus facilitated their adoption. In the case of sorghum, conventional breeding so-called "standard" or "formal" breeding) has only explored and exploited a fraction of the available diversity of genetic resources, focusing mainly on the productive potential, resistance to biotic stress and grain quality for limited uses. However, issues such as adaptation to complex cropping systems and specific usage by particular target groups (women, local processors, etc.) are yet to find a place in these conventional breeding programs. In order to address these shortcomings, participatory sorghum breeding programs introduced farmers to many different varieties with new agro-morphological characteristics or combinations of such characteristics. This diversity can be in the form of traditional or improved exotic varieties, or even of forgotten or little-known local varieties.

A key point of the participatory breeding process is to allow farmers to assess these different varieties in their environmental conditions and according to cropping practices specific to their production systems. The selection criteria are thus related to grain yield as well as to various other uses of different parts of the plant. The assessment of this new diversity provides an opportunity for an exchange of know-how and knowledge between farmers and plant-breeder researchers. Breeders can accordingly fine-tune their perceptions of farmers' selection criteria while farmers can improve opportunities for innovation that will meet their needs.

The significance of photoperiodism, one of the new criteria being taken under consideration in breeding programs, is worth exploring. As a character trait in traditional varieties, it allows the control of flowering in relation to variations in day length. In Sudano-Sahelian Africa, the length of the agricultural season depends to a large extent on the arrival of the first rains. This date depends on the latitude, and varies greatly from year to year, unlike the end date of the rainy season which is more constant. In order to address these climatic variations, farmers have, over time, bred pearl millet and sorghum varieties that are sensitive to photoperiodism; plants that mature at the right time, that is, at the end of the rainy season when the family workload is less. This criterion is an advantage, especially in the context of climate change.

Box 17.1 presents an example of sorghum varieties obtained via participatory breeding approaches in Mali, Burkina Faso and Nicaragua. An interesting case here is the Coludo Nevado ("white tail") released variety which was disseminated in Nicaragua. In the course of a participatory survey conducted at the beginning of this program, farmers indicated the desired targets for their photoperiodic sorghums, which are generally sown with maize as part of a strategy to minimize climate risk. The aim was to reduce the plant height while at the same time retaining the compact panicles of the traditional cultivars, which are considered to be very productive, and improving both grain quality for the preparation of tortillas and straw quality for forage. Coludo Nevado was among the many African photoperiodic varieties which were evaluated with the involvement of these farmers in the next stage. Except for grain quality, Coludo Nevado exhibited none of the above characteristics. However, as a result of several years of in situ experimentation, farmers were able to uncover other qualities, such as its high plasticity in relation to soil types, drought tolerance, and easier harvesting despite its height (due to drooping panicles), as well as high, stable yield. Consequently, the Coludo Nevado variety was rapidly adopted in the northern region of the country.

Box 17.1. Participatory breeding, a method of dialogue and mutual learning

Kirsten vom Brocke and Gilles Trouche

The process of participatory breeding involves a continuous and long-term interaction between researchers and farmers. It leads to the joint production of varietal ideotypes that allows the breeding of new varieties adapted to changing family needs.

Participatory selection of breeding characteristics and joint experiments conducted in the fields of farmers has resulted in the success of three common varieties listed below.

Variety	Country or region	Breeding criteria and objectives set initially*	Key characteristics determining adoption of the variety
Soumba	Mali	Combining high yield potential and short height of caudatum varieties with the grain quality of guinea varieties (hardness, mold tolerance, quality of the porridge).	Stems resistant to breaking and less attacked by birds because grains are "hidden" by long glumes. Allows late harvest: Adaptation to harvest schedules, especially for individual fields belonging to women. Straw of good forage quality.
Coludo Nevado	Nicaragua	Shortening the height of photoperiodic traditional varieties while increasing the size and improving the quality of the grain and fodder value of straw.	Despite the criteria of a good height and a loose panicle being a priori rejected and having straw of low forage value, the variety is highly valued for its hardiness and drought tolerance, its productivity which is equivalent to varieties with compact panicles, its ease of harvest as stems bend when mature, and its excellent grain quality.
Gnossiconi	Burkina Faso	Abandoned by farmers 40 years ago due to its earliness which led to it being attacked by birds.	Earliness, regularity of yield and hardiness are essential criteria nowadays in the context of greater climatic variability, resulting in the readoption of the variety.

* By the sorghum breeder or a participatory diagnostic at the beginning of the project.

Most local varieties grown in Mali belong to the *guinea* race. They are tall, have loose and drooping panicles with vitreous grains. Varieties of the *caudatum* race, originating from Central and East Africa, have a short to medium height with rather compact and upright panicles and less vitreous grains. The adoption of improved varieties of the *caudatum* race is very limited despite their potential for higher yields. The Soumba variety was obtained from crosses of these two races with a phenotype more aligned to *caudatum*. In the context of declining cotton production, it was adopted by farmers in the Dioïla region of south-central Mali both for family consumption and for exchange. . The main reason for Soumba's adoption is that it combines certain characteristics that allow it to be successfully integrated into changing cropping systems. Owing to its good resistance to lodging, due to its strong stems, and its resistance to bird damage due to its long glumes that hide the grain, it can be left longer in the field without fear of damage until other harvesting work has been completed. This is an important feature for family farms which rarely use external labor. Moreover, the fact that the plant remains green when the grain matures is a well appreciated quality for animal nutrition that should not be overlooked when breeding sorghum varieties. Breeders had not considered these criteria when they created this variety. The flexible harvest date and the commercial appeal of this variety's large grains are two other advantages that have convinced women who sell their surplus production in the market to grow this variety in their fields.

The local *Gnossiconi* variety has been maintained in the gene bank of INERA (Environment and Agricultural Research Institute) in Saria for 40 years. It was reintroduced to its region of origin in Burkina Faso in 2002 through a participatory breeding program, and was subsequently adopted by farmers for the very reason it was abandoned 40 years ago: its earliness. Farmers explain that this characteristic was a major drawback in former times when more favorable rainfall patterns allowed the growing of late variety sorghums that were more productive than *Gnossiconi*, which was susceptible to bird attacks because of its early flowering. At present, in a context of less favorable and more irregular rainfall, farmers have readopted the variety because of its earliness, stability and productivity. The fact that a group of farmers in this area has adopted these early varieties greatly reduces the risk of bird damage, a risk that is aggravated when plots are few and far between.

This cooperation between farmers and researchers to create new varietal ideotypes was taken further in some participatory programs in Nicaragua and Burkina Faso. These programs have produced a new generation of varieties that are being disseminated in these countries.

17.4 Participatory Modeling Applied to Seed Systems: The Example of Mali

In spite of various public policies to support the establishment of national seed systems for disseminating varieties across West African countries over the last 30 years, the traditional seed system remains the primary route for the movement of seeds (Bazile and Abrami 2008) accounting for nearly 90 % of seed transfers (Delaunay et al. 2008). In large areas of Mali, there has been no adoption of improved varieties without local germplasms (Yapi and Debrah 1998), demonstrating the failure of the dissemination of "improved" varieties since the 1960s. Cropping systems continue, however, to incorporate a wide diversity of local species and varieties.

In such a context, new ways must be devised to manage this varietal diversity and disseminate seeds in order to address the needs of and changes in family farming. Current research describes farmers' seed systems either on the basis of genetic diversity models or according to business models in order to take into account the risks of agricultural production. These two approaches can be combined to describe the dynamics of family farming biodiversity in changing environments. Innovations in modeling research are, in fact, focused on linkages between national seed systems and traditional seed systems in order to better understand the various issues concerning a dynamic conservation (Wood and Lenne 1997), through a combination of the genetic and economic dimensions. New civil society actors (farmer organizations, cooperatives, associations, NGOs, etc.) are working together for a reconfiguration of seed systems, requiring us to re-evaluate traditional geographical and social networks (Subedi et al. 2003). Companion modeling approaches can help analyze the complexity of these dynamics.

Companion modeling is an inter-disciplinary approach which encourages a participatory management of renewable resources (Barreteau et al. 2013; Le Page et al. 2013). It results either in a collective representation of a problem in managing a shared resource, or in changes in organizational techniques (Bousquet et al. 1993). Effort can be directed to studying the dynamics of biodiversity resulting from interactions between populations and their environment (Étienne et al. 2003). The work by Vejpas et al. (2004) on rice in Thailand, and that of Bazile and Abrami (2008) on sorghum in Mali have shown a new method of applying the multi-agent modeling approach to the management of varietal diversity in family farming systems. This approach is used either for understanding knowledge of local management practices or for a simulation of scenarios. In the latter perspective, it is a matter of testing new strategies on family farms or assessing the impact of public policies (Bazile et al. 2012; Belem et al. 2011).

The representation of a situation and its dynamics are formalized in multi-agent systems (MAS).[3] However, the use of role-playing games (RPG)[4] based on the same conceptualization as MAS facilitates the representation of a complex situation in a controlled situation. The role playing game is created on the basis of the same assumptions as those of the model.

---

[3] MAS originate from the field of distributed artificial intelligence and are used to solve problems of coordination of independent heterogeneous agents. MAS are suitable for simulating different forms of coordination, especially changes in management rules and the overlapping effects of individual strategies and collective rules.

[4] MAS use role-playing games which correspond to the representation of a complex situation in a controlled space. The Seed-Div role-playing game was created on the basis of assumptions and is a model or archetype of reality. It allows one to step back from the real world and thus serves as an intermediary with the reality of the situations observed in order to discuss with actors the actions they have undertaken or to confront them with new situations.

In addition to examples of participatory breeding at the producer level – such as the sorghum example described above –, participatory modeling based on multi-level interactions of cereal seed systems of farmers, traders and institutions was developed in West Africa to support actors of seed systems in managing agrobiodiversity dynamically. This section of the chapter examines this experiment and its generic and operational lessons learned in Mali.

From 2004 to 2007, six workshops were held based on methodologies of participatory modeling, each of which were attended by 20–30 farmers and members from farmer organizations and NGOs in Mali and Niger. Specific role-playing games (each game with a question based on an hypothesis) were used which led participants to arrive at a shared vision of seed systems based on five criteria:

- a characterization of farm types according to the diversity found in their cropping systems;
- a classification of varieties in functional groups for exchanges;
- a description of three principal individual behaviors to manage varieties (farmer experimenter, imitator or conservator);
- an improved understanding of the decision-making processes involved in selecting a variety and the formulation of rules for conducting tests on a farm;
- a characterization of different seed supply sources and conditions to access them in these different supply chains.

One of the major lessons learnt from this dialogue between actors is that family farmers prefer to maintain their local system because its greater diversity leads to greater security and productivity for them in comparison to the national seed system. Families access a "formal" seed system only when it helps them meet new requirements and supply objectives. In order to develop, new seed systems must first validate the role of key farmers in seed exchanges and help them adjust to a framework of collective rules on a larger scale – different from that of their farm – but consistent at the scale of their community.

The Seed-Div role-playing game was co-constructed in a 3-day workshop held to encourage participants to come up with alternative systems for collectively managing pearl millet and sorghum seeds. This role-playing game was then used to simulate operating rules of different types of entities (village association, seed cooperative, farmer organization) to achieve seed conservation, multiplication and distribution objectives. The shared experience of the role-playing game allowed results of different management systems to be compared and discussed and to continue the simulations to adapt operating rules to targeted goals in order to actually create these institutions. Seed-Div is now used for training in France and elsewhere largely due to its generic nature. At the same time, research is continuing to formalize the MAS approach and role-playing game in seed systems (Box 17.2).

Box 17.2. Application of companion modeling for agrobiodiversity: the case of the IMAS project.

Group of researchers coordinated by Didier Bazile

A generic system was developed as part of the IMAS project (Impact of the modalities of seed access on the dynamics of agricultural genetic diversity, ANR 2008-2012) in order to link seed systems. It pertained to two contrasting situations: the *in situ* conservation of the diversity of traditional cereals (pearl millet and sorghum) in Mali and the revival of quinoa cultivation in Chile in association with *ex situ* conservation.

Two major findings emerged from the use of MAS and role-playing game:

– The development of a MAS application allowed the analysis of the dynamics of biodiversity using simulations of scenarios based on links with the market, the implementation of agricultural policies and the impact of climate change;

– The first permanent national roundtable on quinoa in Chile was created within ODEPA (Ministry of Agriculture in Chile) bringing together different public and private actors to boost regional and national groups (Bazile *et al.*, 2012).

The modeling is participatory because it was created jointly by many actors (farmers, FOs, NGOs, research groups, seed growers, etc.). It simulates favorable mechanisms for the conservation, maintenance and use of varietal diversity in family farming systems.

The simulations carried out in workshops contribute to a gradual increase of shared knowledge with the help of a model that is regularly re-evaluated by all actors. The researchers' representations of the system are subject to collective review, thus providing the farmers involved with a sense of active participation in the research process.

In addition to the analysis of agrobiodiversity determinants at the level of the family farm and the characterization of varietal dynamics in different agrarian contexts (agricultural, environmental and socio-institutional), this approach has provided a framework for discussion on how to monitor and problematize the impact of the introduction of an improved variety in a self-production situation which is, after all, the norm in family farming.

These participatory dynamics have engendered significant changes in seed systems in terms of access of family farming systems to quality seeds, by describing, highlighting and promoting the critical role of FOs (Box 17.3). In addition to its efficiency, the local dissemination of seeds through the FO network makes it possible to link the formal state systems (national seed system, NSS) and the traditional system, when the latter cannot cater to the demand on its own (Coulibaly et al. 2008). The activities of the FOs are not only based on an agroecological vision of farming surfaces, but also on locally respected organizations, thus recognizing the importance of the links of kinship between family farms. This articulation between different systems allows the emergence of a comprehensive seed system that includes and interlinks seed networks of farmers and of the State at the national level, with the FO responsible for local-level transitions.

Box 17.3. Seeds from the Association of Professional Farmer Organizations in Mali.

Didier Bazile

In the late 1990s, FOs wanted to come out from under the control of the Malian Textile Development Company (CMDT). CMDT had established village associations to monitor the distribution of agricultural inputs (including seeds) and harvests in the cotton growing region. In 1997, the Association of Professional Farmer Organizations in Mali (AOPP) identified the following problems at a national workshop with all FOs of the country present: disappearance of certain varieties, reduced seed quality (especially pearl millet and sorghum), decreased rainfall, lower yields, high cost of fertilizers and difficulty of access to credit.

AOPP suggested that its members initiate a discussion on certified seeds. This led to the creation of the grain commission which, in turn, developed a network, starting in 1999, of experimenter farmers, known as *Si fileli kela*. Every year, about 20 local FOs can nominate 15 of their member farmers to receive training. These farmers are then responsible to conduct comparative tests between their traditional seeds and certified seeds from the national seed system (NSS). This continuous training has created an extensive network of more than a thousand experimenters, or "testers" as they call themselves. AOPP requests the NSS technicians it selects to get the farmer himself to provide training on the use of certified seeds and the method of on-farm tests. AOPP buys certified seeds and distributes them to farmers. The farmers compare their best local variety with the improved variety proposed by the trainers for the region under consideration by cultivating it on a quarter hectare on their farms.

From 2000 to 2004, seed production (pearl millet, maize, groundnuts and sorghum) spread from Mandé to Bélédougou, in Tominian and in Seno in the Dogon area. The production of basic seeds started in 2005. Seven seed cooperatives with a suitable legal status were created under AOPP for this purpose (three in the Office du Niger area and four in upland areas for dry crops). In 2011, the program was extended to the regions of Koulikoro (four cooperatives) and Ségou (one cooperative). The focus is currently on ten major crops: pearl millet, sorghum, maize, rice, cowpea, fonio, groundnuts, sesame, okra and hibiscus. The members of seed cooperatives provide R1 or R2 seed that are packed in accordance with recognized and adapted standards with the AOPP label. Awareness is raised and certified seeds are distributed with the help of the AOPP networks and through radio announcements.

The support provided by AOPP to family farming systems in Mali demonstrates its dynamism in addressing the inadequacies of the formal seed evaluation and dissemination system. Local FOs now systematically review tests conducted after each cropping season. The quality of this network, developed with limited resources, shows the ability of family farming systems to organize themselves in order to develop their own monitoring and capitalization tools.

Different conditions need to be met in order for collaboration in a global and multi-actor seed system to be more effective. Companion modeling favors the implementation and sustainability of these conditions.

The production and dissemination of first and second generation seeds is currently undertaken primarily by decentralized cooperatives of the government seed system, which operates in parallel with the system of FOs and their network of experimenter farmers. Seed producers of the state system are supervised by NSS technicians and receive financial support for their crops. Although the quantity of seeds produced is significant, they represent very few varieties per locality. As a result, cooperatives have to deal with large quantities of unsold seeds. On the other hand, as the choice of the variety can vary greatly according to the annual climate, or even during replanting, the stocks held by NSS cooperatives prove inadequate to effectively meet unanticipated demand, which usually concerns a wide range of varieties.

Members of the AOPP network, with its experimenter farmers and seed cooperatives, encourage the dissemination of varieties beyond the limited scale of the family or the village. Simulations of the participatory model show us that aligning this system with the national system could lead to a balance between local production and demand. AOPP maintains tight control over its operations for breeding, distribution and marketing of seeds to achieve prices that are lower than those of the NSS, while providing suitable high quality seeds to family farms. It could take over the NSS's responsibilities. Conversely, the technical staff of the NSS could rely more on member-farmers of FOs for breeding and seed certification. Thus, each actor has a role to play but is also constrained. The institutional anchoring of AOPP at the national level legitimizes a process to disseminate improved varieties in the country that maintains a close and strong relationship with locally anchored family farming systems. Although FOs are actively involved in supplying improved seeds to their members, it is not easy for them to undertake nationwide operations. They also encounter technical difficulties occasionally. The NSS commands a wide network of technicians and enjoys institutional legitimacy. The varietal diversity maintained in family farming systems constitutes an extremely rich and diverse pool of genetic resources. This wealth often goes unnoticed by farmers themselves and is not always accessible to researchers and seed breeders for breeding programs. Moreover, family farms do not have the means to monitor the development of traditional varieties and plan a sound conservative management strategy at a scale that goes beyond their farm or, at best, the village.

Companion modeling allows actors to share each other's roles and constraints. It allows a systemic representation of the seed system, its different objectives and expected performance, both for the purposes of conservation and for increasing production and adapting to product valorizations, between consumption and commercialization. This sharing creates favorable conditions that encourage deliberations on strategic priorities and changes that could be made in the production and distribution of seeds. In Mali, for example, farmers noted that the workshops had helped them understand for the first time what the researchers wanted and did, which then allowed them to better explain their activities and roles in the system.

17.5 For Participatory and Multi-actor Research on Support

The new paradigm that integrates agricultural research and environmental protection assumes that research assimilates this development to create new knowledge by basing itself on interactions between mankind, societies and the environment. This widening of knowledge requires the creation of new analytical frameworks that incorporate the diversity of productions. A high biodiversity exists within these productions, to which is linked know-how that could provide new indicators for changes in biodiversity under different contexts.

In addition to the need for a multi-disciplinary approach, agricultural research must open itself to accommodate development and civil society actors in a more meaningful way if it has to have any hope of dealing with the agricultural system as a complex object. An agricultural system can no longer be considered as being contained only within the limits of a plot with production factors that are isolated from each other, foremost among which are cultivated species or varieties.

It is difficult to arrive at any genericity because of the many biodiversity management methods practiced in family farming systems, an aspect that results in criticism of participatory methods. However, as examples in this chapter show, innovative participatory methodologies are being developed which encourage dialogue between various actors to better lay out seed selection criteria and discuss links to markets and agricultural policies and the impacts of climate change. This dialogue also leads to a more accurate and comprehensive understanding of actors and issues of cultivated biodiversity, helping us understand the intrinsic characteristics of family farming models and their relationship with their living environment. From this viewpoint, the participatory approach, despite its limitations, seems essential and its discussion should not be limited to an activist discourse.

Support for family farming systems presented through these case studies is still in its nascent stages, but it points to a growing research effort that extends its approach beyond a mere management of the diversity of genetic resources, and which seeks to be practical, collaborative and inventive at the same time. This welcome reactivity and responsibility of research helps us reflect on the evolution of agricultural technical models as a whole with the help of debates on various perspectives of ecological intensification. This reflection is the subject of the next chapter.

Chapter 18
Lessons and Perspectives of Ecological Intensification

François Affholder, Laurent Parrot, and Patrick Jagoret

18.1 What is Ecological Intensification?

Consistently with Cassman's (1999) original wording, ecological intensification is commonly defined as the imperative to attain high productivity per surface area unit and per time unit with a concomitant "ecological" commitment to protect the environment. For most authors who subscribe to this concept the principle of mobilizing ecosystem processes that support and regulate primary production is the key to overcoming this challenge (Egger 1987; Breman and Sissoko 1998; Affholder et al. 2008; Chevassus-au-Louis and Griffon 2008; Bonny 2011; Doré et al. 2011; Bommarco et al. 2013; Hochman et al. 2013). By accepting this sense, we are justified in using the expressions "ecological intensification" and "ecologically intensive agriculture" interchangeably, with the latter expression suggesting more explicitly the forceful mobilization of ecological processes for high yields, and not simply the search for a combination of increased intensification and low environmental impact.

Ecological intensification is thus the opposite of artificialization of the environment, which was the basis for the massive increase in agricultural production in the twentieth century. This artificialization of the environment greatly reduced competition for resources between living organisms present in natural ecosystems and plants grown and livestock reared by the farmers. They did this mainly through a drastic reduction in the number of "competitors" with the help of pesticides and physical manipulation of the soil. They also relied on supplying in great quantities resources that are likely to be limiting factors, such as water and nutrients through irrigation and fertilizers. The paradigm of ecological intensification, on the other hand, harnesses ecological processes coupled with a controlled, precise and parsimonious use of "external inputs." Such inputs are seen as resources that are not

---

F. Affholder (✉) • L. Parrot • P. Jagoret
Persyst, Cirad, Montpellier, France
e-mail: francois.affholder@cirad.fr; laurent.parrot@cirad.fr; patrick.jagoret@cirad.fr

provided by the local ecosystem but which are likely to improve its ability to provide services. It specifically encourages facilitation between living species, i.e., the processes by which species make resources more easily available to other associated species, so that the yield through such association is greater than the sum of the yields obtained with the same species cultivated separately. The scientific rationale of ecological intensification relies on agroecology, the science of interactions between plants, animals, human beings and the environment within agricultural systems (Dalgaard et al. 2003).

Two broad types of agroecological levers have a strong influence on the structure of the agricultural systems in which they are implemented. They can be regarded as ecological engineering archetypes that have to be mobilized for ecological intensification.

The first agroecological lever is the optimal use of resources such as solar radiation, water and nutrients through a spatial and temporal organization of species. The typical example is agroforestry (Box 18.1), a cultivation system that encourages a simultaneous or sequential association of trees, annual crops or livestock to obtain goods and services useful to human beings (Torquebiau 2000).

More generally, many plant species growing in association convert more sunlight into biomass than when these species are grown separately in similar densities. The geometry of aboveground and root systems of associated species and their relative dynamics over a period of time is indeed such that these species use a greater amount of light for photosynthesis, without a commensurately increased competition between them for water and nutrients. An organized association of various species can also help reduce erosion and improve synchronization between the nutritional needs of plants and mineralization of soil organic matter in a way that the association reduces the amount of water and nutrient lost from the portion of soil explored by plant roots. Moreover, if a share of the additional biomass is returned to the soil, it can lead to an increase in – and a longer term stabilization of – the amount of nutrients available in the soil. In this case, we witness facilitations between associated species, a process of a particular magnitude if one or more of these species are legumes directly or indirectly making atmospheric nitrogen available to other species (see, for example, Baldé et al. 2011; Rusinamhodzi et al. 2012; Jamont et al. 2013). An example of the application of this lever to livestock production is provided by the case of a community of fish species maintained in a pond with two aims: to obtain a yield of marketable fish and to purify nitrate-contaminated sewage water from an intensive pig farm (Mikolasek et al. 2009a).

Box 18.1. Agroforestry in cocoa cultivation, a credible alternative for an uncertain future.

Patrick Jagoret

There is an increasing debate on the conventional model of cocoa cultivation in the tropics. The world of research and development has long been recommending farmers to completely clear a forest area and then to grow cocoa as a pure crop or under partial shade with an intensive use of chemical inputs (mineral fertilizers and pesticides). This model does achieve high cocoa yields in the first years but very often it does not match the strategies of farmers. The farmers generally do not have the financial means to support such a model in the context of highly volatile global prices. Indeed, the very existence of farms that depend on this speculation is threatened when cocoa prices take a downturn. Furthermore, the cocoa yield decreases after 10–20 years of continued cropping without sufficient fertilization. This may lead to the abandonment of cocoa plots in favor of new plantations on forest clearings (Ruf 1995). This cocoa cultivation model has now reached its agronomic limit (lack of sustainability), environmental limit (deforestation) and social limit (threat to the existence of cocoa farms).

In contrast, there is a wide range of cocoa agroforestry systems providing numerous ecosystem services that have been extensively studied in recent years. In Cameroon, contrary to what is usually observed, Jagoret et al. (2011) described notable cocoa agroforestry systems that have sustained yields over a long term (over 60 years) without recourse to chemical fertilizers. The same authors have shown that an appropriate management of fruit and forest species associated with cocoa trees allows farmers to reduce the use of pesticides to control mirids and black pod disease, thus demonstrating that the spatial interaction between individuals of different species greatly promotes natural processes that control cocoa pests. This biological control is an alternative to conventional pest control methods based on the use of chemical inputs. Furthermore, agroforestry practices adopted by farmers in central Cameroon significantly improve soil fertility and encourage sustainable cultivation of cocoa in unfavorable areas (Jagoret et al. 2012).

The key factors of the remarkable longevity of these cocoa agroforestry systems seem to be the rehabilitation practices in cocoa stands and the management of associated trees. Cocoa stand density, whose age is indeed less than is apparent, remains stable over time, and its vegetative growth is offset by removing redundant associated trees. The sustainability of this cocoa system is also related to its flexibility, which offers margins for maneuver to farmers in organizing their activities, including off-farm activities, and in developing their systems towards their own production objectives.

The second type of agroecological lever is the stimulation of soil biological activity by decreasing mechanical operations on the soil and increasing the amount of biomass returned to it, coupled with a simultaneous reduction of periods when the soil surface is exposed to direct sunlight, wind, runoff and the direct impact of the high kinetic energy of raindrops. This stimulation of soil biological activity helps improve ecosystem regulation services such as flood control, water quality, and carbon sequestration (see, for example, Blanchart et al. 2007). It also improves mineral and water balances and reduces their temporal variability which, in turn, enhances production functions (Scopel et al. 2012).

Conservation agriculture is a typical example of how these levers are used (Box 18.2). Moreover, the combination of these two types of levers helps improve pest control. The activation of telluric biological processes and a planned introduction or maintenance of a specific diversity in agroecosystems, based on different spatio-temporal deployment modalities, can effectively reduce pest pressure on useful plants. Several processes have been identified that function on their own or in combination (Ratnadass et al. 2013a). The spatial and temporal separation of host and non-host plants, either through a mixed cultivation of species or varieties, through crop rotations or through landscape configurations, limits the spread of pathogens and pests since their targets are "diluted" in space and time. In addition, some plants release chemical mediators that have a repellent or depressive effect on certain pests or plants and the introduction of such plants in associations can thus be beneficial. The abundant availability of biomass in aboveground and subterranean ecosystem compartments leads to a biological richness promoting the presence of organisms which fight pests and pathogens (see, for example, Ratnadass et al. 2006). Finally, the quality of the mineral nutrition of plants, enhanced by the presence of abundant and biologically diverse organic matter, is in itself conducive to an improved physiological resistance to pests and diseases (Chaps. 12 and 16).

Box 18.2. Conservation agriculture in family farming.

François Affholder

Conservation agriculture is a cropping method founded on three principles: the reduction in or elimination of plowing to stimulate biological activity; the maintenance of dead (mulch) or living vegetation cover on the soil which helps protect it from climate-induced erosion and provides biomass that stimulates soil biological activity; and finally the maintenance of species diversity through crop rotations or associations. This approach thus mobilizes all the agroecology levers. Moreover, it has also demonstrated its ability to produce grain and biomass yields that are comparable to intensive agriculture, particularly in the case of the major annual cereal and legume crops.

Conservation agriculture is currently commonly practiced on entrepreneurial farms or on some specialized and relatively well-capitalized family farms (Chapters 2 and 4) of a relatively large size, a sufficiently high room for financial maneuver, and with motorized mechanization. Conservation agriculture already covers more than one hundred million hectares all over the world (Derpsch *et al.*, 2010) and is expanding progressively under different names and forms including, for example, some types of "Techniques Culturales Simplifiées" ("simplified cropping techniques") developed in France.

The savings on plowing costs on these farms are of the same order of magnitude as the new costs induced by this management technique, such as those resulting from higher herbicide requirements or lower yields due to increased pest pressure, at least in the initial years after conversion from conventional management. In addition, although farmers who adopt these techniques are sometimes exposed to moderate short-term risks of reduced income, they accept them for two reasons: first, the possibility of better performance over the longer term, and secondly, they feel their efforts to reduce environmental impacts of agriculture contributes to restore trust between citizens and farmers (Goulet and Vinck, 2012).

To date, in this context, and for some major types of environments, conservation agriculture when actually implemented have made relatively little use of specific biodiversity. However, it does provoke an increase in soil biological activity and its regulatory functions, with a marked reduction of soil erosion, decreased greenhouse gas emissions due to reduced combustion of fossil fuels, and increased carbon sequestration and organic nitrogen in the soil that will eventually reduce the use of chemical fertilizers (Scopel *et al.*, 2012). On the other hand, it rather results in an increase in the use of pesticides and, in particular, of herbicides.

However, short-term financial constraints are currently the primary reason that prevents widespread adoption of conservation agriculture by the majority of family farms in the world. This is especially true when cropping systems currently practiced are not mechanized or do not use inputs. In such cases, the retention of mulch and the abandon of plowing initially lead to an increased risk of competition for nutrients from weeds and soil microorganisms. This risk can only be mitigated by the use of extra plowing, herbicides and mineral fertilizers, additional expenses which are not adequately compensated for, in the short term at least, by savings made on other costs. Families with income levels at a few Euros a day cannot easily adopt these methods, even if it is a question of using relatively small, even very small, quantities of inputs, or of a moderate increase in workload.

There are, in theory and in practice, many ways to use and combine the different agroecological levers mentioned depending on the ecosystem services to be promoted, the environment in which to apply them and the external resources to be mobilized. An important feature of ecological intensification is that it relies on ecological engineering to design appropriate combinations of these levers, i.e., combinations which effectively comply with ecological principles and which are adapted to the local ecosystem and to the objectives and constraints of the farm wanting to implement them, with its specific technical skills (Bergen et al. 2001).

Since ecological intensification can be implemented in many different ways, the range of its agronomic and environmental performances is very wide. Few studies have been carried out to compare performances of eco-intensive systems with "conventional" intensive systems because even relatively old eco-intensive systems, like cocoa agroforestry, have been little studied by agronomists, often due to the pre-conceived idea that they lead to low yields of commercially interesting species.

Nevertheless, we can note that a transition from intensive to eco-intensive agriculture frequently results in decreased yields of the marketable product which, in turn, can bring down the farmer's income in the initial years. This is because competition within the ecosystem initially overcomes facilitations in the balance of effects on the provisioning services that are valued by the market or on which farmers and their families directly depend for their food security. The feasibility of ecological intensification is thus based on the assumption that long-term improvement of ecosystem services is likely to offset, in one way or another, a short-term decrease in supply. This assumption has so far not been demonstrated in a general manner.

However, there are many systems based on the principles described above, in which the main marketable product's yield is greater than, or equal to, that of conventional systems, despite the reduced use of external resources (hence the "eco-efficient" qualification of these systems). This is especially true, for example, for certain rubber production agroforestry systems. Moreover, if we take into account not only the main product but include all other usable products, the performance of eco-intensive systems can clearly be higher than that of the intensive pure monocropping system to which they can be compared.

However, this type of comparison ignores issues of collecting products – often of low value – from widely dispersed sources across territories as against those obtained from specialized production regions. There also exist ecologically intensive systems where the reduction of working time and fuel consumption, mainly from the discontinuation of plowing, makes these systems economically viable to producers despite sometimes slightly lower yields. This can be observed in the example of a maize/soybean rotation under direct seeding by entrepreneurial agriculture in Brazil and Argentina, and more generally in the "Techniques Culturales Simplifiées" ("simplified crop management") systems developed in France or in conservation agriculture in moto-mechanized agriculture (Box 18.2). However, the long-term sustainability of these systems remains to be assessed.

Finally, there are also documented cases where conventional intensive systems deplete resources to the point that cultivation areas are relocated in a cyclical manner (for example, cocoa systems, see Box 18.1).

18.2 How to Transition to Ecological Intensification?

A reader from industrialized regions would readily suppose this entails a transition from intensive systems, and it would be simply appropriate to introduce diversity and complexity, and to reduce the use of industrial inputs such as pesticides and fertilizers. In this scenario, the agronomic objective would be to maintain or increase yields (based on available room for improvement) while reducing environmental impacts (Carberry et al. 2013) or even to accept lower yields if this is possible without a reduction in farm income. However, for extensive or less intensive farming methods,[1] very prevalent in tropical family farming (Chaps. 2, 3 and 4), it is a matter of intensification without abusing local or global (fertilizer, fuel, etc.) ecosystem resources and without any loss of biodiversity.

Obviously there are a multitude of intermediate cases between these two extremes: many different starting points, of course, but also possible "arrival points" depending, most notably, on the hierarchies to establish between the various ecosystem services expected from ecologically intensive agriculture.

Many cases of "conventional" intensification of family farming have been observed over the past decade, especially in emerging countries when changes in markets create an opportunity for both the increased use of inputs and a greater market integration of the farms (Chap. 6). This type of intensification typically brings with it new risks for farmers. First, intensive agriculture activities are more sensitive to climatic hazards than are extensive activities (Affholder 1997). Second, the specialization of cultivation systems which often accompanies conventional intensification renders farm income more sensitive to the inherent hazards of the main activity. And, finally, farm income becomes very sensitive to price fluctuations, especially to those of inputs and products.

Despite these drawbacks, conventional intensification has expanded through veritable agricultural revolutions. Indeed, it has been adopted on a massive scale. It takes just a few years to implement when its risks appear to be significantly offset by a sufficiently remunerative market, a significant increase in average farm incomes and in the resulting living standards of rural families (Bainville et al. 2005). Once this conventional intensification is well-established, the issue of ecological intensification clearly becomes similar to that of intensive agriculture in developed countries (how to transition from intensive agriculture to ecologically

---

[1] Or at least for farming systems where yields and the use of exogenous inputs are both low – which does not prevent them from being very intensive in labor, in knowledge or in the mobilization of ecological processes.

intensive agriculture?). However, research should be oriented to studying opportunities and means to quickly reorient such trajectories towards ecological intensification and, consequently, also examine why the trajectories currently followed remain conventional, despite widespread awareness of the unsustainable nature of the systems they lead to.

There are also situations where, due to a lack of market access to sell agricultural products and non-availability of land, family farms must increase their yield and performance to meet the increased food requirements per available unit of agricultural land resulting from population growth. They must make the best possible use of intensification without the use of external inputs. For example, cultivation systems with very high labor-intensity per unit area have been observed. Activities that are particularly labor-intensive include weed and pest management and the transfer of fertility to the cultivated area via biomass flows (see, for example, Andriarimalala et al. 2013). While such situations are found in some remote areas of emerging countries – where they are, however, becoming increasingly rare –, they are very common in sub-Saharan Africa and Madagascar. It should be noted that in all types of farming systems without external inputs, regardless to the intensification level of crops, farmers use the agroecological engineering levers mentioned above. And these farmers have a deep knowledge of the ecosystems they exploit and of these ecosystems' responses to their actions (Chap. 6).

In such a scenario, where population pressure increases even as market integration opportunities remain low, intensification typically means ecological intensification, albeit without the use of external inputs, and often with production levels per unit surface area significantly lower than those of conventional intensive agriculture. Yet, these situations are likely to lead, sooner or later, to environmental damage that is often extremely difficult to reverse, and to a decline in production that may compel a part of the population to migrate (Demont et al. 2007). This reminds us that there is a limit to intensification without recourse to inputs external to the agroecosystem. Research must keep on tackling the question on the ways to extend this limit. However, on a more political level, it must also attempt to demonstrate the importance of helping these farmers integrate with the market well before environmental or social crisis situations are reached.

Finally, in countries of the South, instances of the implementation of ecological intensification are still uncommon in family farms that are integrated into markets. Peri-urban agriculture is one example, where the high availability of organic matter is favorable to ecological intensification (Box 18.3). Cropping systems based on conservation agriculture have been adopted by a small number of mechanized family farms highly integrated into the Brazilian market, which, however, is in contrast to their widespread adoption by entrepreneurial farms in the same region. There may also be movements to adopt these systems in Southeast Asia and Madagascar. However, apart from cases where development projects strongly support such cropping systems over extended periods of time, the adoption of these systems by family farms in the South remains the exception rather than the rule (Giller et al. 2009).

Box 18.3. Ecological intensification in urban areas and waste management.

Laurent Parrot

A growing share of agriculture in sub-Saharan Africa is increasingly being practiced in highly anthropized environments. We have already seen in Chapter 6 the numerous economic, social and environmental functions this agriculture fulfills, as also the limited recognition and inadequate support it benefits from. Different possible ecological intensification options have been identified and are being studied to help this agriculture develop and to reduce the risks of pollution in towns from the excessive use of fertilizers and pesticides (de Bon *et al.*, 2010).

Thus, the use of compost in urban areas in Cameroon has been studied and shows a wide range of situations concerning soil fertilization (Parrot *et al.*, 2009; Sotamenou and Parrot, 2013). A third of the farmers in swampy lowland areas uses compost solely or in combination with chemical fertilizers. Raw material is generally abundant. For example, nearly 70% of municipal solid waste in the Cameroonian city of Yaoundé is composed of organic matter.

Factors that make compost an attractive alternative include cultivation of high-value crops (vegetables in particular), land tenure security (which can allow investments to be stretched over several years) and short distances between habitats and plots (which makes compost easier to use).

The existence of associations that provide artisanal compost can offset the lack of public support and directly encourage environmentally intensive practices. Moreover, while the potential is good in urban lowlands, areas on slopes, where mechanized equipment cannot be used and where there is no construction, also offer possibilities for land tenure security and proximity to dwellings.

18.3 Assessment of Ecological Intensification in Family Farming

Farmers generally find the "ecological" intensification approach less economically attractive in the short-term than the "conventional" intensification because the former often produces lower yields, is more labor-intensive and requires greater knowledge of the local cultivation ecosystem. It also runs the risk of failure, especially in the initial years of operation when changing from one cultivation system to another (Affholder et al. 2010; Penot et al. 2012; Scopel et al. 2012) or during transmission of the farm from one generation to the next, as has been observed in old cocoa agroforestry plantations in central Cameroon (Jagoret et al. 2011). An *a posteriori* reconstruction of the technical trajectory of these old cocoa plantations shows that there is a disruption in their technical management in most cases following the death of the farmer who was the only one with the expertise to manage associated communities and cocoa populations (regeneration). This phase of disruption, a period of varying length, is usually followed by a recovery phase during which the inheritor, who had often been away from the village until then, learns the technical skills involved.

These difficulties can be moderated in the long term but the low financial flexibility of many family farmers, especially in the South, normally forces them to build their strategy around short-term income.

At the scale of the cropping or livestock system, when an ecologically intensive activity appears to be more profitable than the currently practiced system, it may not necessarily be so at the farm scale, when the opportunity cost of resources at the time they are used is taken into account. It is thus quite possible that non-farm activities (for example, those linked to urbanization, land pressure, tourism development, etc.) offer a better return on the labor or money invested than an eco-intensive farming method, whereas this is not the case for the "conventional" method, despite it being less profitable per unit surface area (Affholder et al. 2010).

Ecologically intensive approaches may also lead to changes in the access or use of resources managed not by individuals but collectively by rural communities. This can greatly curb the individual enthusiasm towards these techniques or generate new tensions in the collective use of these resources. Thus, for example, service plants used during the pasture grazing period for cattle can lead farmers who cultivate them to build fences around the fields.

Eco-intensive techniques are therefore generally unlikely to find a place on farms, especially non-mechanized ones, as long as compensation for farmers' labor depends solely on the value of products, and so long as it is remains insufficient to allow farmers to invest in the long-term sustainability of their production systems.

One of the major advantages of family farms vis-à-vis ecological intensification is their ability to apprehend the ecosystem in its complexity and heterogeneity (Chap. 6). This seems partly due to the relatively small size of family farms compared to other types of farms. It is also due to the fact that in family farms, technical management is carried out by family members with a certain freedom of initiative and experimentation, as compared to employees of entrepreneurial farms, expected to follow instructions and technical directions received from their superiors. But it can also be the result of the passing on of experience and knowledge gathered on the field by one generation to the next. Nevertheless, this aspect must not be idealized since such a transmission of family or community knowledge to the children clashes with knowledge they obtain from their schools. It is thus not uncommon for young people, who have spent their time in urban schools far away from their family farm, to have no particular expertise in agriculture at the time of inheriting the family farm (Chap. 4).

Another advantage of family farming is its ability to draw value from ecosystem services other than those related to provisioning, for example by developing an accommodation service suited to tourists attracted by a protected environment. It can also draw value more directly from secondary provisioning services that also generate other types of services useful to the production system, but that would not be economically sound to bring to the market, for example, the use as fuel or construction material of wood obtained from agroforests or hedgerows, or the family use of medicinal or dye plants. A study conducted in central Cameroon showed that farmers attribute value to a wide range of ecosystem services (Box 18.4).

Box 18.4. How to assess performances of agroforestry systems without getting completely baffled?

Patrick Jagoret

Complex agroforestry systems illustrate the difficulties of assessing eco-intensive agricultural systems. How to assess their overall productivity? Which unit of measurement to adopt to evaluate different productions?

In order to estimate the yield of a species in mixed cropping, it is necessary to understand its behavior as a pure crop. However, except for a few major commercial species (citrus fruits, oil palm, etc.), such information is not available for species used in tropical agroforestry (see, for example, Lamanda, 2004). On the other hand, the land equivalent ratio (LER) method – which permits a theoretical comparison of the productivity of several associated species on a plot with that of the same species as a pure crop – quickly shows its limitations when it comes to studying complex associations because of the many interactions involved (Malézieux et al., 2009).

Various products are obtained from species present in agroforestry systems, some of which are consumed by the household itself while others are marketed and serve as a source of farm income. Some products are part of the pharmacopoeia (leaves, bark) while others are used for construction (timber). The nature of the harvested parts also varies depending on the species: fruits for fruit trees, sap for oil palm, and bark or leaves for some medicinal species.

A simple addition of these products would be of little significance or relevance. Production cycles also vary according to the species: some have a cycle that spans the full year over the production period, while production in others is concentrated over a shorter and more defined period of time. Furthermore, the size of the production of a particular species, although regular, may be subject to variations due to weather conditions or genetic origin. The harvest cycle also varies according to species and the parts harvested. In some cases, the crop can be completely or partially harvested due to various factors such as variations in prices paid to farmers (cocoa, citrus fruits for instance) or household requirements (indigenous fruit, medicinal products, timber).

It is therefore difficult to find an optimal period over which to measure all production types and choose a common time unit to express it. It is also difficult to define a common measurement unit to incorporate different types of functions such as commercial production, the satisfaction of household needs, maintenance or development of social relationships, improvement of soil fertility, maintenance of shade for cocoa plants, etc. (Fig.18.1).

Fig.18.1 Illustration of the multi-functionality of cocoa.

In an attempt to overcome these constraints, farmers and researchers in central Cameroon have jointly identified seven main uses of the different species in cocoa agroforestry systems. They then assigned to each species, depending on its use(s) and significance, a number of tokens representing their value. With a hundred tokens being distributed to each farmer, the totaling of scores indicates, to some extent, the usage profile of cocoa farmers.

The score of the "sale of products" function for different species in the agroforestry system represented 33% of the total value, while the "medicinal product" function and "home consumption" represented 17% and 16% respectively of the total value according to the farmers' estimates. The remaining four uses, "shade," "fertility," "timber and firewood" and "social" together accounted for the remaining 34%.

More generally, the future of ecological intensification depends to a great extent on the methods identified for attributing value to ecosystems, in addition to the value of their short-term provisioning function (see, for example, Rapidel et al. 2011), and for protecting farmers against the risk of production losses incurred before improvements of ecosystem regulation services could lead to a long-term reduction of these risks. Moreover, the way to take advantage of these methods could a priori depend not so much on the farm belonging or not to family farming as on its financial room for maneuver, its links to communication and information networks, and the nature of its insertion into markets. However, its familial nature can, in the long-term, strengthen technical management systems targeted at a high level of environmental services, provided that such farms are properly supported (Chap. 10).

New compromises must therefore be built between the objectives of production, farmer incomes and the preservation of the environment. It follows that a reflection on ecological intensification can lead to a redrawing of power relations between stakeholders of agricultural production, from the farm to the dining table – or to other places where agricultural products are consumed (Méndez and Bacon 2013).

18.4 What Research for Which Ecological Intensification?

The first challenge for applied and support research is to enhance our capacity to identify technical solutions based on the principles of ecological intensification best suited to a broad range of biophysical and socio-economic environments. This leads to a simultaneous adoption of agronomic, environmental (depending on risks to water quality, the health of farmers and consumers, changes in soils, biodiversity) and economic indicators (profitability, risk, impact on income, and elaboration of product cost prices) which help compare innovations by taking into account the various criteria of sustainability. The methods to be developed (choice of indicators, their values, aggregation methods between indicators and changes of scale) must be acceptable to the various stakeholders involved in the building of the new compromises mentioned in this chapter.

A second challenge is to clarify public policy decision-making to facilitate the emergence of these new compromises and consequently the institutional, economic and financial arrangements that could resolve conflicts between short-term constraints of producers and environmental issues. It is a matter of contributing to the definition of policies suited to the different scenarios mentioned above in terms of transition towards ecological intensification, with or without support for increasing productivity or yield. It is also a matter of supporting the production of environmental services through public investments and improvements in basic services, technology and capital markets, and an increase in human and social capital. New agronomic approaches must be considered. They will have to be founded largely, but not exclusively, on local resources. They will also have to be pragmatic and be based on deliberative and local approaches to problems. This does not mean that past experiences will be completely ignored to build new so-called universal models but that the effort has to be based instead on current perceptions, representations, knowledge and principles of all stakeholders as they actually exist in real situations (Maris 2010; Wegner and Pascual 2011; Francis et al. 2013; Parks and Gowdy 2013).

Finally, it is essential to assess the extent to which agriculture is progressing towards ecological intensification, and continuously identify factors that favor and hinder this transition. However, because of the need to adapt the way ecological intensification principles are used to existing biophysical and socioeconomic contexts, these principles cannot be seen as a "ready-to-use" list of technologies available to farmers "off the shelf" that they can simply select or reject. The concepts of innovation and of its design and adoption themselves will have to be reviewed in depth (Chaps. 14 and 15).

These issues require research to adopt a systemic and interdisciplinary perspective across spatial and temporal scales and involving multiple stakeholders (Giller et al. 2011). It has to address several methodological challenges faced by science in the twenty-first century:

- challenges specific to systemic approaches. These concern issues related to the complexity, diversity and dynamics of the systems considered, especially the difficulty of mastering the large number of interacting factors through experiments, thus requiring an increased use of modeling;
- challenges specific to the integration of scales and disciplines. An interdisciplinary effort is constrained by various misunderstandings between researchers using the same words for different concepts (Naiman 1999) or having different perceptions of methodological issues and thus of hierarchies between variables to consider in addressing a common problem. And most importantly, such an effort is hampered by power issues between disciplines which have asymmetrical roles in addressing a problem. This, for example, is the case when some disciplines are expected to provide expertise whereas others are harnessed for producing new scientific knowledge (MacMynowski 2007). Necessary changes in scale, for example, between the plot and the farm or between the farm and the territory, also involve changes in hierarchies between variables in the determinism of ecological and economic processes. Methods pertaining to modeling, designing of environmental mechanisms, and the formulation of theories need to be renewed in order to better address these challenges.

More generally, the challenges faced by research efforts on systemic and interdisciplinary approaches to ecological intensification reside – for biophysical sciences that are still influenced by the idea of a unique reality, independent of science and accessible to science – in the recognition of a degree of subjectivity, which increases with the complexity of the systems considered. In the social sciences, new experimental approaches are emerging which, contrastingly, aim at bringing more objective arguments to theoretical constructions (Banerjee and Duflo 2009; Vakulabharanam 2013). These approaches remain, however, subject to controversies. Finally, for science as for society in general, ecological intensification reinforces the need for problems to be addressed by the entire community of stakeholders concerned by them, where each group of stakeholders recognizes the value of learning from the others.

General Conclusion

The overview presented in this book clearly demonstrates that family farming encompasses several very different realities. It is hard to pin down, both as a concept as well as for more practical purposes. Furthermore, since family farming is not a statistical item, it does not appear in any national or international databases. Consequently, the concept of family farming is used to support various political and activist actions and discourses, discourses which may vary depending on whether the context is local, national or international. Indeed, the lack of a clear definition for family farming paradoxically strengthens its concept and its recent spread in local and international arenas. Expectedly, this very ambiguity generates substantial debate and criticism. Thus, proponents of alternative agricultural models, radically different from the dominant, conventional and widely supported productivist model, do not feel very comfortable with the concept of family farming.

In the usual and common representation, industrial or commercial agriculture is characterized by the strategic role of financial capital, the legal status and division of labor and a business-like approach to farming. Some family farms, however, can exhibit very high levels of capitalization and embrace the modernization standards based on artificializing production systems. Thus, highly capitalized farmers who participated in the major events marking the International Year of Family Farming and who promote family farming are very different from those farmers who do not subscribe to conventional modernization, such as, for example, those belonging to international peasant movements like Via Campesina. Such very disparate groups do not share the same rationale, do not subscribe to the same beliefs, nor have the same concerns and, hence, express very different goals. And yet, they come together under the family farming banner.

© Éditions Quæ, 2015
J.-M. Sourisseau (ed.), *Family Farming and the Worlds to Come*,
DOI 10.1007/978-94-017-9358-2

1. Family Farming: A Means to Understand Agricultural Issues

From the point of view of researchers and when defining family farming through the inextricable (organic) linkages that exist between the family and the farm within the agricultural process and through the exclusive recourse to permanent family workers, it makes sense to put family farmers in a separate analytical category. The results presented in this book demonstrate that, in spite of the diversity of family farming systems, such an approach throws light on the various agricultural strategies in a more global perspective of reproduction of families and their territories. They demonstrate too that performance has to go beyond productivity and has to also integrate social and environmental dimensions. Indeed, looking through the prism of "family" helps us move away from conventional productivist agricultural intensification and its performance which are strictly focused on volumes produced and productivity per unit of area or labor, on an emphasis on physical capital over labor and, increasingly, on rapid returns on investment in the case of the most commercial forms. It therefore becomes possible, by changing perspective, to fully include environmental, social and territorial sustainabilities, since from the point of view of family farmers, the costs and benefits for their services to society, whether recognized or not by the market, are also costs and benefits for themselves.

A more precise definition of family farming and its use by the research community also imparts meaning to multidisciplinary approaches, especially for action research. The chapters of this book suggest a possible family-and-farm approach which assesses the performance of production systems in terms of knowledge of the natural environment, biodiversity management, understanding of energy issues, and the management of health risks. The rationale behind this family-and-farm combination helps us rescale from the farm level to the global level through various intermediate sizes and types of territories – as illustrated, in particular, by the case of invasive pests and diseases of animals and plants. While practices and levels of capitalization may vary, the available analytical frameworks can converge to improve interactions, not only between different scientific disciplines but also those between both research and development communities.

Finally, one cannot consider family farming without taking into account all the activities undertaken by rural families – even those outside the strictly agricultural sphere – in the economy of one or more territories. To incorporate such a family pluriactivity and multilocality in rural and territorial development support systems, it is necessary to "de-sectorize" our reasoning. Agricultural production has to be thought of as one means, among others, of the livelihood of families, irrespective of whether this production is sold or not on a market. What emerges, therefore, is a vision of a diversified rural economy, one that goes beyond agriculture alone, and which radically changes the principles and practices of accompanying and supporting agricultural and non-agricultural sectoral dynamics.

In addition to emphasizing the importance and relevance of these changes in approach, comparative and interdisciplinary perspectives on family farming brought together in this book invite us to draw many lessons for targeted research, which we have presented in detail according to major themes. These cross perspectives ultimately help develop more holistic approaches to agricultural foresight, possible agricultural futures, and public policies that could or should be associated with them, and in considering new avenues opened by and for research.

2. Family Farming: A Social Project but Relying on Which Agricultural Models?

The diversity of world agriculture and the impressive gaps in productivity between different agricultural models result from long-term processes which have been supported by large-scale public policies. The technical dimension was and still remains at the heart of these processes. Today, in industrialized countries as well as when development assistance to agriculture in low-income countries is being considered, the priority is still to specialize and professionalize farmers and maximize production and labor productivity – with the result that the means of production become ever more concentrated. However, this orientation has resulted in economic, social and environmental imbalances that are hardly sustainable at either the local or the global level. The risk is of a future in which agricultural goals are reduced to those of financial profitability, exacerbating the existing asymmetries and gaps in productivity between different types of farms. And current financializing processes only reinforce such risks. Reliance on financial capital in agriculture is no longer confined to the upstream and downstream segments of the agricultural value chain. Financial capital is now firmly and increasingly involved in production itself through land acquisitions, segmentation and control of certain technical operations in the production cycle or of primary processing.

Placing family farming at the center of the debate, despite its highly composite character, may crystallize a challenge to the dominant model based on labor productivity, still presented as the only credible path to development. The research results presented in this book emphasize the sustainability of complex family-based activity systems, in which non-market relations dominate. Our conclusions highlight their multifaceted contributions in addressing challenges of economic and social development and of global change. They also show the flexibility inherent in family farms with a view to improving existing family-based models and promoting new ones.

Development has to be considered in terms of economic and social progress while remaining anchored at the territorial level. This paradigm shift calls for considering activities involving the use of the natural environment as providing functions other than solely those of producing agricultural products or raw materials. It aims for the restoration of ecological balances and social ties through new representations of farm performance. It encourages the generation and maintenance of jobs and self-employment in the context of difficult economic transitions for the least developed countries. It promotes the development and strengthening of linkages between farms at the local, national and international scales. Family farming systems, in all their diversity, can be potential vectors of this encompassing social project – which goes beyond agricultural production and its market orientations –, by rethinking the production of food primarily in terms of family food security. A truly vast scope of work is therefore required to consider and design models for tomorrow, capable of meeting the challenges of the future.

3. Towards Targeted, Adapted, Flexible, Multi-dimensional and Inclusive Public Policies

Is family farming the farming of the future? It certainly can be if we take advantage of its intrinsic strengths in social and territorial anchoring, its knowledge of environments and natural resources that it can bring to bear, and it ability to adapt and respond to changes. But, as highlighted in the conclusions of various chapters, it will only be so if it receives suitable public support. The productivist agricultural model that dominates today could not have arrived at its most extreme forms without favorable policy and institutional environments that allowed the implementation of incentivizing policies and appropriate support. In the same way, new models of sustainable development in which family farms will play a central role cannot be developed in isolation; they too will require enabling conditions and incentives.

Even though this book does not aim to make policy recommendations in favor of family farming, we can still point out some general principles so that the family dimension of farming is better taken into account in the accompaniment of and support that is provided to the agricultural sector and rural territories.

Firstly, the diversity of national and subnational situations calls for a diversity of public policy. The definition of family farming – a prerequisite to targeting aid – as well as the identification of its major constraints and the levers to overcome them, depend on national histories and natural and institutional environments of each context. It is therefore important to recast strategic thinking to take account of these histories and their specific implications. In order to challenge the standard model of farming in the world and to consider alternative macro-economic transitions, each national or sub-regional entity must assert its choices of territorial and sectoral development (including agricultural). One crucial aspect concerns the livelihood of the global population which continues to grow and, more specifically, the possible sources of income and employment. Intrinsically labor-inclusive, family farming is undoubtedly a solution for providing employment to young people entering the labor market, especially in sub-Saharan Africa and South Asia, where populations are growing the fastest and where the alternatives offered by other sectors remain woefully inadequate. However, such a solution will be viable only if agriculture is made more attractive, which means more remunerative activities (prices, management of risks), less arduous labor (small-scale mechanization) and the recourse to value-enhancing new technologies (especially those of information and communication). Thus, the models of agricultural development to promote and, consequently, the policies that need to be implemented must necessarily take into account the degree of diversification of economies, which largely determines available flexibility in terms of activities.

The instruments of public policy must also take into greater account the main strengths of family farming and, in particular, its flexibility and capacity to adapt to change. Targeted policies must match the flexibility of forms of organization and must themselves be flexible and rapidly adaptable to changes in context. This calls for a reflection on the temporalities and conditionalities of public policy in order to encourage more inventivizing actions than those found in the standard toolbox of agricultural modernization and specialization. In the same vein, the pluriactivity of rural families and the embedding of agricultural strategies in more complex activity systems remain structural. These features also benefit from the adaptability and specificities of family farming and call for the decompartmentalization of support mechanisms. And yet, territorial development policies are at present struggling to actually take advantage of the beneficial linkages between sectors of activity, being content most often to merely juxtapose various sectoral policies. But for any support and accompaniment mechanism for family farming to be effective, it has to rely on these linkages. This applies in particular to the recasting of technical messages to take into account the use of time of the rural workforce, specific local knowledge and the purpose of each household activity. Such a recasting challenges traditional extension and advisory practices, and more broadly those of agronomy.

Finally, some of the experiences described and analyzed in this book emphasize the perspectives offered by: (i) composite public policies which combine grants for agricultural production with guaranteed prices; (ii) public procurement reserved for family farming; and (iii) support extended to rural diversification activities coupled with social welfare measures and investment in public goods (infrastructure for markets, primary processing and transport, for example). This strategy of increasing the number of instruments is meant specifically to target certain rural populations. It therefore comes at a cost, but it also increases the chances of activating the right levers and removing the most significant constraints, which are rarely confined solely to the agricultural sector. Such a strategy, requiring strong involvement of the State as well as local communities in rural areas, effectively leads to positive results in reducing inequality, while limiting the risk of creating "poverty traps".

4. A Unifying Field of Research

Family farming, as a subject both of study as well as of development-oriented research, is analyzed in this book from multiple angles and in different contexts. This synthesis not only reflects traditional trajectories of research but also opens up new avenues that we believe are worth exploring – of which three seem to be particularly interesting.

First, the efforts to define – and differentiate between – different forms of agricultural production should be pursued. Family farming will benefit from a stable definition and a more precise characterization. Indeed, the issue is more important than may appear and reaches far beyond a seemingly academic debate. The figures available today to assess the significance and weightage of family farming in agricultural production, food security, environmental conservation and rural employment suffer from a lack of explicit data and sometimes rough estimates. These data are not based on standardized and validated statistics because "family farming" is not a separate category in agricultural statistics. And yet, accurate figures are necessary in order to discuss objectively the issues surrounding family farming and to implement and evaluate public policies in its favor. As we have shown, the criterion of labor offers the best perspectives in this regard, but it is certainly not sufficient. It has to be complemented by other criteria, such as the level of substitution of labor by physical capital, dependence on activities upstream and downstream of production, territorial anchoring and intra-family dynamics. Furthermore, it is as important to better differentiate between family farming and entrepreneurial or family business farming as it is to characterize the diversity of family forms themselves. Research therefore has an important role to play in undertaking analyses that can contribute to international debates and the formulation of national policies, and even help in redesigning a statistical system that takes family farming into account.

Second, research teams working on plants, animals, natural resources, agricultural practices and agro-ecological systems will find that including the family dimension and its specificities in their work can be a source of innovation. These innovations bear on research protocols, methods for interpretation and analysis, and improved effectiveness of targeted development-oriented studies. They pertain primarily to the inclusion in their analyses of the link between the family and the farm, and thus the multidimensional criteria mobilized by family farmers, in order to understand their technical itineraries, their choices in the end-use of their production and the impact of their agricultural activities. These innovations also focus on possible improvements to the way farmers use their natural and institutional environments, which must fit into the overall goals of their activity systems and be in concordance with their own understanding of environmental, economic and social risks.

A third line of research addresses the crucial issue of future agricultural models in a context of rapid transformation of world agricultures marked by structural asymmetries. There is little doubt that the industrialization movement will continue at least for the next few decades, depending on the speed of global changes (climate, energy, water resources, etc.). The future of family farming thus depends not only on its relationships with other forms of agriculture, but also in its ability to evolve into configurations that can better compete with them. Research must explore how different forms of agriculture can be better linked at the scale of local and national territories. In our efforts to define family farming, we have noted the existence of family business farming, which borrows characteristics from both entrepreneurial and family forms but which obeys its own rationality and contributes in its own original way to territorial construction. New evolutionary forms of farming are sure to emerge and research must participate in their development by strengthening and proposing hybrid forms based on the specificities of family farming and by monitoring their performance.

To us, family farming clearly appears to be the future of agriculture, but the particular forms that it will take are yet to be discovered. Strengthening of public debates, both within nations and internationally – to which the International Year of Family Farming has amply contributed –, should promote collective awareness of what is possible. The future of agriculture is far from being already determined; each one of us has to contribute to its construction.

Boxes

Box 1.1. Some conventions used in this book.

Box 4.1. Pastoral camps in the Sahel: a way of life beyond the farm. *Christian Corniaux*

Box 4.2. Agricultural families in Office du Niger: an organization at the heart of rice production and social reproduction. *Jean-François Bélières and Jean-Michel Sourisseau*

Box 4.3. The shepherds of the Ferlo: evidence of the monetization of the pastoral economy. *Véronique Ancey*

Box 4.4. Unexpected inclusions and exclusions: effects of coffee certifications in Costa Rica. *Nathalie Cialdella*

Box 4.5. Unexpected inclusions and exclusions: effects of coffee certifications in Costa Rica. *Nicole Sibelet*

Box 4.6. The diversification of activities: an old strategy brought up to date by the Cameroonian cocoa farmers. *Philippe Pédelahore*

Box 4.7. Agricultural families in Nicaragua: social cohesion in agricultural work and anchoring to the *terroir* maintained through multi-localization. *Sandrine Fréguin-Gresh*

Box 4.8. Mobility in southern Africa: heading towards a break with family farming? *Sara Mercandalli*

Box 4.9. Mobility of young herdsmen in the Ferlo in Senegal: between strategies of security, emancipation and changing lifestyles. *Claire Manoli and Véronique Ancey*

Box 4.10. The long-term building up of patrimony by small Indonesian farmers. *Éric Penot*

Box 4.11. The conversion of black miners into farmers in South Africa. *Sandrine Fréguin-Gresh*

Box 4.12. The partial failure of Nicaraguan land reform: the imposition "from above" of forms of land tenure at odds with rural social relationships. *Pierre Merlet*

Box 4.13. The Mexican land reform: a family farming project. *Emmanuelle Bouquet and Éric Léonard*

Box 5.1. Processing of palm oil. *Sylvain Rafflegeau*

Box 5.2. Paternalist agricultural capitalism in Indonesia. *Stéphanie Barral*

Box 5.3. The experiment of *alianzas* in Colombia. *Sylvain Rafflegeau*

Box 5.4. The financialization of South African agriculture. *Ward Anseeuw*

Box 6.1. The densification of trees in Sahelian landscapes. *Régis Peltier*

Box 6.2. Family farming systems in forested and periforested Central Africa: the legacy of shifting slash-and-burn cultivation. *Jean-Noël Marien*

Box 6.3. Ecosystem services and payments for ecosystem services. *Denis Pesche*

Box 6.4. The farmer's cow in the Amazon. What a program! *Soraya Abreu Carvalho, René Poccard-Chapuis, Amaury Burlamaqui Bendahan, Jonas Bastos da Veiga, Jean-François Tourrand*

Box 6.5. Indonesian agroforests, a biodiversity not always desired. *Laurène Feintrenie*

Box 6.6. Herdsmen in the Sahel. *Abdrahmane Wane and Christian Corniaux*

Box 6.7. Territorial changes and adaptation of the shifting cultivation system by the Amerindians of Guyana. *Isabelle Tritsch, Valéry Gond, Philippe Karpe*

Box 6.8. Family farming, forest conservation and transition to a green economy in the Brazilian Amazon. *Marie-Gabrielle Piketty, Isabel Drigo, Émilie Coudel, Joice Ferreira, Plinio Sist*

Box 7.1. The place and roles of family farms in the structuring of rural territories in the Sudano-Sahelian zone of West Africa. *Jean-François Bélières*

Box 7.2. Multifunctionality in New Caledonia. *Jean-Michel Sourisseau*

Box 7.3. The localized agri-food system (LAS): a key to understanding the dynamics of family farming in rural zones. *Claire Cerdan*

Box 7.4. Private enterprise of former South African cooperatives. *Ward Anseeuw*

Box 7.5. Family farmers and foreign investors: going beyond clichés in Office du Niger. *Amandine Adamczewski*

Box 7.6. Supplying milk to Greater Cairo. *Véronique Alary, Christian Corniaux, Salah Galal*

Box 7.7. Socio-spatial arrangements of intra-urban family farmers in Bobo-Dioulasso. *Ophélie Robineau*

Box 8.1. "Natural" palm groves: a particularity of family farming in West Africa. *Sylvain Rafflegeau*

Box 8.2. Small African, Haitian and Malagasy family orchards supply international markets with products of recognized quality. *Magalie Lesueur-Jannoyer, Michel Jahiel, Jean-Yves Rey, Éric Malézieux*

Box 9.1. Cooperatives in independent Africa. *Pierre-Marie Bosc*

Box 9.2. The contribution of the National Federation of Coffee Growers of Colombia in providing public goods. *Pierre-Marie Bosc*

Box 9.3. Fall and rise of cooperatives: the revival of Pima cotton in Peru. *Michel Dulcire*

Box 9.4. Establishment of supranational networks of producer organizations in Africa. *Pierre-Marie Bosc*

Box 10.1. Poverty, employment and agriculture: the case of Cameroon. *Laurent Parrot*

Box 10.2. Food markets in West Africa. *Nicolas Bricas*

Box 10.3. Income transfer programs. *Philippe Bonnal*

Box 10.4. Anti-poverty policies in China. *Jacques Marzin*

Box 10.5. Guaranteeing minimum prices for cereals to stimulate investment in family farming in developing countries. *Franck Galtier*

Box 11.1. Integrating family farming into a national biofuel production chain. The Brazilian example. *Marie-Hélène Dabat, Denis Gautier, Laurent Gazull, François Pinta*

Box 11.2. Reconciling food and energy production at the farm level: a major challenge for the participation of family farming in the bioenergy market. *Marie-Hélène Dabat, Denis Gautier, Laurent Gazull, François Pinta*

Box 11.3. Energy for food conservation and processing: the case of drying mangoes in Burkina Faso. *Marie-Hélène Dabat, Denis Gautier, Laurent Gazull, François Pinta*

Box 12.1. An example of an environment where different health issues come face to face: the association of duck breeding and rice cultivation. *Julien Cappelle, Pascal Bonnet, Alain Ratnadass*

Box 12.2. Veterinary compliance on family farms. *Muriel Figuié, Aurélie Binot, Marisa Peyre, François Roger*

Box 12.3. Breeders managing risk within the "surveillance territories": example of avian influenza in southeast Asia. *Aurélie Binot, Muriel Figuié, Marie-Isabelle Peyre, François Roger*

Box 13.1. Central Africa, rich in untapped natural resources. *Laurène Feintrenie*

Box 13.2. When a rush hides another: land, water and capital in Office du Niger (Mali). *Jean-Yves Jamin, Thomas Hertzog et Amandine Adamczewski*

Box 13.3. The new REDD+ mechanism to fight against deforestation: a benefit for ecological intensification of family farming? *Alain Karsenty*

Box 14.1. A few emblematic projects of experiments of action research in partnership. *Olivier Mikolasek, Éric Sabourin, Éric Vall*

Box 14.2. An example of an intermediate object linked to the family nature of the farm: the Cikeda model. *Nadine Andrieu and Aristide Semporé*

Box 14.3. Forms and changes of the disengagement. *Éric Vall, Eduardo Chia*

Box 15.1. Towards privatization of agricultural advice: implications for dairy producers in the Mantaro Valley. *Guy Faure, Kary Huamanyauri Méndez, Ivonne Salazar, Michel Dulcire*

Box 15.2. Management advice for family farms in northern Cameroon. *Michel Havard, Anne Legile, Patrice Djamen Nana*

Box 15.3. Experience of PASEA in Cuba. *Jacques Marzin, Teodoro Lopez Betancourt*

Box 16.1. Chlordecone in the West Indies. *Magalie Lesueur-Jannoyer*

Box 17.1. Participatory breeding, a method of dialogue and mutual learning. *Kirsten Vom Brocke, Gilles Trouche*

Box 17.2. Application of companion modeling for agrobiodiversity: the case of the IMAS project. Group of researchers coordinated by *Didier Bazile*

Box 17.3. Seeds from the Association of Professional Farmer Organizations in Mali. *Didier Bazile*

Box 18.1. Agroforestry in cocoa cultivation, a credible alternative for an uncertain future. *Patrick Jagoret*

Box 18.2. Conservation agriculture in family farming. *François Affholder*

Box 18.3. Ecological intensification in urban areas and waste management. *Laurent Parrot*

Box 18.4. How to assess performances of agroforestry systems without getting completely baffled? *Patrick Jagoret*

Liste des auteurs

Abreu Carvalho Soraya	UFPA – Belém, Brésil	carvalhosoraya@ymail.com
Adamczewski Amandine	Cirad – ES – Montpellier	amandine.adamczewski@cirad.fr
Affholder François	Cirad – Persyst – Montpellier	francois.affholder@cirad.fr
Alary Véronique	Cirad – ES – Égypte	veronique.alary@cirad.fr
Ancey Véronique	Cirad – ES – Montpellier	veronique.ancey@cirad.fr
Andrieu Nadine	Cirad – ES – Colombie	nadine.andrieu@cirad.fr
Anseeuw Ward	Cirad – ES – Afrique du Sud	ward.anseeuw@cirad.fr
Barral Stéphanie		phanette.barral@gmail.com
Bastide Philippe	Cirad – Persyst – Montpellier	philippe.bastide@cirad.fr
Bastos da Veiga Jonas	IDESP – Belém, Brésil	jonas.veiga@superig.com.br
Bazile Didier	Cirad – ES – Montpellier	didier.bazile@cirad.fr
Bélières Jean-François	Cirad – ES – Madagascar	jean-francois.belieres@cirad.fr
Bertrand Benoît	Cirad – Bios – Montpellier	benoit.bertrand@cirad.fr
Binot Aurélie	Cirad – ES – Thaïlande	aurelie.binot@cirad.fr
Bonnal Philippe	Cirad – ES – Montpellier	philippe.bonnal@cirad.fr
Bonnet Pascal	Cirad – ES – Montpellier	pascal.bonnet@cirad.fr
Bosc Pierre-Marie	Cirad – ES – Montpellier	pierre-marie.bosc@cirad.fr
Bouquet Emmanuelle	Cirad – ES – Montpellier	emmanuelle.bouquet@cirad.fr
Bricas Nicolas	Cirad – ES – Montpellier	nicolas.bricas@cirad.fr
Vom Brocke Kirsten	Cirad – Bios – Montpellier	kristen.vom_brocke@cirad.fr
Burlamaqui Bendahan Amaury	Cirad – ES – Montpellier	amaury.burlamaqui@cirad.fr
Cerdan Claire	Cirad – ES – Montpellier	claire.cerdan@cirad.fr

(continued)

Charmetant Pierre	Cirad – Bios – Montpellier	pierre.charmetant@cirad.fr
Chia Eduardo	Cirad – ES – Montpellier	eduardo.chia@cirad.fr
Cialdella Nathalie	Cirad – ES – Montpellier	nathalie.cialdella@cirad.fr
Clavel Danièle	Cirad – Bios – Montpellier	danièle.clavel@cirad.fr
Corniaux Christian	Cirad – ES – Montpellier	christian.corniaux@cirad.fr
Coudel Émilie	Cirad – ES, Montpellier	emilie.coudel@cirad.fr
Dabat Marie-Hélène	Cirad – ES – Montpellier	dabat@cirad.fr
Daviron Benoît	Cirad – ES – Montpellier	benoit.daviron@cirad.fr
Djamen Nana Patrice	African Conservation Tillage Network (ACT) – Burkina Faso	djamenana@yahoo.fr
Drigo Isabelle	Nexus Social and Environmental Consultancy	isabel.drigo@gmail.com
Dulcire Michel	Cirad – ES – Montpellier	michel.dulcire@cirad.fr
Faure Guy	Cirad – ES – Montpellier	guy.faure@cirad.fr
Feintrenie Laurène	Cirad – ES – Montpellier	laurene.freintrenie@cirad.fr
Ferreira Joice	EMBRAPA-CPATU, Belém, Brésil	joice.ferreira@embrapa.br
Figuié Muriel	Cirad – ES, Montpellier	muriel.figuie@cirad.fr
Fréguin-Gresh Sandrine	Cirad – ES – Nicaragua	sandrine.freguin@cirad.fr
Galal Salah	Université Ain Shams – Égypte	sgalal@gmail.com
Galtier Franck	Cirad – ES – Montpellier	franck.galtier@cirad.fr
Gautier Denis	Cirad – ES – Burkina Faso	denis.gautier@cirad.fr
Gazull Laurent	Cirad – ES – Montpellier	laurent.gazull@cirad.fr
Gond Valéry	Cirad – ES, Montpellier	valery.gond@cirad.fr
Havard Michel	Cirad – ES – Burkina Faso	michel.havard@cirad.fr
Hertzog Thomas	Cirad – ES – Montpellier	thomas.hertzog@cirad.fr
Huamanyauri Mendez Kary	Universidad Nacional Agraria La Molina –Pérou	canewame@hotmail.com
Jagoret Patrick	Cirad – Persyst – Montpellier	patrick.jagoret@cirad.fr
Jahiel Michel	Cirad – Persyst – Madagascar	michel.jahiel@cirad.fr
Jamin Jean-Yves	Cirad – ES – Montpellier	jean-yves.jamin@cirad.fr
Karpe Philippe	Cirad – ES, Cameroun	philippe.karpe@cirad.fr
Karsenty Alain	Cirad – ES – Montpellier	alain.karsenty@cirad.fr
Legile Anne	AFD – Paris	legilea@afd.fr
Léonard Éric	IRD – Montpellier	eric.leonard@ird.fr
Lescot Thierry	Cirad – Persyst – Montpellier	thierry.lescot@cirad.fr
Lesueur-Jannoyer Magalie	Cirad – Persyst – Montpellier	magalie.jannoyer@cirad.fr
Lopez Betancourt Teodoro	Université agraire de La Havane – Cuba	teodoro@unah.edu.cu
Losch Bruno	Cirad – ES – Montpellier	bruno.losch@cirad.fr
Malézieux Éric	Cirad – Persyst – Montpellier	eric.malezieux@cirad.fr
Manoli Claire	ESA d'Angers	c.manoli@groupe-esa.com
Marien Jean-Noël		jean-noel.marien@cirad.fr
Marzin Jacques	Cirad – ES – Montpellier	Jacques.marzin@cirad.fr

(continued)

Mercandalli Sara		Sara_mercandalli@hotmail.com
Merlet Pierre	AGTER – Nogent sur Marne	michel.merlet@agter.org
Mikolasek Olivier	Cirad – Persyst – Montpellier	olivier.mikolasek@cirad.fr
Molia Sophie	Cirad – ES – Montpellier	sophie.molia@cirad.fr
Moumouni Ismail	Université de Parakou – Bénin	mmismailfr@yahoo.fr
Parrot Laurent	Cirad – Persyst – Montpellier	laurent.parrot@cirad.fr
Pédelahore Philippe	Cirad – ES – Montpellier	philippe.pedelahore@cirad.fr
Peltier Régis	Cirad – ES – Montpellier	regis.peltier@cirad.fr
Penot Éric	Cirad – ES – Montpellier	eric.penot@cirad.fr
Pesche Denis	Cirad – ES – Montpellier	denis.pesche@cirad.fr
Peyre Marie-Isabelle	Cirad – ES – Vietnam	marisa.peyre@cirad.fr
Piketty Marie-Gabrielle	Cirad – ES – Montpellier	marie-gabrielle.piketty@cirad.fr
Pinta François	Cirad – Persyst – Burkina Faso	francois.pinta@cirad.fr
Piraux Marc	Cirad – ES – Brésil	marc.piraux@cirad.fr
Poccard-Chapuis René	Cirad – ES – Brésil	rené.poccard-chapuis@cirad.fr
Prades Alexia	Cirad – Persyst – Montpellier	alexia.prades@cirad.fr
Rafflegeau Sylvain	Cirad – Persyst – Montpellier	sylvain.rafflegeau@cirad.fr
Ratnadass Alain	Cirad – Persyst – Montpellier	alain.ratnadass@cirad.fr
Rey Jean-Yves	Cirad – Persyst – Sénégal	jean-yves.rey@cirad.fr
Robineau Ophélie	Cirad – ES – Montpellier	ophelie.robineau@cirad.fr
Roger François	Cirad – ES, Thaïlande	francois.roger@cirad.fr
Sabourin Éric	Cirad – ES – Brésil	eric.sabourin@cirad.fr
Sainte-Beuve Jérôme	Cirad – Persyst – Montpellier	jerome.sainte-beuve@cirad.fr
Salazar Ivonne	Universidad Nacional Agraria La Molina –Pérou	sri@lamolina.edu.pe
Semporé Aristide	Cirdes – Bobo-Dioulasso, Burkina Faso	aristide.sempore@cirad.fr
Sibelet Nicole	Cirad – ES – Costa-Rica	nicole.sibelet@cirad.fr
Sist Plinio	Cirad – ES, Montpellier	plinio.sist@cirad.fr
Sounigo Olivier	Cirad – Bios – Cameroun	olivier.sounigo@cirad.fr
Sourisseau Jean-Michel	Cirad – ES – Montpellier	jean-michel.sourisseau@cirad.fr
Temple Ludovic	Cirad – ES – Montpellier	ludovic.temple@cirad.fr
Toillier Aurélie	Cirad – ES – Burkina Faso	aurelie.toillier@cirad.fr
Torquebiau Emmanuel	Cirad – Persyst – Montpellier	emmanuel.torquebiau@cirad.fr
Tourrand Jean-François	Cirad – DGDRS – Montpellier	jean-francois.tourrand@cirad.fr
Tritsch Isabelle	Ecofog – Guyane française	isabelle.tritsch@gmail.com
Trouche Gilles	Cirad – Bios – Montpellier	gilles.trouche@cirad.fr
Valette Élodie	Cirad – ES – Montpellier	elodie.valette@cirad.fr
Vall Éric	Cirad – ES – Montpellier	eric.vall@cirad.fr
Wane Abdrahmane	Cirad – ES – Sénégal	abdrahmane.wane@cirad.fr

References

Abrunhosa, L., Serra, R., & Venâncio, A. (2002). Biodegradation of ochratoxin A by fungi isolated from grapes. *Journal of Agricultural and Food Chemistry, 50*, 7493–7496.

Adamczewski, A., Tonneau, J. P., Coulibaly, Y., & Jamin, J. Y. (2013a). Dynamiques sociales induites par les concessions de terres dans la zone Office du Niger au Mali. *Études rurales, 191*, 37–61.

Adamczewski, A., Jamin, J. Y., Burnod, P., Boutout Ly, E. H., & Tonneau, J. P. (2013b). Terre, eau et capitaux: Investissements ou accaparements fonciers à l'Office du Niger ? *Cahiers Agricultures, 22*, 22–32.

Affholder, F. (1997). Empirically modelling the interaction between intensification and climatic risk in semiarid regions. *Field Crops Research, 52*, 79–93.

Affholder, F., Jourdain, D., Morize, M., Quang, D. D., & Ricome, A. (2008). Éco-intensification dans les montagnes du Vietnam. Contraintes à l'adoption de la culture sur couvertures végétales. *Cahiers Agricultures, 17*, 289–296.

Affholder, F., Jourdain, D., Quang, D. D., Tuong, T. P., Morize, M., & Ricome, A. (2010). Constraints to farmers' adoption of direct-seeding mulch-based cropping systems: A farm scale modeling approach applied to the mountainous slopes of Vietnam. *Agricultural Systems, 103*, 51–62.

Affholder, F., Poyedebat, C., Corbeels, M., Scopel, E., & Tittonell, P. (2013). The yield gap of major food crops in family agriculture in the tropics: Assessment and analysis through field surveys and modelling. *Field Crops Research, 143*, 106–118.

Agence Bio. (2012). L'agriculture biologique, ses acteurs, ses produits, ses territoires, http://www.agencebio.org/la-bio-en-chiffres-historique. Retrieved 17 Mar 2014.

Aïach, P. (2010). *Les inégalités sociales de santé* (280 pp.). Paris: coll. Sociologiques, Economica Anthropos.

Alami, S., Clavel, D., Maffezzoli, C., Bertrand, B. (2013). De l'invention technique à l'innovation sociale: Quels rôle et responsabilité de la recherche dans l'accompagnement du changement ? Éclairages à travers la production locale de nouvelles variétés hybrides de café en Amérique centrale. In C. Boutillier, F. Djellal, & D. Uzunidis (Eds.), *L'innovation. Analyser, anticiper, agir* (pp. 251–270), Réseau de recherche sur l'innovation, coll. Business and Innovation, 5. Paris: Peter Lang International Editions.

Alarcón, E., & Ruz, E. (2011). *Institucionalidad de la extensión rural y las relaciones público-privadas en América Latina*. Santiago de Chile: RIMISP.

Albaladéjo, C., & Casabianca, F. (1997). Éléments pour un débat autour des pratiques de recherche-action. La recherche action. Ambitions, pratiques, débats. *Études et recherches sur les systèmes agraires et le développement, 30*, 127–149.

Alpha, A., Bricas, N., & Fouilleux, E. (2013, June). La difficile mise en œuvre d'une action publique intersectorielle en matière de sécurité alimentaire et de nutrition en Afrique. In *First international conference in developing countries: New approaches to an old challenge*, Grenoble.

Ameur, F., Hamamouche, M. F., Kuper, M., & Benouniche, M. (2013). La domestication d'une innovation technique: La diffusion de l'irrigation au goutte-à-goutte dans deux *douars* au Maroc. *Cahiers Agricultures, 22*, 311–318.

Anadon, M. (dir.). (2007). *La recherche participative: Multiples regards* (232 pp.). Quebec: University of Quebec Press.

Ancey, G. (1975). *Niveaux de décision et fonctions d'objectifs en milieu rural africain*. Paris: Amira.

Anderson, J. R., & Feder, G. (2004). Agricultural extension: Good intentions and hard realities. *World Bank Research Observer, 19*, 41–60.

Andriarimalala, J. H., Rakotozandriny, J. N., Andriamandroso, A. L. H., Penot, E., Naudin, K., Dugue, P., Tillard, E., Decruyenaere, V., & Salgado, P. (2013). Creating synergies between conservation agriculture and cattle production in crop-livestock farms: A study case in the lake Alaotra region of Madagascar. *Experimental Agriculture, 49*, 352–365.

Andrieu, N., Chia, E., & Vall, E. (coord.). (2011). Recherche et innovations dans les exploitations de polyculture-élevage d'Afrique de l'Ouest. Quelles méthodes pour évaluer les produits de la recherche ? *Revue d'élevage et de médecine vétérinaires des pays tropicaux*, numéro thématique, *64*, 1–4.

Andrieu, N., Dugue, P., Le Gal, P. Y., Rueff, M., Schaller, N., & Sempore, A. (2012). Validating a whole farm modelling with stakeholders: Evidence from a West African case. *Journal of Agricultural Science, 4*, 159–173.

Anseeuw, W. (2004). *La réforme agraire en Afrique du Sud. Le maintien d'une ségrégation agricole post-Apartheid*. Sarrebruck: Éditions universitaires européennes.

Anseeuw, W., Alden Wily, L., Cotula, L., & Taylor, M. (2012). *Land rights and the rush for land: Findings of the global commercial pressures on land research project* (84 pp.). ILC/

Argyris, C., & Schön, D. (1978). *Organizational learning: A theory of action perspective* (344 pp.). Reading: Addison Wesley.

Arrighi, G. (1994). *The long twentieth century: Money, power and the origins of our times*. London: Verso.

Asare, R., Afari-Sefa, V., Gyamfi, I., Okafor, C., & Mva Mva, J. (2010). Cocoa seed multiplication: An assessment of seed gardens in Cameroon, Ghana and Nigeria. STCP Working Paper Series, 11.

Augé, M. (1973). L'illusion villageoise, limites sociologiques et politiques du "développement" villageois en Côte d'Ivoire. *Archives internationales de sociologie de la coopération et du développement, 34*, 240–251.

Avelino, J., Romero-Gurdián, A., Cruz-Cuellar, H. F., & Declerck, F. A. J. (2012). Landscape context and scale differentially impact coffee leaf rust, coffee berry borer, and coffee rootknot nematodes. *Ecological Applications, 22*, 584–596.

Avenier, M.-J., & Schmitt, C. (2007). *La construction de savoirs pour l'action* (245 pp.). Paris: L'Harmattan.

Azhar, M., Lubis, A. S., Siregar, E. S., Alders, R. G., Brum, E., McGrane, J., Morgan, I., & Roeder, P. (2010). Participatory disease surveillance and response in Indonesia: Strengthening veterinary services and empowering communities to prevent and control highly pathogenic avian influenza. *Avian Diseases, 54*(Suppl. 1), 749–753.

Babin, R., Ten Hoopen, G. M., Cilas, C., Enjalric, F., Yede, G. P., & Lumaret, J. P. (2010). Impact of shade on the spatial distribution of *Sahlbergella singularis* in traditional cocoa agroforests. *Agricultural and Forest Entomology, 12*, 69–79.

Bainville, S., Affholder, F., Figuié, M., & Madeira, N. J. (2005). Les transformations de l'agriculture familiale de la commune de Silvânia: Une petite révolution agricole dans les Cerrados brésiliens. *Cahiers Agricultures, 14*, 103–110.

Bairoch, P. (1989). Les trois révolutions agricoles du monde développé: Rendements et productivité de 1800 à 1985. *Économies, sociétés, civilisations, 2*, 317–353.

Baldé, A. B., Scopel, E., Affholder, F., Corbeels, M., Silva, F. A. M. D., Xavier, J. H. V., & Wery, J. (2011). Agronomic performance of no-tillage relay intercropping with maize under smallholder conditions in Central Brazil. *Field Crops Research, 124*, 240–251.

Baldy, C., & Stigter, C. J. (1993). *Agrométéorologie des cultures multiples en régions chaudes, CTA/Inra, 246 p.*

Banerjee, A. V., & Duflo, E. (2009). L'approche expérimentale en économie du développement. *Revue d'Economie Politique, 119*, 691–726.

Barbault, R. (2008, October 28–30). Pourquoi des biodiversités? *Rencontres de Valdeblore*, http://www.mercantour.eu/valdeblore2008/images/actes/j1/intervention_barbault.pdf. Retrieved 17 Mar 2014.

Baron, S., Goutard, F., Nguon, K., & Tarantola, A. (2013). Use of a text message-based pharmacovigilance tool in Cambodia: Pilot study. *Journal of Medical Internet Research, 15*(8):e68 (8 pp.). doi:10.2196/jmir.2477. Retrieved 17 Mar 2014.

Barral, S. (2012). Plantations de palmiers à huile en Indonésie et déprolétarisation. *Paris, EHESS, Études rurales, 190*, 63–76.

Barral, S. (2013). Capitalisme agraire en Indonésie: Les marchés du travail et de la terre comme déterminants des rapports salariaux dans les plantations de palmier à huile. *Revue de la régulation, 13*.

Barreteau, O., Bots, P., Daniell, K. A., Etienne, M., Perez, P., Barnaud, C., Bazile, D., Becu, N., Castella, J. C., Daré, W., & Trébuil, G. (2013). Participatory approaches. In B. Edmonds & R. Meyer (Eds.), *Simulating social complexity: A handbook* (pp. 197–234). Heidelberg/Allemagne: Springer.

Barthélémy, D., Delorme, H., Losch, B., Moreddu, C., & Nieddu, M. (Eds.). (2003, March 21–22). *La multifonctionnalité de l'activité agricole et sa reconnaissance par les politiques publiques: Actes du colloque international de la Société française d'économie rurale*. Dijon: Educagri.

Barthez, A. (1984). *Famille, travail et agriculture* (82 pp.). Paris: Economica, Inra.

Bazile, D., & Abrami, G. (2008). Des modèles pour analyser ensemble les dynamiques variétales du sorgho dans un village malien. *Cahiers Agricultures, 17*(2), 203–209.

Bazile, D., & Soumaré, M. (2004). Gestion spatiale de la diversité variétale en réponse à la diversité écosystémique: Le cas du sorgho (*Sorghum bicolor* (L) Moench) au Mali. *Cahiers Agricultures, 13*, 480–487.

Bazile, D., Martinez, E. A., Hocdé, H., & Chia, E. (2012). Primer encuentro nacional de productores de quínoa de Chile: Una experiencia participativa del proyecto internacional IMAS a través de una prospectiva por escenarios usando una metodología de "juego de roles". *Tierra Adentro (Chile), 97*, 48–54.

Belem, M., Bousquet, F., Müller, J. P., Bazile, D., & Coulibaly, H. (2011, September 19–23). A participatory modeling method for multi-points of view description of a system from scientist's perceptions: Application in seed systems modeling in Mali and Chile. In *7th European Social Simulation Association conference* (ESSA 2011) (12 p.), Montpellier.

Bélières, J. -F., Bonnal, P., Bosc, P. -M., Losch, B., Marzin, J., & Sourisseau, J. -M. (2013). *Les agricultures familiales du monde. Définitions, contributions et politiques publiques* (306 pp.). Montpellier: Cirad, AFD, MAAF, MAE.

Bélières, J.-F., Hilhorst, T., Kebe, D., Keïta, M. S., Keïta, S., & Sanogo, O. (2011). Irrigation et pauvreté: Le cas de l'Office du Niger au Mali. *Cahiers Agricultures, 20*, 144–149.

Benfekih, L., Foucart, A., & Petit, D. (2011). Central Saharan populations of *Locusta migratoria cinerascens* (*Orthoptera: Acrididae*) in irrigated perimeters: Is it a recent colonisation event? *Annales de la Société entomologique de France (N.S.). International Journal of Entomology, 47*(1–2), 147–153. doi:10.1080/00379271.2011.10697706.

Benouniche, M., Kuper, M., Poncet, J., Hartani, T., & Hammani, A. (2011). Quand les petites exploitations adoptent le goutte-à-goutte: Initiatives locales et programmes étatiques dans le Gharb au Maroc. *Cahiers Agricultures, 20*, 40–47.

Bentley, J., & Baker, P. (2000). The Colombian coffee growers' federation: Organised, successful smallholder farmers for 70 years. Agricultural Research and Extension Network (AgREN), Papers 100.

Bergen, S. D., Bolton, S. M., & Fridley, L. J. (2001). Design principles for ecological engineering. *Ecological Engineering, 18*, 201–210.

Bergeret, P., & Dufumier, M. (2002). Analyser la diversité des exploitations agricoles. In *Mémento de l'agronome* (pp. 411–432). Paris: Cirad-Gret, Ministère des Affaires étrangères.

Bertrand, R. (2011). *L'Histoire à parts égales. Récits d'une rencontre Orient-Occident* (664 pp.). Paris: Seuil.

Bertrand, B., Étienne, H., Cilas, C., Charrier, A., & Baradat, P. (2005). Coffea arabica hybrid performance for yield, fertility and bean weight. *Euphytica, 141*(3), 255–262.

Bertrand, B., Montagnon, C., Georget, F., Charmetant, P., & Étienne, H. (2012). Création et diffusion de variétés de caféiers Arabica: Quelles innovations variétales ? *Cahiers Agricultures, 21*(2–3), 77–88.

Bessière, C., Giraud, C., & Renahy, N. (2008). Famille, travail, école et agriculture. *Revue d'é tudes en agriculture et environnement, 88*(3), 5–19.

Biénabe, E., Berdegué, J., Peppelenbos, L., & Belt, J. (2011). *Reconnecting markets. Innovative global practices in connecting small-scale producers with dynamic food markets.* Gower: IIED.

Binot, A., Goutard, F., Duboz, R., Peyre, M., Cappelle, J., & Roger, F. (2012, October 29–31). Managing animal health risks in a changing environment: Going beyond technical tools for risk communication between heterogeneous stakeholders. In *ICT-Asia Regional Seminar 2012.*

Binswanger, H., & Rosenweig, M. (1986). Behavioural and material determinants of production relations in agriculture. *Journal of Development Studies, 22*(3), 503–539.

Birner, R., Davis, K. E., Pender, J., Nkonya, E., Anandajayasekeram, P., Ekboir, J., Mbabu, A., Spielman, D. J., Horna, D., Benin, S., & Cohen, M. (2009). From best practice to best fit: A framework for designing and analyzing pluralistic agricultural advisory services worldwide. *Journal of Agricultural Education and Extension, 15*, 341–355.

Blanchard, M., Chia, E., Koutou, M., & Vall, E. (2012). Recherche-action-en-partenariat, une démarche de réconciliation entre recherche et société. In E. Coudel, H. Devautour, C. Soulard, G. Faure, & B. Hubert (Eds.), *Apprendre à innover dans un monde incertain: Concevoir les futurs de l'agriculture et de l'alimentation* (208 pp.). Versailles: Quæ.

Blanchart, E., Bernoux, M., Sarda, X., Siqueira, N. M., Cerri, C. C., Piccolo, M. D. C., Douzet, J.-M., & Scopel, E. (2007). Effect of direct seeding mulch-based systems on soil carbon storage and macrofauna in Central Brazil. *Agriculturae Conspectus Scientificus, 72*, 81–87.

Blandin, P. (2009). *De la protection de la nature au pilotage de la biodiversité* (122 pp.). Versailles: Éditions Quæ.

Bocci, R., & Chable, V. (2008). Semences paysannes en Europe: Enjeux et perspectives. *Cahiers Agricultures, 17*, 216–221.

Boltanski, L., & Chiapello, E. (1999). *Le nouvel esprit du capitalisme.* Paris: Gallimard.

Bommarco, R., Kleijn, D., & Potts, S. G. (2013). Ecological intensification: Harnessing ecosystem services for food security. *Trends in Ecology and Evolution, 28*, 230–238.

Bonilla, H. (2008). La cuestión agraria en el Perú después de la Reforma Agraria. *Revista socialismo y participación, 105*, 155–162.

Bonin, M., Cattan, P., Dorel, M., & Malezieux, E. (2006). L'émergence d'innovations techniques face aux risques environnementaux. Le cas de la culture bananière en Guadeloupe: Entre solutions explorées par la recherche et évolution des pratiques. In J. Caneill (Ed.), *Agronomes et innovations* (pp. 123–135). Paris: L'Harmattan.

Bonnal, P. (2010). Production des politiques et compromis institutionnels autour du développement durable. Rapport de fin de projet. Montpellier, Projet ANR Propocid, no. ANR-06-PADD-016, 39 p.

Bonnal, P., & Kato, K. (2009). *Analise comparativa das politicas de desenvolvimento territorial, relatorio final, CPDA-IICA,* 110 p.

Bonnal, P., Bosc, P. -M., Diaz, J. M., & Losch, B. (2004). Multifuncionalidad de la agricultura y Nueva Ruralidad ¿Reestructuración de las políticas públicas a la hora de la globalización? In E. Pérez Correa, & M. A. Farah Quijano (Eds.), *Desarollo rural y nueva ruralidad en América Latina y la Unión Europea* (pp. 19–41). Bogota: Pontificia Universidad Javeriana, Facultad de Estudios Ambientales y Rurales, Departamento de Desarrollo Rural y Regional, Maestría en Desarrollo Rural.

Bonnet, P., & Lesnoff, M. (2009). Decision making, scales and quality of economic evaluations for the control of contagious bovine pleuropneumonia (CBPP): The use of economic analysis methods in combination with epidemiological and geographical models to help decision making for CBPP control in Ethiopia. In J. A. Rushton (Ed.), *The economics of animal health and production* (pp. 279–285). Wallingford: CABI Publishing.

Bonnet, P., Tibbo, M., Workalemahu, A., & Gau, M. (2001, août 30–31). Rift Valley Fever an emerging threat to livestock trade and food security in the Horn of Africa: A review. In *Proceedings of the 9th national conference of the Ethiopian Society of Animal Production (ESAP)* (13 pp.). Addis-Abeba: Éthiopie.

Bonnet, P., Carmeille, A., Dini Ibrahim, Didier Laurent, S., Sundar Das, S., & Bayou Aberra, S. (2003). Patterns of diffusion of an institutional innovation in pastoral areas of Ethiopia. The community based animal health worker and its geographical and social spread in Afar region. In *10th international symposium for veterinary epidemiology and economics*. Viña del Mar, Chile: ISVEE.

Bonnet, P., Thiaucourt, F., & Bendali, F. (2005). "PPCB Média", "CBPP Media", une compilation de documents de politique sanitaire sur la PPCB (péripneumonie contagieuse bovine) avec supports vidéo et photos sur la maladie. Cirad-EMVT: Montpellier, 1 (2004–2005).

Bonnet, P., Bedane, B., Bheenick, K.J., Juanes, X., Girardot, B., Coste, C., Gourment, C., Wanda, G., Madzima, W., Oosterwijk, G., & Erwin, T. (2010). The LIMS Community and its collaborative Livestock Information Management System for managing livestock statistics and sharing information in the SADC region (Southern African Development Community). In *Scientific and technical information and rural development* (9 pp.). Montpellier: Agropolis International.

Bonneuil, C., & Thomas, F. (2012). *Semences: Une histoire politique* (2016 pp.). Éditions Charles Léopold Mayer.

Bonny, S. (2011). L'agriculture écologiquement intensive: Nature et défis. *Cahiers Agricultures, 20*(6), 451–462.

Bontoulougou, J., Oulé, J. M., Pellissier, J. P., & Tallet, B. (2000). La participation des acteurs, un exercice difficile. Leçons de l'expérience d'un plan de lutte contre la trypanosomose animale africaine dans la vallée du Mouhoun (Burkina Faso). *Natures, sciences, sociétés, 8*(1), 33–43.

Bosc, P. -M., Mercoiret, M. R., & Sabourin, E. (2004). Agricultures familiales, action collective et organisation. In E. Sabourin, M. Antonna, & E. Coudel (Eds.), *Séminaire transversal MOISA: Action collective* (7 pp.). Montpellier: Cirad.

Bosc, P.-M., Eychenne, D., Hussein, K., Losch, B., Mercoiret, M.-R., Rondot, P., & Macintosh-Walker, S. (2001). *The role of Rural Producers Organisations (RPOs) in the World Bank rural development strategy. Background study*. Washington, DC: World Bank.

Bosc, P.-M., Berthomé, J., Losch, B., & Mercoiret, M. (2002). Le grand saut des organisations de producteurs agricoles africaines. De la protection sous tutelle à la mondialisation. *Recma, 285*, 47–62.

Bosc, P.-M., Dabat, M.-H., & Maître, D.'h. É. (2010). Quelles politiques de développement durable au Mali et à Madagascar ? *Économie rurale, 320*, 24–38.

Boserup, E. (1965). *The conditions of agricultural growth. The economics of agriculture under population pressure*. London: Allen and Unwin.

Boucher, J. (1990). *Théorie de la régulation et rapport salarial*. Quebec: coll. Études théoriques, CRISES, Centre de recherche sur les innovations sociales.

Boullenger, G., Le Bellec, F., Girardin, P., & Bockstaller, C. (2008). Évaluer l'impact des traitements des agrumes sur l'environnement: Adaptation d'I-Phy, indicateur environnemental d'effet de l'utilisation des produits phytosanitaires, à l'agrumiculture guadeloupéenne. *Phytoma-La défense des végétaux, 617*, 22–25.

Bouman, B. A. M., Lampayan, R. M., & Tuong, T. P. (2007). *Water management in irrigated rice: Coping with water scarcity* (54 pp.). Los Baños, Laguna: IRRI Publications. http://dspace.irri.org:8080/dspace/handle/10269/266. Retrieved 14 Mar 2014.

Bouquet, E., & Colin, J.-P. (2009). L'État, l'*ejido* et les droits fonciers: Ruptures et continuités du cadre institutionnel formel au Mexique. In J.-P. Colin, P.-Y. Le Meur, & E. Léonard (Eds.), *Les politiques d'enregistrement des droits fonciers: Du cadre légal aux pratiques locales* (pp. 299–332). Paris: Karthala.

Bourdieu, P. (Ed.). (1993). *La misère du monde* (948 pp.). Paris: coll. Libre examen, Le Seuil.

Bourdieu, P., & Sayad, A. (1964). *Le déracinement. La crise de l'agriculture traditionnelle en Algérie*. Paris: Éditions de Minuit.

Bousquet, F., Cambier, C., Mullon, C., Morand, P., Quensiere, J., & Pavé, A. (1993). Simulating the interaction between a society and a renewable resource. *Journal of Biological Systems, 1* (2), 199–214.

Boussard, J. M. (1986). Hétérogénéité technique et structurelle dans les exploitations agricoles. *Économie rurale, 10*(176), 3–10.

Bouzidi, Z., Abdellaoui, E. H., Faysse, N., Billaud, J. P., Kuper, M., & Errahj, M. (2011). Dévoiler les réseaux locaux d'innovation dans les grands périmètres irrigués. Le développement des agrumes dans la plaine du Gharb au Maroc. *Cahiers Agricultures, 20*, 34–39.

Bowyer, C., & Kretschmer, B. (2011). *Anticipated indirect land use change associated with expanded use of biofuels and bioliquids in the EU – An analysis of the national renewable energy action plans*. London: Institute for European Environmental Policy.

Braudel, F. (1979). *Civilisation matérielle, économie et capitalisme. XVe-XVIIIe siècle*. Paris: Armand Colin.

Braudel, F. (1986). *L'identité de la France: Les hommes et les choses*. Paris: Arthaud-Flammarion.

Braudel, F. (1993). *Grammaire des civilisations* (1re ed. 1963, 625 pp.). Paris: Flammarion.

Brauman, K. A., Siebert, S., & Foley, J. A. (2013). Improvements in crop water productivity increase water sustainability and food security: A global analysis. *Environmental Research Letters, 8*(2). doi:10.1088/1748-9326/8/2/024030.

Breman, H., & Sissoko, K. (1998). *L'intensification agricole au Sahel* (1000 pp.). Paris: Économie et développement, Karthala.

Brévault, T., Heuberger, S., Zhang, M., Ellers-Kirk, C., Ni, X., Masson, L., Li, X., Tabashnik, B. A., & Carrière, Y. (2013). Potential shortfall of pyramided transgenic cotton for insect resistance management. *Proceedings of the National Academy of Sciences of the USA, 110*, 5806–5811.

Bricas, N., Tchamda, C., & Thirion, M. C. (2013). Consommation alimentaire en Afrique de l'Ouest et centrale: Les productions locales tirées par la demande urbaine, mais les villes restent dépendantes des importations de riz et de blé. *Déméter, Économie et stratégies agricoles, 2014*, 125–142.

Brossier, J., Devèze, J.-C., & Kleene, P. (2007). Qu'est-ce que l'exploitation agricole familiale en Afrique ? In M. Gafsi, P. Dugué, J.-Y. Jamin, & J. Brossier (Eds.), *Exploitations agricoles familiales en Afrique de l'Ouest et du Centre*. Montpelllier: Quæ-CTA.

Byarugaba, F., Grimaud, P., Godreuil, S., & Etter, E. (2010). Risk assessment in zoonotic tuberculosis in Mbarara, the main milk basin of Uganda. *Bulletin of Animal Health and Production in Africa, 58*(2), 125–132. Retrieved 17 Dec 2010.

Cabidoche, Y. M., & Lesueur-Jannoyer, M. (2012). Contamination of harvested organs in root crops grown on chlordecone-polluted soils. *Pedosphere, 22*(4), 562–571.

Cano, G. J., Balcazar, A., Castillo, J., Giraldo, J. C., Arcila, A., & Rodriguez, C. (2006). Alianzas estrategicas en palma de aceite en Colombia: estudio de caracterizacion. *Palmas, 27*, 47–63.

Carberry, P. S., Liang, W.-L., Twomlow, S., Holzworth, D. P., Dimes, J. P., McClelland, T., Huth, N. I., Chen, F., Hochman, Z., & Keating, B. A. (2013). Scope for improved eco-efficiency varies among diverse cropping systems. *Proceedings of the National Academy of Sciences, 110*, 8381–8386.

Caron, P., Craufurd, P., Martin, A., Mc Donald, A., Abedini, W., Afiff ,S., Bakurin, N., Bass, S., Hilbeck, A., Jansen, T., Lhaloui, S., Lock, K., Newman, J., Primavesi, O., Sengooba, T., Ahmed, M., Ainsworth, E. A., Ali, M., Antona, M., Avato, P., Barker, D., Bazile, D., Bosc, P. M., Bricas, N., Burnod, P., Cohen, J.I., Coudel, E., Dulcire, M., Dugué, P., Faysse, N., Farolfi, S., Faure ,G., Goli, T., Grzywacz, D., Hocdé, H., Imbernon, J., Ishii-Eiteman, M., Leakey, A., Leakey, C., Lowe, A., Marr, A., Maxted, N., Mears, A., Molden, D.J., Müller, J.P., Padgham, J., Perret, S., Place, F., Raoult-Wack, A.L., Reid, R., Riches, C., Scherr, S.J., Sibelet, N., Simm, G., Temple, L., Tonneau, J.P., Trébuil, G., Twomlow, S., & Voituriez, T. (2009). Impacts of AKST on development and sustainability goals. In B. D. McIntyre, H. R. Herren, J. Wakhungu, & R. T. Watson (Eds.), *Agriculture at a crossroads. International Assessment of Agricultural Knowledge, Science and Technology for Development (IAASTD): Global report* (pp. 145–253). Washington, DC: Island Press.

Caron, A., de Garine-Wichatitsky, M., & Morand, S. (2011). Parasite community ecology and epidemiological interactions at the wildlife/domestic/human interface: Can we anticipate emerging infectious diseases in their hotspots? [Abstract]. *EcoHealth, 7*, S24. Retrieved 3 Aug 2011.

Caron, A., Miguel, E., Gomo, C., Makaya, P. V., Pfukenyi, D., Foggin, C., Hove, T., & de Garine-Wichatitsky, M. (2013). Relationship between burden of infection in ungulate populations and wildlife/livestock interfaces, *Epidemiology and infection, 141*(7), 1522–1535. 10.1017/S0950268813000204. Retrieved 17 Mar 2014.

Cassman, K. G. (1999). Ecological intensification of cereal production systems: Yield potential, soil quality, and precision agriculture. *National Academy of Sciences Colloquium, 96*, 5952–5959.

Chambers, R., & Conway, G. (1991). *Sustainable rural livelihoods: Practical concepts for the 21st century* (33 pp.). London: Institute of Development Studies.

Chambers, R., Pacey, A., & Thrupp, L. A. (1989). *Farmer first. Farmer innovation and agricultural research*. Londres: Intermediate Technology Publication.

Chang, H.-J. (2002). *Kicking away the ladder: Development strategy in historical perspective*. London: Anthem Press.

Chang, H. -J. (2009). Rethinking public policy in agriculture. Lessons from distant and recent history. *Policy assistance series* (Vol. 7). Rome: FAO.

Charmes, J., Couty, P., & Winter, G. (1985). Rapports Nord-Sud: Pour des stratégies de développement plus réalistes et mieux informées. In *Économies en transition* (pp. 51–83). Paris: Orstom.

Chauveau, J. -P., & Yung, J. -M. (eds). (1995). Innovation et sociétés. Quelles agricultures ? Quelles innovations ? II. Les diversités de l'innovation. In *Actes du Séminaire d'économie rurale, 14*, 13–16 septembre 1993. Montpellier: Cirad.

Chen, S., & Ravallion, M. (2012). *More relatively-poor people in a less absolutely-poor world* (42 pp.). Policy Research Working Paper 6114. Washington, DC: The Word Bank.

Chevalier, V., Thiongane, Y., & Lancelot, R. (2009a). Endemic transmission of Rift Valley Fever in Senegal. *Transboundary and Emerging Diseases, 56*(9–10), 372–374. 10.1111/j.1865-1682.2009.01083.x. Retrieved 17 Mar 2014.

Chevalier, V., Tran, A., Gerbier, G., Olive, M.-M., Gély, M., Bonnet, P., & Roger, F. (2009b). *Rift Valley Fever outbreaks and control in the Middle East* (pp. 28–29). Paris: OIE Report.

Chevassus-au-Louis, B., & Bazile, D. (2008). Cultiver la diversité, Éditorial. *Cahiers Agricultures, 17*(2), 77–78.

Chevassus-au-Louis, B., & Griffon, M. (2008). La nouvelle modernité: Une agriculture productive à haute valeur écologique. *Déméter, Économie et stratégies agricoles, 14*, 7–48.

Cheyns, E., & Rafflegeau, S. (2005). Family agriculture and the sustainable development issue: Possible approaches from the African oil palm sector: The example of Ivory Coast and Cameroon. *Oléagineux, corps gras, lipides, 12*, 111–120.

Chia, E. (2004). Principes, méthodes de la recherche en partenariat: Une proposition pour la traction animale. *Revue d'Élevage et de Médecine Vétérinaire des Pays Tropicaux, 57*(3–4), 233–240.

Chia, E., Barlet, B., Tomedi Eyango Tabi, M., Pouomogne, V., & Mikolasek, O. (2008, July 6–10). Co-construction of a local fish culture system: Case study in Western Cameroon. In B. Dedieu (Ed.), *Empowerment of the rural actors. A renewal of farming systems perspectives. 8th European IFSA Symposium* (12 p.). Clermond-Ferrand: IFSA.

Chiffoleau, Y. (2005). Learning about innovation through networks: The development of environment-friendly viticulture. *Technovation, 25,* 1193–1204.

Chillet, M., Castelan, F. P., Abadie, C., Hubert, O., & de Lapeyre de Bellaire, L. (2013). Necrotic leaf removal, a key component of integrated management of *Mycosphaerella* leaf spot diseases to improve the quality of banana: The case of Sigatoka Disease. *Fruits, 68,* 271–277.

Chombard de Lauwe, J., Poitevin, J., & Tirel, J.-C. (1969). *Nouvelle gestion des exploitations agricoles.* Paris: Dunod.

Cialdella, N. (2005). *Stratégies d'élevage dans les projets familiaux en milieu aride. Usage des ressources locales pour gérer l'incertain, cas de la Jeffara (sud-est tunisien)* (PhD). AgroParisTech.

Cirad. (2012). Projet FFEM (Traitement environnemental de la lutte antiacridienne en Afrique de l'Ouest et du Nord-Ouest): Définition de méthodologies d'exploitation de l'imagerie satellitaire pour des applications directement opérationnelles. Document de travail.

Cirad-FAO. (2012). *Atlas des évolutions des systèmes pastoraux au Sahel* (36 pp.). Rome: FAO-Cirad.

Cirad-Tera. (1998). Agricultures familiales. Atelier 2–3 February 1998. Montpellier: Cirad.

Cissokho, M. (2009). *Dieu n'est pas un paysan* (295 pp.). Grad: Présence africaine.

Cittadini, R. (2010). Food safety and sovereignty, a complex and multidimensional problem. *Voces en el Fenix.* Buenos Aires: University of Buenos Aires. http://www.vocesenelfenix.com

Clarke, L., & Bishop, C. (2002, July 30). Farm power – present and future availability in developing countries. Invited overview paper presented at the *Special session on agricultural engineering and international development in the third millennium.* Chicago: ASAE Annual International Meeting/CIGR World Congress.

Clasadonte, L., de Vries, E., Trienekens, J., Arbeletche, P., & Tourrand, J. F. (2013). Network companies: A new phenomenon in South American farming. *British Food Journal, 115*(6), 850–864. http://www.emeraldinsight.com/journals.htm?issn=0007-070x&volume=115&issue=6&articleid=17090541. Retrieved 17 Mar 2014.

Cochet, H. (2011). *L'agriculture comparée* (160 pp.). Versailles: Quæ.

Colin, J. -P., Le Meur, P. -Y., & Léonard, E. (Eds.). (2009). *Les politiques d'enregistrement des droits fonciers. Du cadre légal aux pratiques locales* (534 pp.). Paris: Karthala.

Collier, P., & Dercon, S. (2013). African agriculture in 50 years: Smallholders in a rapidly changing world? In *Stanford symposium on global food policy and food security in the 21st century* (16 pp.). Oxford: Center of Food Security and the Environment.

Commons, J. R. (1934). *Institutional economics. Its place in the political economy.* MacMillan, réédition 1990, Transaction Publishers, 915. New York: Macmillan Company.

Conan, A., Goutard, F. L., Sorn, S., & Vong, S. (2012). Biosecurity measures for backyard poultry in developing countries: A systematic review. *BMC Veterinary Research, 8*(240).

Convention of Biological Diversity (CBD). (2013). http://www.cbd.int/convention/articles/default.shtml?a=cbd-08. Retrieved 17 Mar 2014.

Cook, S. M., Khan, Z. R., & Pickett, J. A. (2007). The use of push-pull strategies in integrated pest management. *Annual Review of Entomology, 52,* 375–400.

Copans, J. (1987). Classes, État, marches. Une crise conceptuelle opportune. *Politique africaine,* (26), 2–14.

Corniaux, C. (2006). Gestion du troupeau et droit sur le lait: Prise de décision et production laitière au sein de la concession sahélienne. *Cahiers Agricultures, 15*(6), 515–522.

Coudel, E., Piketty, M. -G., Gardner, T., Viana, C., Ferreira, J., Morello, T. F., Parry, L., Barlow, J., & Antona, M. (2012, June 16–19). Environmental compliance in the Brazilian Amazon: Exploring motivations and institutional conditions. In *12th Biennial conference of the International Society for Ecological Economics (ISEE 2012 conference) "Ecological Economics and Rio+20: Challenges and Contributions for a Green Economy"*, Rio de Janeiro (34 p.). http://www.isee2012.org/anais/pdf/261.pdf. Retrieved 22 Mar 2014.

Coulibaly, H. (2011). *Le rôle des Organisations paysannes dans la conservation in situ des variétés locales de céréales au Mali: Articulation des réseaux semenciers formel étatique et traditionnel paysan. Cas des mils et sorghos* (Thèse de doctorat) (294 pp.). Université Paris X-Nanterre.

Coulibaly, H., Bazile, D., Sidibé, A., & Abrami, G. (2008). Les systèmes d'approvisionnement en semences de mils et sorghos au Mali: Production, diffusion et conservation des variétés en milieu paysan. *Cahiers Agricultures, 17*(2), 199–202.

Coulomb, P., Delorme, H., Hervieu, B., Jollivet, M., & Lacombe, P. (dir.). (1990). *Les agriculteurs et la politique*. Paris: Presses de la Fondation nationale des sciences politiques.

Courleux, F. (2011). Augmentation de la part des terres agricoles en location: échec ou réussite de la politique foncière ? *Économie et statistique, 444–445*.

Crozier, M., & Friedberg, E. (1977). *L'acteur et le système. Les contraintes à l'action collective* (500 pp.). Paris: Le Seuil.

Dalgaard, T., Hutchings, N. J., & Porter, J. R. (2003). Agroecology, scaling and interdisciplinarity. *Agriculture, Ecosystems and Environment, 100*, 39–51.

Dallaire, R., Muckle, G., Rouget, F., Seurin, S., Monfort, C., Multigner, L., Bataille, H., Kadhel, P., Thomé, J. P., Jacobson, S. W., Boucher, O., & Cordier, S. (2012). Cognitive, visual and motor development of infants exposed to chlordecone in Guadeloupe. *Environmental Research, 118*, 79–85.

Dao The Anh, Moustier, P., & Dao The Tuan. (2008). New challenges for the development of cooperatives in Vietnam. In *SFER seminar* (SFER ed.) (p. 19), Paris.

Darghouth Medimegh, A. (1992). *Droits et vécu de la femme en Tunisie* (206 pp.). Paris: L'Hermès, Edilis.

David, A., Hatchuel, A., & Laufer, R. (2001). *Les nouvelles fondations des sciences de gestion. Éléments d'épistémologie de la recherche en management* (215 pp.). Paris: Vuibert.

Daviron, B. (2002). Small farm production and the standardization of tropical products. *Journal of Agrarian Change, 2*(2), 162–184.

Davis, K. E. (2006). Farmer field schools: A boon or bust for extension in Africa? *Journal of International Agricultural and Extension Education, 13*, 91–97.

de Bon, H., Parrot, L., & Moustier, P. (2010). Sustainable urban agriculture in developing countries. A review. *Agronomy for Sustainable Development, 30*(1), 21–32.

De Garine-Wichatitsky, M., Caron, A., Kock, R., Tschopp, R., Munyeme, M., Hofmeyr, M., & Michel, A. (2013). A review of bovine tuberculosis at the wildlife-livestock-human interface in sub-Saharan Africa. *Epidemiology and Infection, 141*(7), 1342–1356.

de La Roque, S., Michel, J. F., Cuisance, D., de Wispelaere, G., Augusseau, X., Solano, P., Guillobez, S., & Arnaud, M. (2001). *Le risque trypanosomien, du satellite au microsatellite, une approche globale pour une décision locale* (151 p.). Cirad: Montpellier.

De Meyer, M., Robertson, M. P., Mansell, M., Ekesi, S., Tsuruta, K., Mwaiko, W., Vayssières, J. F., & Peterson, A. (2010). Ecological niche and potential geographic distribution of the invasive fruit fly *Bactrocera invadens* (*Diptera Tephritidae*). *Bulletin of Entomological Research, 100*(1), 35–48.

Dedieu, B., Gibon, A., Ickowicz, A., & Tourrand, J. F. (2010). Transformation des élevages extensifs et des territoires ruraux. *Cahiers Agricultures, 19*(2), 81–83.

Deguine, J. P., Rousse, P., & Atiama-Nurbel, T. (2012). Agroecological crop protection: Concepts and a case study from Reunion. In L. Larramendy, & S. Soloneski (Eds.), *Integrated pest management and pest control: Current and future tactics* (pp. 63–76). Intech Publisher.

Deguine, J. P., Augusseau, X., Insa, G., Jolet, M., Le Roux, K., Marquier, M., Rousse, P., Roux, E., Soupapoullé, Y., & Suzanne, W. (2013). Gestion agroécologique des Mouches des légumes à La Réunion. *Innovations Agronomiques, 28*, 59–74.

Deheuvels, O., Avelino, J., Somarriba, E., & Malezieux, E. (2012). Vegetation structure and productivity in cocoa-based agroforestry systems in Talamanca, Costa Rica. *Agriculture, Ecosystems and Environment, 149*, 181–188.

Deininger, K. W., & Byerlee, D. (2011). *Rising global interest in farmland: Can it yield sustainable and equitable benefits?* (263 pp.) Washington, DC: World Bank Publications.

Delabouglise, A., Antoine-Moussiaux, N., Binot, A., Vu Dinh, T., Nguyen, V. K., Duboz, R., & Peyre, M. (2012). Methodological framework for a participatory study to evaluate the socio-economic factors impairing the efficacy of animal health surveillance systems. In *13th international symposium on veterinary epidemiology and economics* (327 pp.), Maastricht.

Delaunay, S., Tescar, R. P., Oualbégo, A., Vom, B. K., & Lançon, J. (2008). La culture du coton ne bouleverse pas les échanges traditionnels de semences de sorgho. *Cahiers Agricultures, 17*(2), 189–194.

Deléage, E., & Sabin, G. (2012). Modernité en friche. Cohabitation de pratiques agricoles. *Ethnologie Française, 42*(4), 667–676.

Delphy, C. (Ed.). (2013). *L'ennemi principal. 1. Économie politique du patriarcat* (262 pp.). Paris: Syllepse.

Demeke, M., Dawe, D., Teft, J., Ferede, T., & Bell, W. (2012). Stabilizing price incentives for staple Grain producers in the context of broader agricultural policies. *Debates and Country Experiences*, FAO, ESA Working Paper, 12-05. http://www.fao.org/docrep/016/ap536e/ap536e.pdf. Retrieved 19 Mar 2014.

Demont, M., Jouve, P., Stessens, J., & Tollens, E. (2007). Boserup versus Malthus revisited: Evolution of farming systems in northern Cote d'Ivoire. *Agricultural Systems, 93*, 215–228.

Derpsch, R., Friedrich, T., Kassam, A., & Hongwen, L. (2010). Current status of adoption of no-till farming in the world and some of its main benefits. *International Journal of Agricultural and Biological Engineering, 3*, 1–25.

Desvaux, S., & Figuié, M. (2011). Formal and informal surveillance systems: How to build bridges? *Épidémiologie et santé animale, 59–60*, 352–355.

Devienne, S. (2013). Régulation de l'accès aux parcours et évolution des systèmes pastoraux en Mongolie. *Études mongoles et sibériennes, centrasiatiques et tibétaines* [online] (pp. 43–44). http://emscat.revues.org/2104. Retrieved 19 Mar 2014.

Djamen, N. P., Djonnewa, A., Havard, M., & Legile, A. (2003). Former et conseiller les agriculteurs du Nord-Cameroun pour renforcer leurs capacités de prise de décision. *Cahiers Agricultures, 12*, 241–245.

Djerma, G. D., & Dabat, M. -H. (2013, November 21–23). Du vide institutionnel au partenariat multiacteurs pour la définition de politiques en faveur des agrocarburants en Afrique de l'Ouest. In *4ᵉ Conférence internationale sur les biocarburants en Afrique. Quel bilan et quelles voies d'avenir pour les biocarburants et les bioénergies en Afrique ?* (10 pp.), Ouagadougou.

Doré, T., Makowski, D., Malézieux, E., Munier-Jolain, N., Tchamitchian, M., & Tittonell, P. (2011). Facing up to the paradigm of ecological intensification in agronomy: Revisiting methods, concepts and knowledge. *European Journal of Agronomy, 34*, 197–210.

Dorin, B. (2012). Agribiom caloric balance sheets. Methodology and data sources detailed in *Agrimonde*. Versailles: Quæ, 2010, 295 p.

Dorin, B., Hourcade, J. -C., & Benoit-Cattin, M. (2013). A world without farmers? The Lewis Path revisited. Working Papers, Cired, 47-2013. http://www.centre-cired.fr/IMG/pdf/CIREDWP-201347.pdf. Retrieved 19 Mar 2014.

Dorward, P., Shepherd, D., & Galpin, M. (2007). The development and role of novel farm management methods for use by small-scale farmers in developing countries. *Journal of Farm Management, 13*, 123–134.

Draper, C., & Conlong, D. E. (2000). Eldana saccharina and predator populations in sugarcane fields of rural and commercial growers. *Proceedings of the South Africa Sugarcane Technologists Association, 74*, 228.

Drigo, I., Piketty, M. G., Driss, W., & Sist, P. (2013). Cash income from community-based forest management: Lessons from two case studies in the Brazilian Amazon. *Bois et Forêts des Tropiques, 315*(1), 40–49.

Duby, G. (1962). *L'économie rurale et la vie des campagnes dans l'Occident médiéval*. Paris: Aubier.

Duby, G. (1978). *Les trois ordres ou l'imaginaire du féodalisme*. Paris: Gallimard.

Ducamp Collin, M. N., Arnaud, C., Kagy, V., & Didier, C. (2007). Fruit flies: Disinfestation, techniques used, possible application to mango. *Fruits, 62*, 223–236.

Ducastel, A., & Anseeuw, W. (2011). Le "production grabbing" et la transnationalisation de l'agriculture (sud-)africaine. *Transcontinentales*, (10/11). http://transcontinentales.revues.org/1080. Retrieved 19 Mar 2014.

Ducastel, A., & Anseeuw, W. (2013). Situating investment funds in agriculture. *Farm Policy Journal, 10*(3), 35–43.

Ducatenzeiler, G., & Itzcovitz, V. (2011). La démocratie en Amérique latine. *Revue internationale de politique comparée, 18*, 123–140. doi:10.3917/ripc.181.0123.

Dufour, B., & Frérot, B. (2008). Optimization of coffee berry borer, *Hypothenemus hampei* Ferrari (*Col., Scolytidae*), mass trapping with an attractant mixture. *Journal of Applied Entomology, 132*(7), 591–600.

Dufumier, M. (2010). Agriculture d'abattis-brûlis, fronts pionniers et environnement en Asie du Sud-Est: Le cas du Laos. In B. Wolfer (Ed.), *Agricultures et paysanneries du monde. Mondes en mouvement, politiques en transition* (350 pp.). Versailles: Quæ.

Dugué, M. -J., Pesche, D., & Le Coq, J. -F. (2012). *Appuyer les organisations de producteurs* (144 pp.). Gembloux: coll. Agricultures tropicales en poche, Quæ-CTA, Presses universitaires de Gembloux.

Dulcire, M. (2010). De la passivité à la collaboration. L'évolution des relations entre cacaoculteurs et industriel en Équateur. *Cahiers Agricultures, 19*(4), 249–254.

Dulcire, M. (2012). The organisation of farmers as an emancipatory factor: The setting up of a supply chain of cocoa in São Tomé. *The Journal of Rural and Community Development, 7*(2), 131–141.

Dulcire, M., Vall, E., & Chia, E. (2007). *ATP CIROP, Conception des innovations et rôle du partenariat. Rapport d'ATP* (50 pp.). Montpellier: Cirad.

Dumézil, G. (1968). *Mythes et épopée. 1. L'idéologie des trois fonctions dans les épopées des peuples indo-européens*. Paris: Gallimard.

Duncan, C. A. M. (1996). *The centrality of agriculture: Between humankind and the rest of nature*. Montreal/London: McGill-Queen's University Press.

Durand, F., & Pirard, R. (2008). Quarante ans de politiques forestières en Indonésie, 1967–2007: La tentation de la capture par les élites. *Les cahiers d'outre-mer, 244*, 18.

Duranton, J. -F., Foucart, A., & Gay, P. -E. (2012). *Florule des biotopes du criquet pèlerin en Afrique de l'Ouest et du Nord-Ouest à l'usage des prospecteurs de la lutte antiacridienne*. Rome: FAO-CLCPRO, Cirad copublication.

Duris, D., Mburu, J. K., Durand, N., Clarke, R., Frank, J. M., & Guyot, B. (2010). Ochratoxin A contamination of coffee batches from Kenya in relation to cultivation methods and post-harvest processing treatments. *Food Additives and Contaminants Part A, 27*(6), 836–841.

Duyck, P. F., Rousse, P., Ryckewaert, P., Fabre, F., & Quilici, S. (2004). Influence of adding borax and modifying pH on effectiveness of food attractants for melon fly (*Diptera: Tephritidae*). *Journal of Economic Entomology, 97*, 1137–1141.

Eastwood, R., Lipton, M., & Newell, A. (2010). Farm size. In *Handbook of agricultural economics* (pp. 3323–3397). Burlington: Academic Press.

Efombagn, M. I. B., Sounigo, O., Vefonge, K. D., & Nyassé, S. (2011). Farmer participatory and collaborative approaches to cocoa breeding in Cameroon, Country report. In A. B. Eskes (Ed.),

Collaborative and participatory approaches to cocoa variety improvement. Final report of the CFC/ICCO/Bioversity project on "Cocoa Productivity and Quality Improvement: A Participatory Approach" (2004–2010). Amsterdam/London/Rome: CFC/ICCO/Bioversity International.

Egger, K. (1987). L'intensification écologique conservation (LAE) et amélioration des sols tropicaux par les systèmes agro-sylvo-pastoraux. Aménagements hydro-agricoles et systèmes de production, Montpellier, Cirad-DSA. *Documents systèmes agraires, 2*(6), 129–135.

Ehrlich, P. R., & Mooney, H. A. (1983). Extinction, substitution, and ecosystem services. *Bioscience, 33*(4), 248–254.

Ellis, F. (1993). *Peasant economics. Farm households and agrarian development* (309 pp.). W.s.i. a.a.r. development (2nd ed.). Cambridge: Cambridge University Press.

ENRD (European Network for Rural Development). (2010, April 21–23). Semi-subsistence farming in Europe: Concepts and key issues. Background paper prepared for the seminar *Semi-subsistence farming in the EU: Current situation and future prospects*, Sibiu. http://enrd.ec.europa.eu/app_templates/filedownload.cfm?id=FB3C4513-AED5-E24F-E70A-F7EA236BBB5A. Retrieved 19 Mar 2014.

Eskes, A. B. (Ed.). (2011). *Collaborative and participatory approaches to cocoa variety improvement.* Final report of the CFC/ICCO/Bioversity project on "Cocoa Productivity and Quality Improvement: A Participatory Approach" (2004–2010). Amsterdam/London/Rome: CFC/ICCO/Bioversity International.

Étienne, M., Le Page, C., & Cohen, M. (2003). A step-by-step approach to building land management scenarios based on multiple viewpoints on multi-agent system simulations. *Journal of Artificial Societies and Social Simulation, 6*(2). http://jasss.soc.surrey.ac.uk/6/2/2.html. Retrieved 19 Mar 2014.

Étienne, H., Bertrand, B., Montagnon, C., Bobadill, A., Landey, R., Dechamp, E., Jourdan, I., Alpizar, E., Malo, E., & Georget, F. (2012). Un exemple de transfert de technologie réussi dans le domaine de la micropropagation: La multiplication de *Coffea arabica* par embryogenèse somatique. *Cahiers Agricultures, 21*(2–3), 115–124.

Ezzine de Blas, D., Börner, J., Violato-Espada, A. L., Nascimento, N., & Piketty, M. G. (2011). Forest loss and management in land reform settlements: Implications for REDD governance in the Brazilian Amazon. *Environmental science and policy, 14*(2), 188–200.

FAO. (2000). *The energy and agriculture nexus*. Environment and Natural Resources Working Paper no. 4. Rome: FAO.

FAO. (2007). *A system of integrated agricultural censuses and surveys: Vol. 1. World programme for the census of agriculture 2010. FAO Statistical Development Series*, (11), Rome. ftp://ftp.fao.org/docrep/fao/008/a0135e/a0135e00.pdf. Retrieved 19 Mar 2014.

FAO. (2008). The state of food and agriculture 2008. Biofuels: Prospects, risks and opportunities. Rome: Food and Agriculture Organization of the United Nations. ftp://ftp.fao.org/docrep/fao/011/i0100e/i0100e.pdf. Retrieved 21 Mar 2014.

FAO. (2010). 2000 World census of agriculture. Main results and metadata by country (1996–2005). *FAO Statistical Development Series*, (12), 350, Rome.

FAO. (2012, April). *World agriculture watch. Monitoring structural changes in agriculture informing policy dialogue* (48 pp.). Methodological framework. Version 2.6. Rome: FAO.

FAO. (2013a). The state of food insecurity in the world 2012. http://www.fao.org/docrep/016/i3027e/i3027e00.htm. Retrieved 19 Mar 2014.

FAO. (2013b). *Climate-smart agriculture sourcebook* (570 pp.), Rome: FAO.

Farolfi, S., Müller, J. P., & Bonte, B. (2010). An iterative construction of multi-agent models to represent water supply and demand dynamics at the catchment level. *Environmental Modelling and Software, 25*, 1130–1148.

Faure, G., & Kleene, P. (2004). Lessons from new experiences in extension in West Africa: Management advice for family farms and farmers' governance. *Journal of Agricultural Education and Extension, 10*, 37–49.

Faure, G., Gasselin, P., Triomphe, B., Temple, L., & Hocdé, H. (éd. scientifiques). (2010). *Innover avec les acteurs du monde rural: La recherche-action en partenariat* (221 pp.). Paris: coll. Agriculture tropicale en poche, Quaé, CTA, Presse agronomique de Gembloux

Faure, G., Desjeux, Y., & Gasselin, P. (2011). Revue bibliographique sur les recherches menées dans le monde sur conseil en agriculture. *Cahiers Agricultures, 20*(5), 327–342.

Favareto, A. (2011, November 14–16). Le Programme national de production et d'utilisation du biodiesel (Brésil). De l'opportunité à la réalité sociale. In *3e Conférence internationale sur les biocarburants en Afrique. Les biocarburants: Quels potentiels pour l'Afrique ?* (15 pp.), Ouagadougou.

Faye, B., Alexandre, G., Bonnet, P., Boutonnet, J.-P., Cardinale, E., Duteurtre, G., Loiseau, G., Montet, D., Mourot, J., & Regina, F. (2011). Élevage et qualité des produits en régions chaudes. *Productions Animales, 24*, 77–88.

Faysse, N., Taher, S. M., & Errahj, M. (2012). Local farmers' organisations: A space for peer-to-peer learning? The case of milk collection cooperatives in Morocco. *The Journal of Agricultural Education and Extension, 18*(3), 285–299.

Feder, G., Murgai, R., & Quizon, J. B. (2004). The acquisition and diffusion of knowledge: The case of pest management training in Farmer Field Schools, Indonesia. *Journal of Agricultural Economics, 55*, 221–243.

Feintrenie, L. (2013). Opportunities to responsible land-based investment practices in Central Africa. In *World Bank International Conference on Land and Poverty* (pp. 8–11). Washington, DC: World Bank.

Feintrenie, L., & Levang, P. (2009). Sumatra's rubber agroforests: Advent, rise and fall of a sustainable cropping system. *Small-Scale Forestry, 8*(3), 323–335.

Feintrenie, L., & Rafflegeau, S. (2012, septembre 25–28). Oil palm development: Risks and opportunities based on lessons learnt from Cameroon and Indonesia. In *Memorias XVIIth Conferencia Internacional sobre Palma de Aceite y expopalma 2012*. Cartagena: Santafé de Bogota, FEDEPALMA, slide show, 1 slide show (23 slides) 17.

Feintrenie, L., Schwarze, S., & Levang, P. (2010). Are local people conservationists? Analysis of transition dynamics from agroforests to monoculture plantations in Indonesia. *Ecology and Society, 15*(4), 37 [online], http://www.ecologyandsociety.org/vol15/iss4/art37. Retrieved 19 Mar 2014.

Fig, D. (2011). Agrocarburants au Mozambique: Entre espoirs et déboires. *Alternatives Sud, 18*, 77–90.

Figuié, M., & Peyre, M. -I. (2010, November 22–24). L'incertitude dans la gouvernance internationale des zoonoses émergentes. *In colloque: Colloque Agir en situation d'incertitude. Quelles constructions individuelles et collectives des régimes de protection et d'adaptation en agriculture ?* (6 pp.). Montpellier: Cirad-Inra.

Francis, C., Breland, T. A., Østergaard, E., Lieblein, G., & Morse, S. (2013). Phenomenon-based learning in agroecology: A prerequisite for transdisciplinarity and responsible action. *Agroecology and Sustainable Food Systems, 37*(1), 60–75.

Freire, P. (1973). *Extensión o comunicación? La comunicación en el medio rural* (108 pp.). México: Siglo Veintiuno Ed.

Friedmann, H. (1978). World market, State and family farm: Social bases of household production in the era of wage labor. *Comparative Studies in Society and History, 20*, 545–586.

Friedmann, H. (2013). Farming households in 1973 and today: One path for agriculture or many paths for farming? Processed.

FruiTrop, 178. (2010).

Fuller, T. L., Gilbert, M., Martin, V., Cappelle, J., Hosseini, P., Njabo, K. Y., Abdel, A. S., Xiao, X., Daszak, P., & Smith, T. B. (2013). Predicting hotspots for influenza virus reassortment. *Emerging Infectious Diseases, 19*(4), 581–588.

Furuno, T. (2001). *The power of duck: Integrated rice and duck farming, Tasmania, Australia*. Sisters Creek: Tagari Publications.

Gabas, J.-J., & Losch, B. (2008). La fabrique en trompe-l'œil de l'émergence. In *L'enjeu mondial. Les pays émergents* (C. Jaffrelot, dir.) (pp. 25–40). Paris: Presses de Sciences Po-L'express.

Galtier, F. (2012). *Gérer l'instabilité des prix alimentaires dans les pays en développement. Une analyse critique des stratégies et des instruments* (306 pp.) Paris: Cirad-AFD. http://www.afd.fr/webdav/site/afd/shared/PUBLICATIONS/RECHERCHE/Scientifiques/A-savoir/17-A-Savoir.pdf. Retrieved 19 Mar 2014)

Gari, G., Bonnet, P., Roger, F., & Waret-Szkuta, A. (2011). Epidemiological aspects and financial impact of lumpy skin disease in Ethiopia. *Preventive Veterinary Medicine, 102*(4), 274–283. 10.1016/j.prevetmed.2011.07.003. Retrieved 19 Mar 2014.

Gasson, R. (1986). Part time farming strategy for survival? *Sociologia Ruralis, 26*(3–4), 364–376.

Gastellu, J.-M. (1980). Mais où sont donc ces unités économiques que nos amis cherchent tant en Afrique ? *Cahiers Orstom, série Sciences humaines, 17*(1–2), 3–11.

Gautier, D., Benjaminsen, T. A., Gazull, L., & Antona, M. (2012). Neoliberal forest reform in Mali: Adverse effects of a World Bank "Success". *Society and Natural Resources, 26*(6), 702–716.

Gellner, E. (1989). *Nations et nationalisme* (208 pp.). Paris: Payot.

Gentil, D. (1986). *Les mouvements coopératifs en Afrique noire. Interventions de l'État ou Organisations paysannes* (269 pp.). Paris: UCI, L'Harmattan.

Gerbier, G., Bonnet, P., Tran, A., & Roger, F. (2006). Descriptive and spatial epidemiology of Rift valley fever outbreak in Yemen 2000–2001. *Annals of the New York Academy of Sciences, 1081*, 240–242.

Geres-Iram. (2012). Note de positionnement pour un soutien aux expérimentations sur les agrocarburants paysans, 2 pp. http://www.geres.eu/fr/points-de-vue/124-pour-un-soutien-aux-experimentations-sur-les-agrocarburants-paysans. Retrieved 19 Mar 2014.

Gervais, M., Servolin, C., & Weil, J. (1965). *Une France sans paysans*. Paris: Seuil.

Gifford, R. C. (1985). Mécanisation agricole et développement: Directives pour l'élaboration d'une stratégie. *Bulletin des services agricoles* (p. 45). Rome: FAO.

Gil, P., Servan de Almeida, R., Molia, S., Chevalier, V., Traoré, H. A., Samake, K., & Albina, E. (2009, May 12–15). Isolation and molecular characterization of virulent newcastle disease viruses in Mali in 2007 and 2008. In *3th epizone meeting on epizootic animal diseases: Towards shared genetic information*, Antalya.

Gilbert, M., Xiao, X., Chaitaweesub, P., Kalpravidh, W., Premashthira, S., Boles, S., & Slingenbergh, J. (2007). Avian influenza, domestic ducks and rice agriculture in Thailand. *Agriculture, Ecosystems and Environment, 119*, 409–415.

Giller, K. E., Witter, E., Corbeels, M., & Tittonell, P. (2009). Conservation agriculture and smallholder farming in Africa: The heretics' view. *Field Crops Research, 114*, 23–34.

Giller, K. E., Corbeels, M., Nyamangara, J., Triomphe, B., Affholder, F., Scopel, E., & Tittonell, P. (2011). A research agenda to explore the role of conservation agriculture in African smallholder farming systems. *Field Crops Research, 124*, 468–472.

Giordano, M., & Villholth, K. G. (Eds.). (2007). *The agricultural groundwater revolution: Opportunities and threats to development*. Wallingford/Colombo: CABI/IWMI.

Giraud, P. -N. (1996). *L'inégalité du monde. Économie du monde contemporain* (352 pp.). Paris: Gallimard.

Godelier, M. (2004). *Métamorphoses de la parenté* (678 pp.). Paris: Fayard.

Goebel, F. R., & Way, M. (2009). Crop losses due to two sugarcane stem borers in Réunion and South Africa. *Sugar Cane International, 27*(3), 107–111.

Goergen, G., Vayssières, J. F., Gnanvossou, D., & Tindo, M. (2011). Bactrocera invadens (Diptera: Tephritidae), a new invasive fruit fly pest for the afrotropical region: Host plant range and distribution in West and Central Africa. *Environmental Entomology, 40*(4), 844–885.

Goody, J. (2006). *The theft of history* (342 pp.). Cambridge: Cambridge University Press.

Gordon Childe, V. (1949). *L'Aube de la civilisation européenne* (1re ed. 1925, 384 pp.). Paris: Payot.

Goulet, F., & Vinck, D. (2012). Innovation through subtraction: Contribution to a sociology of "detachment". *Revue Française de Sociologie, 53*(2), 195–224.

Gouyon, A., De Foresta, H., & Levang, P. (1993). Does jungle rubber deserve its name? An analysis of rubber agroforestry systems in Southeast Sumatra. *Agroforestry Systems, 22*(3), 181–206.

Grataloup, C. (2007). *Géohistoire de la mondialisation. Le temps long du monde* (256 pp.). Paris: Armand Colin.

Grau, R., Kuemmerle, T., & Macchi, L. (2013). Beyond 'land sparing versus land sharing': Environmental heterogeneity, globalization and the balance between agricultural production and nature conservation. *Current Opinion in Environmental Sustainability, 5*(5), 477–483. 10.1016/j.cosust.2013.06.001. Retrieved 20 Mar 2014.

Grechi, I., Hilgert, N., Sauphanor, B., Senoussi, R., & Lescourret, F. (2010). Modelling coupled peach tree-aphid population dynamics and their control by winter pruning and nitrogen fertilization. *Ecological Modelling, 221*, 2363–2373.

Grechi, I., Sane, C. A. B., Diame, L., De Bon, H., Benneveau, A., Michels, T., Huguenin, V., Malézieux, E., Diarra, K., & Ray, J.-Y. (2013). Mango-based orchards in Senegal: Diversity of design and management patterns. *Fruits, 68*(6), 447–466.

Griffon, M., (2006). *Nourrir la planète*, Éditions Odile Jacobs, 456 p.

Guétat-Bernard, H. (2007). Développement, mobilités spatiales, rapports de genre : une lecture des dynamiques des ruralités contemporaines (Inde du sud, Ouest Cameroun, Amazonie brésilienne). HDR, Université de Toulouse-Le Mirail.

Gumuchian, H., & Pecqueur, B. (2007). *La ressource territoriale* (254 pp.). Paris: Édition Anthropos.

Guyard, S., Apithy, L., Bouard, S., Sourisseau, J. M., Passouant, M., Bosc, P. M., & Bélières, J. F. (2013). *L'agriculture en tribu. Poids et fonctions des activités agricoles et de prélèvement* (4 pp.). Pouembout: Enquête IAC, IAC-Cirad. http://www.iac.nc/

Hainzelin, E. (2013). *Cultivating biodiversity to transform agriculture* (261 pp.). Versailles: Quæ.

Hatton, T. J., & Williamson, G. W. (2005). *Global Migration and the World Economy. Two centuries of Policy and Performance*. Cambridge: The MIT Press.

Hautdidier, B., & Gautier, D. (2005). What local benefits does the implementation of rural wood markets in Mali generate? In M. A. F. Ros-Thonen, & A. J. Dietz (Eds.), *African forests between nature and livelihood resources: Interdisciplinary studies in conservation and forest management* (pp. 191–220). London: Edwin Mellen Press.

Havard, M., & Djamen, N. P. (2010). Réforme de l'accompagnement des producteurs au Nord-Cameroun: Leçons d'un partenariat entre recherche-développement-producteurs. *Agridape, 2*(3), 14–16.

Havard, M., & Sidé, S. C. (2013, November 21–23). Les dynamiques de mécanisation de la production et de la transformation agricoles en Afrique de l'Ouest. In *4e Conférence internationale sur les biocarburants en Afrique. Quel bilan et quelles voies d'avenir pour les biocarburants et les bioénergies en Afrique ?* (10 pp.), Ouagadougou.

Hayami, Y. (1996). The peasant economic modernization. *American Journal of Agricultural Economics, 78*(5), 1157–1167.

Hayami, Y. (2002). Family farms and plantations in tropical development. *Asian Development Review, 19*(2), 67–89.

Hayami, Y. (2010). Plantations agriculture. *Handbook of Agricultural Economics, 4*, 3305–3322.

Heong, K. L., Chien, H. V., Escalada, M. M., & Trébuil, G. (2013). Réduction de l'usage des insecticides dans la riziculture irriguée d'Asie du Sud-Est: De l'écologie expérimentale au changement des pratiques à grande échelle. *Cahiers Agricultures, 22*, 378–384.

Hertzog, T., Adamczewski, A., Molle, F., Poussin, J. C., & Jamin, J. Y. (2012). Ostrich-like strategies in Sahelian sands? Land and water grabbing in the Office du Niger, Mali. *Water Alternatives, 5*(2), 304–321.

Hervieu, B., & Purseigle, F. (2011). Des agricultures avec des agriculteurs, une nécessité pour l'Europe. *Projet, 321*, 60–69.

Hill, B. (1993). The 'myth' of the family farm: Defining the family farm and assessing its importance in the European community. *Journal of Rural Studies, 9*(4), 359–370.

HLPE. (2011). *Price volatility and food security: A report by The High Level Panel of Experts on Food Security and Nutrition* (83 pp.). Rome: FAO. http://www.fao.org/fileadmin/user_upload/hlpe/hlpe_documents/HLPE-price-volatility-and-food-security-report-July-2011.pdf. Retrieved 18 Mar 2014.

HLPE. (2013). *Investing in smallholder agriculture for food security. A report by the high level panel of experts on food security and nutrition* (112 pp.). Rome: CFS-HLPE. http://www.fao.org/fileadmin/user_upload/hlpe/hlpe_documents/HLPE_Reports/HLPE-Report-6_Investing_in_smallholder_agriculture.pdf. Retrieved 19 Mar 2014.

Hocdé, H., & Miranda, B. (2000). *Los intercambios campesinos: Más allá de las fronteras… ¡Seamos futuristas!* San Jose: IICA, GTZ, Cirad.

Hocdé, H., Triomphe, B., Faure, G., & Dulcire, M. (2008). Emerging lessons about conducting action-research in partnership with farmers and others stakeholders. *Rural Development News, 2*, 58–63.

Hochman, Z., Carberry, P. S., Robertson, M. J., Gaydon, D. S., Bell, L. W., & McIntosh, P. C. (2013). Prospects for ecological intensification of Australian agriculture. *European Journal of Agronomy, 44*, 109–123.

Hofs, J. L., Gozé, E., Cene, B., Kioye, S., & Adakal, H., (2013, June 3–5). Assessing the indirect impact of Cry1Ac and Cry2Ab expressing cotton (*Gossypium hirsutum* L.) on Hemipteran pest populations in Burkina Faso (West Africa). In J. Romeis, & M. Meissle (Eds.), *Proceedings of the 6th meeting on Ecological Impact of Genetically Modified Organisms (EIGMO) of the IOBC-WPRS Working Group, GMOs in integrated plant production* (pp. 49–54). Berlin: Julius Kühn-Institut.

Hubert, B., & Caron, P. (2009). Imaginer l'avenir pour agir aujourd'hui, en alliant prospective et recherche: L'exemple de la prospective Agrimonde. *Natures sciences sociétés, 17*(4), 417–423.

Huebschle, O. J. B., Ayling, R. D., Godinho, K., Lukhele, O., Tjipura, Z. G., Rowan, T. G., & Nicholas, R. A. J. (2006). Danofloxacin (AdvocinTM) reduces the spread of contagious bovine pleuropneumonia to healthy in-contact cattle. *Research in Veterinary Science, 81*(3), 304–309. 10.1016/j.rvsc.2006.02.005. Retrieved 19 Mar 2014.

IAASTD. (2009). Agriculture at a crossroads. In B. D. McIntyre, H. R. Herren, J. Wakhungu, & R. T. Watson (Eds.), *International assessment of agricultural knowledge, science and technology for development*. Washington, DC: Island Press.

IEA. (2004). *World energy outlook 2004*. Paris: OECD/IEA.

IEA. (2006). *World energy outlook 2006*. Paris: OECD/IEA.

IEA. (2012). *World energy outlook 2012*. Paris: IEA.

IFAD. (2011). *Rural poverty report: New realities, new challenges: New opportunities for tomorrow's generation* (p. 322). Rome: International Fund for Agricultural Development. http://www.ifad.org/rpr2011/report/e/rpr2011.pdf. Retrieved 19 Mar 2014.

IHEDATE. (2011). *Croissance verte et territoires* (215 pp.). Paris.

INS. (2008). Conditions de vie des populations et profil de pauvreté au Cameroun en 2007. Rapport principal de l'ECAM III, Institut national de la statistique, ministère de l'Économie et des Finances, Yaoundé, Cameroun.

IPCC. (2007). Climate change 2007: Synthesis report. In R. K. Pachauri, & A. Reisinger, (Eds.), *Contribution of Working Groups I, II and III to the fourth assessment report of the intergovernmental panel on climate change* (p. 104). Geneva: IPCC.

ISE (International Society of Ethnobiology). (2013). *ISE Newsletter, 5*(1), April, http://ethnobiology.net/docs/ISENewsletter_2012Congress%20issue_no%20photos.pdf. Retrieved 19 Mar 2014.

Jacquet, P., Pachauri, R. K., & Tubiana, L. (2012). *Regards sur la terre. Développement, alimentation, environnement: Changer l'agriculture?* Armand Colin, 355 pp.

Jagoret, P. (2011). *Analyse et évaluation de systèmes agroforestiers complexes sur le long terme: Application aux systèmes de culture à base de cacaoyer au Centre Cameroun* (Doctoral thesis). université de Montpellier III.

Jagoret, P., Ngogue, H. T., Bouambi, E., Battini, J. L., & Nyassé, S. (2009). Diversification des exploitations agricoles à base de cacaoyer au Centre Cameroun: Mythe ou réalité ? *Biotechnologie, Agronomie. Société et Environnement, 13*(2), 271–280.

Jagoret, P., Michel-Dounias, I., & Malézieux, E. (2011). Long-term dynamics of cocoa agroforests: A case study in central Cameroon. *Agroforestry Systems, 81*, 267–278.

Jagoret, P., Michel-Dounias, I., Snoeck, D., Ngogue, H. T., & Malézieux, E. (2012). Afforestation of savannah with cocoa agroforestry systems: A small-farmer innovation in central Cameroon. *Agroforestry Systems, 86*, 493–504.

Jamin, J. -Y. (1994). De la norme à la diversité. L'intensification rizicole face à la diversité paysanne dans les périmètres irrigués de l'ON. PhD, INA PG.

Jamin, J. Y., Bisson, P., Fusillier, J. L., Kuper, M., Maraux, F., Perret, S., & Vandersypen, K. (2005, mai 19). La participation des usagers à la gestion de l'irrigation: Des mots d'ordre aux réalités dans les pays du Sud. In *Colloque Irrigation et développement durable*, Paris. *Les colloques de l'Académie d'agriculture de France, 91*(1), 65–83.

Jamin, J. Y., Bouarfa, S., Poussin, J. C., & Garin, P. (2011). Les agricultures irriguées face à de nouveaux défis. *Cahiers Agricultures, 20*, 10–15.

Jamont, M., Piva, G., & Fustec, J. (2013). Sharing N resources in the early growth of rapeseed intercropped with faba bean: Does N transfer matter? *Plant Soil, 371*, 641–653.

Jarvis, D. I., & Hodgkin, T. (2008). The maintenance of crop genetic diversity on farm: Supporting the Convention on Biological Diversity's Programme of Work on Agricultural Biodiversity. *Biodiversity, 9*(1–2), 23–28.

Jeantet, A. (1998). Les objets intermédiaires dans la conception. Éléments pour une sociologie des processus de conception. *Sociologie du travail, 3*, 291–316.

Jones, W. O. (1968). Plantations. In *International encyclopedia of social sciences*. New York: MacMillan/Free Press.

Joulian, C., Escoffier, S., Le Mer, J., Neue, H. U., & Roger, P. A. (1997). Populations and potential activities of methanogens and methanotrophs in rice fields: Relations with soil properties. *European Journal of Soil Biology, 33*(2), 105–116.

Jourdain, D., Rakotofiringa, A., Quang, D. D., Valony, M. J., Vidal, R., & Jamin, J. Y. (2011). Gestion de l'irrigation dans les montagnes du nord du Vietnam: Vers une autonomie accrue des irrigants ? *Cahiers Agricultures, 20*, 78–84.

Jouve, P., & Mercoiret, M. R. (1987). La recherche-développement: Une démarche pour mettre les recherches sur les systèmes de production au service du développement rural. *Les cahiers de recherche-développement, 16*, 8–15.

Kairu-Wanyoike, S., Kaitibie, S., Taylor, N., Gitau, G., Heffernan, C., Schnier, C., Kiara, H., Taracha, E., & McKeever, D. (2013). Exploring farmer preferences for contagious bovine pleuropneumonia vaccination: A case study of Narok District of Kenya. *Preventive Veterinary Medicine, 110*(3–4), 356–369. 10.1016/j.prevetmed.2013.02.013. Retrieved 20 Mar 2014.

Kautsky, K. (1970). *La Question agraire: Etude sur les tendances de l'agriculture moderne* (1re éd. 1899, traduit de l'allemand par Milhaud E. et Polack C.). Paris: Giard et Bière. réimpression en fac-similé (463 pp.). Paris: François Maspero.

Khan, Z. R., Midega, C. A. O., Bruce, T. J. A., Hooper, A. M., & Pickett, J. A. (2010). Exploiting phytochemicals for developing a 'push-pull' crop protection strategy for cereal farmers in Africa. *Journal of Experimental Botany, 61*, 4185–4196.

Kidd, A. D., Lamers, J. P. A., Ficarelli, P. P., & Hoffmann, V. (2000). Privatising agricultural extension: Caveat emptor. *Journal of Rural Studies, 16*, 95–102.

Kilian, B., Jones, C., Pratt, L., & Villalobos, A. (2006). Is sustainable agriculture a viable strategy to improve farm income in Central America? A case study on coffee. *Journal of Business Research, 59*, 322–330.

Klerkx, L., Grip, K. D., & Leeuwis, C. (2006). Hands off but strings attached: The contradictions of policy-induced demand-driven agricultural extension. *Agriculture and Human Values, 23*, 189–204.

Klerkx, L., Hall, A., & Leeuwis, C. (2009). Strengthening agricultural innovation capacity: Are innovation brokers the answer? *International Journal of Agricultural Resources. Governance and Ecology, 8*(5–6), 409–438.

Kohler, F. (2011). Diversité culturelle et diversité biologique: Une approche critique fondée sur l'exemple brésilien. *Natures Sciences Sociétés, 19*(2), 113–124.

Krausmann, F. (2011). The global metabolic transition: A historical overview. In F. Krausmann (Ed.), *The socio-metabolic transition. Long term historical trends and patterns in global material and energy use*. Social Ecology Working Paper 131. Vienna: Institute of Social Ecology.

Kulikoff, A. (1993). Households and markets: Toward a new synthesis of American agrarian history. *The William and Mary Quarterly*, 3rd serie, *50*, 342–355.

Kuper, M., Bouarfa, S., Errahj, M., Faysse, N., Hammani, A., Hartani, T., Marlet, S., Zairi, A., Bahri, A., Debbarh, A., Garin, P., Jamin, J.-Y., & Vincent, B. (2009). A crop needs more than a drop: Towards a new praxis in irrigation management in North Africa. *Irrigation and Drainage, 58*(S3), S231–S239.

Labarthe, P. (2005). Trajectoires d'innovation des services et inertie institutionnelle. Dynamique du conseil dans trois agricultures européennes. *Géographie, économie, société, 73*, 289–311.

Labatut, J., Aggeri, F., Bibe, B., & Girad, N. (2001). Formes et crises de la coopération dans la production de connaissances pour la gestion des ressources génétiques animales: Une approche par les régimes de sélection. *Revue d'anthropologie des connaissances, 5*(2), 302–336.

Lamanda, N. (2004). *Caractérisation et évaluation agro-écologique des systèmes de culture agroforestiers: Une démarche appliquée aux systèmes de culture à base de cocotiers (Cocos nucifera L.) sur l'île de Malo* (200 pp.). Paris: Institut national agronomique Paris-Grignon.

Lamarche, H. (1991). *L'agriculture familiale. Comparaisons internationales I. Une réalité polymorphe* (303 pp.). Paris: L'Hatmattan.

Lapeyre de Bellaire de, L., Essoh Ngando, J., Abadie, C., Chabrier, C., Blanco, R., Lescot, T., Carlier, J., & Côte, F. (2009). Is chemical control of *Mycosphaerella* foliar diseases of bananas sustainable? *Acta Horticulturae (ISHS), 828*, 161–170.

Lascoumes, P. (2006). Les normes. In L. Boussaguet, S. Jacquot, & P. Ravinet (dir.), *Dictionnaire des politiques publiques* (776 pp.). coll. Références, Presses de Sciences Po.

Laurent, C., & Rémy, J. (2000). L'exploitation agricole en perspective. *Courrier de l'environnement de l'Inra, 41*, 5–22.

Layadi, A., Faysse, N., & Dumora, C. (2011). Les organisations professionnelles agricoles locales, partenaires pour renforcer le dialogue technique ? Une expérience avec des éleveurs de bovins au Maroc. *Cahiers Agricultures, 20*(5), 428–433.

Le Bellec, F., Vélu, A., Le Squin, S., & Michels, T. (2013). Utilisation de l'indicateur I-PHY comme outil d'aide à la décision en verger d'agrumes à la Réunion. Le cas de la Lambda cyhalothrine. *Innovations agronomiques*, (31), 61–73.

Le Page, C., Bazile, D., Becu, N., Bommel, P., Bousquet, F., Étienne, M., Mathevet, R., Souchère, V., Trébuil, G., & Weber, J. (2013). Agent-based modelling and simulation applied to environmental management. In B. Edmonds, & R. Meyer (Eds.), *Simulating social complexity: A handbook* (pp. 499–540). Heidelberg: Springer. 10.1007/978-3-540-93813-2_19. Retrieved 20 Mar 2014.

Leclerc, C., & Coppens d'Eeckenbrugge, G. (2012). Social organization of crop genetic diversity. The G x E x S interaction model. *Diversity, 4*, 1–32.

Lenin, V. I. (1899). The development of capitalism in Russia. In *Collected works* (4th ed., Vol. 3). Moscow: Progress Publishers, 1964. http://www.marxists.org/archive/lenin/works/1899/devel/index.htm. Retrieved 2 Feb 2014.

Léonard, E. (2011). Pluralisme institutionnel et reconfigurations de l'*ejido* au Mexique. De la gouvernance foncière au développement local. *Problèmes d'Amérique latine, 79*, 13–34.

Lesnoff, M., Laval, G., & Bonnet, P. (2002). Demographic parameters of a domestic cattle herd in a contagious-bovine-pleuropneumonia infected area of Ethiopian highlands. *Revue d'élevage et de médecine vétérinaire des pays tropicaux, 55*, 139–147.

Lesnoff, M., Laval, G., & Bonnet, P. (2003, November 17–21). Economic analysis of control strategies for contagious bovine pleuropneumonia (CBPP) outbreaks at herd level: A stochastic epidemiological model for smallholders in Ethiopian highlands. In *10th International symposium for veterinary epidemiology and economics*. Viña del mar, Chile: ISVEE.

Lesnoff, M., Laval, G., Bonnet, P., Lancelot, R., & Thiaucourt, F. (2004). A mathematical model of the effect of chronic carriers on the within-herd spread of contagious bovine pleuropneumonia in an African mixed crop-livestock system. *Preventive Veterinary Medicine, 62*, 101–107. 10.1016/j.prevetmed.2003.11.009. Retrieved 20 Mar 2014.

Lesnoff, M., Bonnet, P., Laval, G., & Abdicho, S. (2005). Effects of antibiotic treatment practices on the within-herd spread of contagious bovine pleuropneumonia: A simulation study in an Ethiopian mixed crop-livestock system. *Ethiopian Veterinary Journal, 9*(1), 59–73.

Lesueur-Jannoyer, M., Cattan, P., Monti, D., Saison, C., Voltz, M., Woignier, T., & Cabidoche, Y. M. (2012). Chlordécone aux Antilles: évolution des systèmes de culture et leur incidence sur la dispersion de la pollution. *Agronomie, environnement et sociétés, 2*(1), 45–58.

Levang, P., & Sevin, O. (1989). Quatre-vingts ans de transmigration en Indonésie (1905–1985). *Annales de géographie, 549*, 28.

Licciardi, S., Assogba-Komlan, F., Sidick, I., Chandre, F., Hougard, J. M., & Martin, T. (2008). A temporary tunnel screen as an eco-friendly method for small-scale growers to protect cabbage crop in Benin. *International Journal of Tropical Insect Science, 27*, 152–158.

Litvine, D., Dabat, M. -H., & Gazull, L. (2013). The influence of proximity on the potential demand for vegetable oil as a diesel substitute: A rural survey in West Africa. *Cahiers de recherche du Creden*, no. 13.09.102, http://www.creden.univ-montp1.fr/downloads/cahiers/CC-13-09-102.pdf. Retrieved 20 Mar 2014.

Liu, M. (1997). *Fondements et pratiques de la recherche action* (351 pp.). Paris: L'Harmattan.

Lizarralde, R. D., Colmenares, B. C., de la Rosa, D., Garcia, S. R., Mantilla, P. G., & Zorro, R. (2012). De las alianzas estrategicas a los negocios inclusivos. *Palmas, 33*(2).

Lobell, D. B., Cassman, K. G., & Field, C. B. (2009). Crop yield gaps: Their importance, magnitudes and causes. *Annual Review of Environment and Resources, 34*, 179–204.

Losch, B. (2012a). Le défi de l'emploi, le rôle de l'agriculture et les impasses du débat international sur le développement. Note préparatoire au colloque *Évolution du marché international du travail, impacts des exclusions paysannes*, Conseil économique, social et environnemental, 16 October 2012. Paris: Cirad-AFD.

Losch, B. (2012b). Relever le défi de l'emploi: L'agriculture au centre. *Perspectives*, 19. Montpellier: Cirad.

Losch, B., & Magrin, G. (2013). Villes et campagnes à la recherche d'un nouveau modèle: Il faut désegmenter les territoires et les politiques. In B. Losch, G. Magrin, & J. Imbernon (dir.), *Une nouvelle ruralité émergente. Regards croisés sur les transformations rurales africaines*. Cirad-Nepad.

Losch, B., Fréguin-Gresh, S., & White, E. (2012). *Structural transformation and rural change revisited: Challenges for late developing countries in a globalizing world.* World Bank-Agence française de développement.

Losch, B., Magrin, G., & Imbernon, J. (Eds.). (2013). *Une nouvelle ruralité émergente: Regards croisés sur les transformations rurales africaines* (46 pp.). Montpellier: Cirad-Nepad.

Louafi, S., Bazile, D., & Noyer, J. L. (2013). Conserving agriculture genetic diversity: Transcending established divides. In É. Hainzelin (Ed.), *Cultivating biodiversity to transform agriculture* (pp. 181–220). Versailles: Quæ/Springer.

MacMynowski, D. P. (2007). Pausing at the brink of interdisciplinarity: Power and knowledge at the meeting of social and biophysical science. *Ecology and Society, 12*, 14.

Madhlopa, A., & Ngwalo, G. (2007). Solar dryer with thermal storage and biomass-backup heater. *Solar Energy, 81*, 449–462.

Magrin, G., & N'Dieye, P. (2007, November 27–29). Biocarburants, aménagement du territoire et politiques agricoles en Afrique: Un éléphant dans un magasin de porcelaine ? In *Conférence internationale Enjeux et perspectives des biocarburants pour l'Afrique*, Ouagadougou, Burkina Faso.

Malézieux, R., Crozat, Y., Dupraz, C., Laurans, M., Makowski, D., Ozier-Lafontaine, H., Rapidel, B., de Tourdonnet, S., & Valantin-Morison, M. (2009). Mixing plant species in cropping systems: Concepts, tools and models. A review. *Agronomy for Sustainable Development, 29*, 43–62.

Maltsoglou, I., Koisumi, T., & Felix, E. (2013). The status of bioenergy development in developing country. *Global Food Security, 2*(2), 104–109, 10.1016/j.gfs.2013.04.002. Retrieved 20 Mar 2014.

Manoli, C., & Ancey, V. (2014). L'ambiguïté des effets de la mobilité des jeunes pasteurs: Entre la sécurisation des conditions de vie pastorales et l'émancipation individuelle au Ferlo, Sénégal. In V. Ancey, G. Azoulay, C. Crenn, D. Dormoy, A. Mangu, & A. Thomashausen (Eds.), *Mobilités et migrations: Figures et enjeux contemporains* (324 pp.). coll. Pluralité des regards et des disciplines, L'Harmattan, Presses universitaires de Sceaux.

Marien, J. N., & Bassaler, N. (2013). Éléments de prospective à l'échéance 2040 pour les écosystèmes forestiers d'Afrique centrale. Rapport de synthèse Comifac, 125 pp.

Marien, J. N., Dubiez, E., Louppe, D., & Larzillière, A. (2013). *Quand la ville mange la forêt: Les défis du bois-énergie en Afrique centrale* (238 pp.). Versailles: Quæ.

Maris, V. (2010). *Philosophie de la biodiversité: Petite éthique pour une nature en péril* (214 pp.). Buchet-Chastel.

Marshall, A. (2011). Terres gagnées et terres perdues. Conséquences environnementales de l'essor de l'agro-industrie dans un désert de Piémont. *Bulletin de l'Institut français des études andines, 40*(2), 375–396.

Martin, T., Assogba-Komlan, F., Houndete, T., Hougard, J. M., & Chandre, F. (2006). Efficacy of mosquito netting for sustainable small holders' cabbage production in Africa. *Journal of Economic Entomology, 99*, 450–454.

Martin, T., Palix, R., Kamal, A., Delétré, E., Bonafos, R., Simon, S., & Ngouajio, M. (2013). A repellent treated netting as a new technology for protecting vegetable crops. *Journal of Economic Entomology, 106*(4), 1699–1706.

Marzin, J., Benoit, S., Lopez, B. T., Cid, L. G., Mercoiret, M. R., & Hocdé, H. (2013). *Una metodología de extensión agraria generalista, sistémica y participativa*. La Habana: ACTAF.

Massardier, G., Sabourin, E., Lecuyer, L., & De Avila, M. L. (2012). La démocratie participative comme structure d'opportunité et de renforcement de la notabilité sectorielle. Le cas des agriculteurs familiaux dans le Programme de développement durable des territoires ruraux au Brésil, territoire Aguas Emendadas. *Participations, 1*(2), 78–102.

Mayaud, J. -L. (1999). *La petite exploitation rurale triomphante. France, XIXe siècle* (278 pp.). Paris: Belin.

Mazoyer, M. (2001). *Protéger la paysannerie pauvre dans un contexte de mondialisation*. Rome: FAO.

Mazoyer, M., & Roudart, L. (1997). *Histoire des agricultures du monde, du néolithique à la crise contemporaine* (546 pp.). Paris: Le Seuil.

Mc Cullough, E. B., Pingali, P., & Stamoulis, K. (2008). *The transformation of agri-food systems. Globalization, supply chains and smallholder farms* (381 pp.). Rome/London: FAO/Earthscan.

McMillan, M., & Rodrik, D. (2011). Globalization, structural change, and productivity growth. NBER Working Paper no. 17143.

Megevand, C. (2013). *Deforestation trends in the Congo Basin: Reconciling economic growth and forest protection*. Washington, DC: World Bank. doi:10.1596/978-0-8213-9742-8.

Meillassoux, C. (1975). *Femmes, greniers et capitaux* (251 pp.). Paris: Maspero.

Meillassoux, C. (2005). La parenté est-elle une affaire de vie ou de survie ? *Actuel Marx, 1*(37), 15–26.

Méndez, V. E., & Bacon, C. M. (2013). Agroecology as a transdisciplinary, participatory, and action-oriented approach. *Agroecology and Sustainable Food Systems, 37*, 3–18.

Mendras, H. (1967). *La fin des paysans*. Paris: SEDEIS, Futuribles.

Mendras, H. (1976). *Sociétés paysannes. Éléments pour une théorie de la paysannerie* (238 pp.). coll. U, Armand Colin.

Mendras, H. (2000). L'invention de la paysannerie. Un moment de l'histoire de la sociologie française d'après-guerre. *Revue française de sociologie, 41*(3), 539–552.

Mercandalli, S. (2013). Le rôle complexe des migrations dans les reconfigurations des systèmes d'activités des familles rurales: La circulation comme ressource ? Localité de Leonzoane, Mozambique 1900–2010. Thèse de doctorat en sciences économiques, Paris, université Paris Sud XI, 497 pp.

Mercer, B., Finighan, J., Sembres, T., & Schaefer, J. (2011). *Protecting and restoring forest carbon in tropical Africa: A guide for donors and funders*. Forests Philanthropy Action Network (FPAN), 322 p, www.forestsnetwork.org.

Mercoiret, M.-R. (2006). Les organisations paysannes et les politiques agricoles. *Afrique contemporaine, 217*, 135–157.

Merlet, P., & Merlet, M. (2010). Legal pluralism as a new perspective to study land rights in Nicaragua. A different look at the Sandinista Agrarian reform. In *Land reforms and management of natural resources in Africa and Latin America conference* (pp. 24–26). Spain: University of Lleida.

Miguel, E., Grosbois, V., Berthouly-Salazar, C., Caron, A., Cappelle, J., & Roger, F. (2013). A meta-analysis of observational epidemiological studies of Newcastle disease in African agrosystems, 1980–2009. *Epidemiology and Infection, 141*(6), 1117–1133.

Mikolasek, O., Khuyen, T. D., Medoc, J.-M., & Porphyre, V. (2009a). The ecological intensification of an integrated fish farming model: Recycling pig effluents from farms in Thai Binh province (North Vietnam). *Cahiers Agricultures, 18*, 235–241.

Mikolasek, O., Barlet, B., Chia, E., Pouomogne, V., & Tomedi, M. (2009b). Développement de la petite pisciculture marchande au Cameroun: La recherche-action en partenariat. *Cahiers Agricultures, 18*(2–3), 270–276.

Millenium Ecosytem Assessment (MEA). (2005a). Millenium ecosytem assessment findings, http://www.unep.org/maweb/documents/document.297.aspx.pdf. Retrieved 20 Mar 2014.

Millennium Ecosystem Assessment (MEA). (2005b). *Ecosystems and human well-being: Our human planet: Summary for decision makers* (109 pp.). Washington, DC: Island Press.

Minet, C., Gil, P., Libeau, G., & Albina, E. (2007, August 20–22). Development of a diva vaccine against peste des petits ruminants by reserve genetic. Does control of animal infectious risks offer a new international perspective ? In *Proceedings of the 12th international conference of the Association of Institutions of Tropical Veterinary Medicine* (Camus, ed., pp. 183–185). Montpellier: Cirad. Accessed 22 Nov 2007.

Missingham, B. D. (2003). *The assembly of the poor in Thailand: From local struggles to national protest movement*. Chiang-Mai: Silkworm Books.

Molia, S. (2012). Deficient reporting in avian influenza surveillance, Mali. *Emerging Infectious Diseases, 18*(4), 691–693. 10.3201/eid1804.111102. Retrieved 20 Mar 2014.

Molia, S., Boly, I. A., Duboz, R., & Fournié, G. (2012, August 20–24). Social network analysis of poultry trade movements in Sikasso county, Mali: Implications for surveillance of avian influenza. In *13th international symposium on veterinary epidemiology and economics*. Book of abstracts. Wageningen: Wageningen Academic Publishers.

Mollard, E., & Walter, A. (Eds.). (2008). *Agricultures singulières* (344 pp.). Paris: IRD Éditions.

Monbeig, P. (1996). Les franges pionnières. In A. Journaux, P. Deffontaines, & M. Delamarre (Eds.), *Géographie générale* (pp. 974–1006). Paris: Encyclopédie de la Pléiade.

Morein, B., Thiaucourt, F., Dedieu, L., & Tulasne, J. J. (1999). *CBPP EU project final scientific report and appendixes to the final scientific report*, Meeting on EU funded Research report. Montpellier: OAU IBAR, Cirad-EMVT, European Union.

Moskow, A. (1999). The contribution of urban agriculture to gardeners, their households, and surrounding communities: The case of Havana, Cuba. In M. Koc, R. MacRae, L. Mougeot, &

J. Walsh (Eds.), *For hunger-proof cities: Sustainable urban food systems*. Ottawa: International Development Research Centre.

Moulin, A. (1992). *Les paysans dans la société française. De la Révolution à nos jours*. Paris: Le Seuil.

Moustier, P. (2010). L'accès des petits producteurs aux filières de qualité au Vietnam. Promouvoir l'information et la coopération. *Perspectives, 5*. http://hal.cirad.fr/hal-00723545/. Retrieved 20 Mar 2014.

Mrema, C. G., Baker, D., & Kahan, D. (2008). Agricultural mechanization in sub-saharan Africa: Time for a new look. FAO Occasional paper 22, 70 pp. ftp://ftp.fao.org/docrep/fao/011/i0219e/i0219e00.pdf. Retrieved 20 Mar 2014.

Muchnik, J., Cañada, J. S., & Salcido, G. T. (2008). Systèmes agroalimentaires localisés: état des recherches et perspectives. *Cahiers Agricultures, 17*(6), 509–512.

Muleke, E. M., Saidi, M., Itulya, F. M., Martin, T., & Ngouajio, M. (2013). The assessment of the use of eco-friendly nets to ensure sustainable cabbage seedling production in Africa. *Agronomy, 3*, 1–12.

Multigner, L., Ndong, J. R., Giusti, A., Romana, M., Delacroix-Maillard, H., Cordier, S., Jégou, B., Thome, J. P., & Blanchet, P. (2010). Chlordecone exposure and risk of prostate cancer. *Journal of Clinical Oncology, 28*, 3457–3462. doi:10.1200/JCO.2009.27.2153.

N'Guessan, R., Corbel, V., Akogbeto, M., & Rowlands, M. (2007). Reduced efficacy of insecticide-treated nets and indoor residual spraying for malaria control in pyrethroid resistance area, Benin. *Emerging Infectious Disease, 13*, 199–206.

Naiman, R. J. (1999). A perspective on interdisciplinary science. *Ecosystems, 2*, 292–295.

Namdar-Irani, M., & Sotomayor, O. (2011). Le conseil agricole au Chili face à la diversité des agriculteurs. *Cahiers Agricultures, 20*(5), 352–358.

Nepad. (2013). *Les agricultures africaines, transformations et perspectives*. Midrand: Nepad.

Neveu, C. (Ed.). (2007). *Cultures et pratiques participatives. Perspectives comparatives* (402 pp.). Paris: coll. Logiques politiques, L'Harmattan.

Nibouche, S., Tibère, R., & Costet, L. (2012). The use of *Erianthus arundinaceus* as a trap crop for the stem borer *Chilo sacchariphagus* reduces yield losses in sugarcane: Preliminary results. *Crop Protection, 42*, 10–15.

Nonato de Souza, H., de Goede, R. G. M., Brussaard, L., Cardoso, I. M., Duarte, E. M. G., Fernandes, R. B. A., Gomes, L. C., & Pulleman, M. M. (2012). Protective shade, tree diversity and soil properties in coffee agroforestry systems in the Atlantic Rainforest biome. *Agriculture, Ecosystems and Environment, 146*, 179–196.

Obosu-Mensah, K. (2002). Changes in official attitudes towards urban agriculture in Accra. *African Studies Quarterly*, (6), 3, http://web.africa.ufl.edu/asq/v6/v6i3a2.htm. Retrieved 20 Mar 2014.

OECD. (2005). *Oslo manual: Guidelines for collecting and interpreting innovation data* (3rd ed., 163 pp.). Luxembourg: OCDE. http://www.oecd.org/sti/oslomanual. Retrieved 20 Mar 2014.

OECD. (2011). *Fostering productivity and competitiveness in agriculture*. OECD Publishing. 10.1787/9789264166820-en. Retrieved 20 Mar 2014.

Oerke, E. C., & Denne, H. W. (2004). Safeguarding production-losses in major crops and the role of crop protection. *Crop Protection, 23*, 275–285.

Oil World. (2013). *Oil world annual* (Eds.). Hamburg: ISTA Mielke GmbH.

Openshaw, K. (2010). Can biomass power development? *Gatekeeper*, 144. London: International Institute for Environment and Development.

Openshaw, K. (2011). Supply of woody biomass, especially in the tropics: Is demand outstripping sustainable supply? *International Forestry Review, 13*(4), 487–499.

Orsini, F., Kahane, R., Nono-Womdim, R., & Gianquinto, G. (2013). Urban agriculture in the developing world: A review. *Agronomy for Sustainable Development, 33*(4), 695–720.

Ostrom, E. (1990). *Governing the commons. The evolution of institutions for collective action* (280 pp.). New York: Cambridge University Press.

Otsuka, K. (2008). Peasants. In S. N. Durlauf, & L. E. Blume (Eds.), *The new Palgrave dictionary of economics*. Basingstoke: Palgrave Macmillan.
Paillard, S., Treyer, S., & Dorin, B. (2011). *Agrimonde. Scénarios et défis pour nourrir le monde* (296 pp.). Versailles: Quæ.
Parks, S., & Gowdy, J. (2013). What have economists learned about valuing nature? A review essay. *Ecosystem Services, 3*, e1–e10.
Parrot, L., Dongmo, C., Ndoumbé, M., & Poubom, C. (2008). Horticulture, livelihoods, and urban transition in Africa: Evidence from South-West Cameroon. *Agricultural Economics, 39*(2), 245–256.
Parrot, L., Sotamenou, J., Dia, K. B., & Nantchouang, A. (2009). Determinants of domestic waste input use in urban agriculture lowland systems in Africa: The case of Yaoundé in Cameroon. *Habitat International, 33*(4), 357–364.
Paugam, S. (1986). Déclassement, marginalité et résistance au stigmate en milieu rural breton. *Anthropologie et sociétés, 10*(2), 23–36.
Paul, M., Wongnarkpet, S., Gasqui, P., Poolkhet, C., Thongratsakul, S., Ducrot, C., & Roger, F. (2011). Risk factors for highly pathogenic avian influenza (HPAI) H5N1 infection in backyard chicken farms, Thailand. *Acta Tropica, 118*(3), 209–216.
Paul, M., Baritaux, V., Wongnarkpet, S., Poolkhet, C., Thanapongtharm, W., Roger, F., Bonnet, P., & Ducrot, C. (2013). Practices associated with highly pathogenic avian influenza spread in traditional poultry marketing chains: Social and economic perspectives. *Acta Tropica, 126*(1), 43–53. 10.1016/j.actatropica.2013.01.008. Retrieved 20 Mar 2014.
Pautasso, M., Aistara, G., Barnaud, A., Caillon, S., Clouvel, P., Coomes, O. T., & Tramontini, S. (2013). Seed exchange networks for agrobiodiversity conservation. A review. *Agronomy for Sustainable Development, 33*(1), 151–175.
Pédelahore, P. (2012). Stratégies d'accumulation des exploitants agricoles: L'exemple des cacaoculteurs du Centre Cameroun de 1910 à 2010. PhD, Toulouse 2.
Peltier, R. (Ed.). (1996). Les parcs à *Faidherbia. Cahiers scientifiques du Cirad-Forêt, 12*, 312 pp., available online since 2012, http://ur-bsef.cirad.fr/publications-et-ressources/ressources-en-ligne/agroforesterie/les-parcs-a-faidherbia. Retrieved 20 Mar 2014.
Peltier, R., Marquant, B., Palou Madi, O., Ntoupka, M., & Tapsou, J. -M. (2013). Boosting traditional management of Sahelian Faidherbia parks? The role of functional diversity for ecosystem services in multi-functional agroforestry. *FUNCITREE final conference*, 23–25 May 2013, Trondheim. http://funcitree.nina.no/Portals/ft/Session%20III%20presentations.pdf. Retrieved 20 Mar 2014.
Penot, É. (2006). Processus d'innovation et crises multiples: Les hévéaculteurs indonésiens dans la tourmente. In J. Caneill (dir.), *Agronomes et innovations, 3e édition des Entretiens du Pradel*. Paris: L'Harmattan.
Penot, É., Macdowall, C., & Domas, R. (2012). Modeling impact of conservation agriculture adoption on farming systems agricultural incomes. The case of lake Alaotra Region, Madagascar (9 pp.). RIME-PAMPA/CA2AFRICA project. Denmark: IFSA.
Perfecto, I., Vandermeer, J., & Wright, A. (2009). *Nature's matrix: Linking agriculture, conservation and food sovereignty* (242 pp.). Oxford: Earthscan.
Perret, S., Farolfi, S., & Hassan, R. (Eds.). (2006). *Water governance for sustainable development: Approaches and lessons from developing and transitional countries* (XXIV-295 pp.). Londres: Earthscan Publications.
Perry, B. D., Randolph, T. F., Ashley, S., Chimedza, R., Forman, T., Morrison, J., Poulton, C., Sibanda, L., Stevens, C., Tebele, N., & Yngström, I. (2003). *The impact and poverty reduction implications of foot and mouth disease control in southern Africa with special reference to Zimbabwe* (137 pp.). Nairobi: ILRI.
Pesche, D. (2007). Dynamique d'organisation des ruraux et renforcement des capacités pour l'élaboration des politiques publiques en Afrique subsaharienne. In H. Delorme, & J. M. Boussard (Eds.), *La régulation des marchés agricoles internationaux. Un enjeu décisif pour le développement* (pp. 158–168). Paris: L'Harmattan.

Petit, M. (1975). Évolution de l'agriculture et caractère familial des exploitations agricoles. *Economie rurale, 106*(1), 45–55.

Peyre, M.-I., Samaha, H., Saad, A., Abd-Elnabi, A., Galal, S., Ettel, T., Dauphin, G., Lubroth, J., Roger, F., & Domenech, J. (2009). Avian influenza vaccination in Egypt: Limitations of the current strategy. *Journal of Molecular and Genetic Medicine, 3*(2), 198–204. Accessed 1 Jan 2010.

Piketty, M. G., Veiga, J. B., Tourrand, J. F., Alves, A. M., Poccard-Chapuis, R., & Thales, M. C. (2005). The determinants of the expansion of cattle ranching in the Eastern Amazon region: Consequences for public policies. *Cadernos de ciência y tecnologia, Embrapa, 22*(1), 221–234.

Piketty, M. G., Drigo, I., Sablayrolles, P., Araujo, E., Pena, D., & Sist, P. (2014). Current barriers threatening income generation from community-based forest management in the Brazilian Amazon. In P. Katila, G. Galloway, W. de Jong, P. Pacheco, G. Mery, & R. Alfaro (Eds.), *Forests under pressure: Local responses to global issues: Vol. 32. IUFRO World Series*. IUFRO.

Pingali, P., Bigot, Y., & Binswanger, H. P. (1988). *La mécanisation agricole et l'évolution des systèmes agraires en Afrique subsaharienne* (204 pp.). Washington, DC: World Bank.

Piou, C., Lebourgeois, V., Benahi, A. S., Bonnal, V., Jaavar, M. E. H., Lecoq, M., & Vassal, J. M. (2013). Coupling historical prospection data and a remotely-sensed vegetation index for the preventative control of Desert locusts. *Basic and Applied Ecology, 14*(7), 593–604.

Pirard, R., & Treyer, S. (2010). Agriculture et déforestation: Quel rôle pour REDD+ et les politiques publiques d'accompagnement ? IDDRI, *Idées pour le débat, 10*. www.iddri.org.

Piraux, M. (2009). Dinâmicas territoriais: Definição e analise: Aplicação no Nordeste do Brasil. In A. G. Da Silva, C. J. Salete Barbosa, & W. M. de Nazareth (Eds.), *Diversificação dos espaços rurais e dinâmicas territoriais no Nordeste do Brasil* (pp. 31–54). João Pessoa: Zarinha Centro de Cultura.

Piraux, M., Assis, W. S., de Rodriguez, V. D. C., Silva, N. N. M., & Wilson, A. J. (2013). Um olhar sobre a diversidade dos Colegiados dos Territórios da Cidadania (Balance of the institutional arrangements in the Joint Committee of Territories of Citizenship). *Novos Cadernos NAEA, 16* (1), 101–124.

Poccard-Chapuis, R. (2004). Les réseaux de la conquête. Rôle des filières bovines dans la structuration de l'espace sur les fronts pionniers d'Amazonie orientale brésilienne. Thèse de doctorat en géographie, Université Paris X-Nanterre, 435 p. + annexes.

Polanyi, K. (1983). *La grande transformation. Aux origines politiques et économiques de notre temps*. Paris: Gallimard, original edition 1944.

Pomeranz, K. (2000). *The great divergence*. Princeton: Princeton University Press.

Ponniah, A., Davis, K. E., & Sindu, W. (2007). Farmer field schools: An alternative to existing extension systems? Experience from Eastern and Southern Africa. *Journal of International Agricultural and Extension Education, 14*, 81–93.

Pratt, A. N., Bonnet, P., Jabbar, M., Ehui, S., & De Haan, C. (2005). *Benefits and costs of compliance of sanitary regulations in livestock markets: The case of rift valley fever in the Somali region of ethiopia* (71 pp.). Nairobi: ILRI.

Prudent, P., Loko, S., Deybe, D., & Vaissayre, M. (2007). Factors limiting the adoption of IPM practices by cotton farmers in Benin: A participatory approach. *Experimental Agriculture, 43* (1), 113–124.

Rafflegeau, S. (2008). *Dynamique d'implantation et conduite technique des plantations villageoises de palmier à huile au Cameroun: facteurs limitants et raisons des pratiques* (148 pp.). (Thèse d'État, Agronomie). Paris: AgroParisTech.

Randrianarivelo, R., Danthu, P., Benoit, C., Ruez, P., Raherimandimby, M., & Sarter, S. (2010). Novel alternative to antibiotics in shrimp hatchery: Effects of the essential oil of *Cinnamosma fragrans* on survival and bacterial concentration of *Penaeus monodon* larvae. *Journal of Applied Microbiology, 109*, 642–650.

Rapidel, B., DeClerck, F., Le Coq, J. -F., & Beer, J. (2011). *Ecosystem services from agriculture and agroforestry: Measurement and payment* (XIX-414 pp.). London: Earthscan Publications.

Rastoin, J. -L. (2008). Les multinationales dans le système alimentaire. *Projet, 307*, 7 pp.

Rastoin, J. -L., & Ghersi, G. (2010). *Le système alimentaire mondial. Concepts et méthodes, analyses et dynamiques* (565 pp.). Versailles: coll. Synthèses, Quæ.

Ratnadass, A., & Djimadoumngar, K. (2001). Les insectes ravageurs des sorghos repiqués ou cultivés en conditions de décrue en Afrique de l'Ouest et du Centre. In J. Comas, & E. Gomez-McPherson (Eds.), AECI*La culture du sorgho de décrue en Afrique de l'Ouest et du Centre* (pp. 65–80). Rome: FAO.

Ratnadass, A., Cissé, B., Diarra, D., Sidibe, B., Sogoba, B., & Thiero, C. A. T. (1999). Faune des stocks de sorgho dans deux régions du Mali et comparaison des pertes infligées aux variétés locales ou introduites pour améliorer le rendement. *Annales de la Société entomologique de France, 35*, 489–495.

Ratnadass, A., Marley, P. S., Hamada, M. A. G., Ajayi, O., Cissé, B., Assamoi, F., Atokple, I. D. K., Beyo, J., Cissé, O., Dakouo, D., Diakité, M., Dossou, Y. S., Le Diambo, B., Vopeyande, M. B., Sissoko, I., & Tenkouano, A. (2003). Sorghum head-bugs and grain molds in West and Central Africa. 1. Host plant resistance and bug-mold interactions on sorghum grains. *Crop Protection, 22*, 837–851.

Ratnadass, A., Michellon, R., Randriamanantsoa, R., & Séguy, L. (2006). Effects of soil and plant management on crop pests and diseases. In N. Uphoff, A. Ball, E. Fernandes, H. Herren, O. Husson, M. Laing, C. Palm, J. Pretty, P. Sanchez, N. Sanginga, & J. Thies (Eds.), *Biological Approaches for Sustainable Soil Systems* (pp. 589–602). Boca Raton: CRC Press.

Ratnadass, A., Cissé, B., Cissé, S., Cissé, T., Hamada, M. A., Chantereau, J., & Letourmy, P. (2008). Combined on-farm effect of plot size and sorghum genotype on sorghum panicle-feeding bug infestation in Mali. *Euphytica, 159*, 135–144.

Ratnadass, A., Fernandes, P., Avelino, J., & Habib, R. (2012a). Plant species diversity for sustainable management of crop pests and diseases in agroecosystems: A review. *Agronomy for Sustainable Development, 32*(1), 273–303.

Ratnadass, A., Razafindrakoto, C. R., Andriamizehy, H., Ravaomanarivo, L. H. R., Rakotoarisoa, H. L., Ramahandry, F., Ramarofidy, M., Randriamanantsoa, R., Dzido, J. L., & Rafarasoa, L. S. (2012b). Protection of upland rice at Lake Alaotra (Madagascar) from black beetle damage (*Heteronychus plebejus*) (*Coleoptera: Dynastidae*) by seed dressing. *African Entomology, 20* (1), 177–181.

Ratnadass, A., Blanchart, E., & Lecomte, P. (2013a). Ecological interactions within the biodiversity of cultivated systems. In É. Hainzelin (Ed.), *Cultivating biodiversity to transform agriculture* (pp. 141–180). Versailles: Quæ/Springer.

Ratnadass, A., Randriamanantsoa, R., Ernest, R. T., Rabearisoa, M., Rafamatanantsoa, E., Moussa, N., & Michellon, R. (2013b). Interaction entre le système de culture et le statut (ravageur ou auxiliaire) des vers blancs (*Coleoptera: Scarabeoidea*) sur le riz pluvial. *Cahiers Agricultures, 22*(5), 432–441. doi:10.1684/agr.2013.0649.

Raton, G. (2013). Les relations villes-campagnes en Afrique de l'Ouest: Une densification à valoriser. In B. Losch, G. Magrin, & J. Imbernon (Eds.), *Une nouvelle ruralité émergente: Regards croisés sur les transformations rurales africaines* (pp. 36–37). Cirad: Montpellier.

Reardon, T., & Timmer, C. P. (2007). Transformation of markets for agricultural output. In R. Evenson, & P. Pingali (Eds.), *Developing countries since 1950: How has thinking changed? Handbook of agricultural economics* (ed. 1), *3*(1), 2807–2855. Elsevier.

Reij, C., Tappan, G., & Smale, M. (2009). Re-greening the Sahel: Farmer-led innovation in Burkina Faso and Niger. IFPRI. http://www.ifpri.org/book-5826/millionsfed/cases/innovation. Retrieved 20 Mar 2014.

Remedio, E. M., & Domac, J. U. (2003). *Socio-economic analysis of bioenergy systems: A focus on employment*. Rome: FAO.

Rémy, J. (2008). "Paysans, exploitants familiaux, entrepreneurs...", de qui parlons-nous? In *Les mondes agricoles en politique* (6 pp.). Paris: Centre d'études et de recherches internationales.

Riaux, J. (2011). Faut-il formaliser les règles de gestion de l'eau ? Une expérience dans le Haut Atlas. *Cahiers Agricultures, 20*, 67–72.

Ribot, J. C. (2004). *Waiting for democracy: The politics of choice in natural resource decentralization.* Washington, DC: World Resources Institute.

Rist, G. (1996). *Le développement. Histoire d'une croyance occidentale.* Paris: Presses de Sciences Po.

Rivera, W. M. (2000). Confronting global market: Public sector agricultural extension reconsidered. *Journal of Extension Systems, 16*, 33–54.

Rivera, W. M., & Alex, G. (2004). Extension system reform and the challenges ahead. *Journal of Agricultural Education and Extension, 10*, 23–36.

Rivier, M., Méot, J.-M., Ferré, T., & Briard, M. (2009). *Le séchage des mangues* (112 pp.). Versailles: Quæ-CTA.

Rivier, M., Méot, J. M., Sebastian, P., & Collignan, A. (2013, November 21–23). Les bioénergies, opportunité pour le développement du secteur agroalimentaire: étude de la filière mangue séchée au Burkina Faso. In *4e Conférence internationale sur les biocarburants en Afrique. Quel bilan et quelles voies d'avenir pour les biocarburants et les bioénergies en Afrique ?* (10 pp.), Ouagadougou

Röling, N., & Jong, F. D. (1998). Learning: Shifting paradigms in education and extension studies. *Journal of Agricultural Education and Extension, 5*, 143–161.

Rondot, P., & Collion, M.-H. (2001). *Organisations paysannes. Leur contribution au renforcement des capacités rurales et à la réduction de la pauvreté. Compte rendu des travaux.* Washington, DC: World Bank.

Roppa. (2013, September 11–14). Déclaration finale de la rencontre de dialogue sur les progrès réalisés dans la mise en œuvre des engagements de Maputo, Monrovia (Liberia).

Rostow, W. W. (1960). *The stages of economic growth: A non-communist manifesto.* Cambridge: Cambridge University Press.

Rouw, A. D. (1995). The fallow period as a weed-break in shifting cultivation (tropical wet forests). *Agriculture, Ecosystems and Environment, 54*, 31–43.

Rowe, J. W. F. (1965). *Primary commodities in international trade.* Cambridge: Cambridge Eng. University Press.

Ruf, F. (1995). *Booms et crises du cacao, les vertiges de l'or brun.* Paris: ministère de la Coopération, Cirad-Sar et Karthala.

Ruf, F. (2011). L'agrumiculture familiale produit une "révolution cacaoyère" » en Indonésie. *Grain de sel, 54–56*, April–December 2011.

Rulli, M. C., Saviori, A., & D'Odorico, P. (2013). Global land and water grabbing. In *Proceedings of the National Academy of Sciences of the USA.* http://www.pnas.org/content/early/2013/01/02/1213163110.abstract. Retrieved 20 Mar 2014.

Rusinamhodzi, L., Corbeels, M., Nyamangara, J., & Giller, K. E. (2012). Maize-grain legume intercropping is an attractive option for ecological intensification that reduces climatic risk for smallholder farmers in central Mozambique. *Field Crops Research, 136*, 12–22.

Sablon, M., Marzin, J., & Lopez, B. T. (2013). *Memorias de los Talleres Nacionales de Profesores de Extensión Agraria.* La Habana: Ed Caminos.

Sabourin, E., Triomphe, B., Lenne, P., Xavier, J. H. V., Oliveira, M., & Scopel, E. (2010, 28 June–1 July). The co-construction of knowledge between researchers and farmers in technical innovation processes: Learning from direct seeding in the Brazilian Cerrados. In *Innovation and sustainable development in agriculture and food, international symposium ISDA 2010,* Montpellier, Abstracts and papers, Cd-Rom.

Sachs, J. D. (2005). *Investing in development: A practical plan to achieve the millenium development goals* (329 pp.), United Nations Millenium Project (UNMP). London: Earthscan.

Sarter, S., Nguyen, H. N. K., Hung, L. T., Lazard, J., & Montet, D. (2007). Antibiotic resistance in Gram-negative bacteria isolated from farmed catfish. *Food Control, 18*, 1391–1396.

Sayago, D., Tourrand, J. F., Bursztyn, M., & Drummond, J. A. (Eds). (2010). *L'Amazonie, un demi-siècle après la colonisation* (271 pp.). Versailles: Quæ.

Schumpeter, J. A. (1935). *Théorie de l'évolution économique. Recherches sur le profit, le crédit, l'intérêt et le cycle de conjoncture* (590 pp.). Paris: Librairie Dalloz.

Scopel, E., Triomphe, B., Affholder, F., Macena da Silva, F. A., Corbeels, M., Xavier, J. H. V., Lahmar, R., Recous, S., Bernoux, M., Blanchart, E., Mendes, I. D. C., & Tourdonnet, S. D. (2012). Conservation agriculture cropping systems in temperate and tropical conditions, performances and impacts. A review. *Agronomy for Sustainable Development, 33*, 113–130.

Sébillotte, M. (2007). Quand la recherche participative interpelle le chercheur. In M. Anadon (dir.), *La recherche participative: Multiples regards* (pp. 49–87). Quebec: Presses de l'Université du Québec.

Seccombe, W. (2005). Les différents types de famille au sein des modes de production (traduit de l'anglais par Luc Benoit). PUF. *Actuel Marx, 37*, 27–42.

Secretariat of the Convention on Biological Diversity. (2010). Global biodiversity outlook 3 – Executive summary (12 pp.), Montreal. http://www.cbd.int/gbo/gbo3/doc/GBO3-Summary-final-en.pdf. Retrieved 14 Mar 2014.

Self, G., Ducamp, M. N., Thaunay, P., & Vayssières, J. F. (2012). The effects of phytosanitary hot water treatments on West African mangoes infested with *Bactrocera invadens* (*Diptera: Tephritidae*). *Fruits, 67*(6), 439–449.

Sélimanovski, C. (2009). Effets de lieu et processus de disqualification sociale. *Espace populations sociétés, 1*, 119–133.

Semporé, A., Andrieu, N., & Bayala, I. (2011). Coconception d'innovations agropastorales assistée par un modèle à l'échelle de l'exploitation. Cas de l'embouche bovine. *Revue d'é levage et de médecine vétérinaire des pays tropicaux, 64*(1–4), 51–60.

Sen, A. (1999). *Development as freedom* (366 pp.). New York: Anchor Book.

Servan de Almeida, R., Fridolin Maminiaina, O., Gil, P., Hammoumi, S., Molia, S., Chevalier, V., Koko, M., Harentsoaniaina, R. A., Traoré, A., Samake, K., Diarra, A., Grillet, C., Martinez, D., & Albina, E. (2009). Africa, a reservoir of new virulent strains of Newcastle disease virus? *Vaccine, 27*(4), 3127–3129. 10.1016/j.vaccine.2009.03.076. Retrieved 20 Mar 2014.

Servolin, C. (1989). *L'agriculture moderne* (289 pp.). Paris: coll. Économie, Le Seuil.

Shanin, T. (1974). The nature and logic of peasant economy. *Journal of Peasant Studies, 1–2*, 186–206.

Shanin, T. (1986). Chayanov message: Illuminations, miscomprehensions and the contemporary "Development theory". In A. V. Chayanov (Ed.), *Introduction to the theory of peasant economy*. Madison: The University of Wisconsin Press.

Shanin, T. (Ed.). (1988). *Peasant and peasant societies* (350 pp.). London: Penguin Books.

Sidibé, A., vom Brocke, K., Coulibaly, H., & Evrard, J. C. (2011). *Production de semences de sorgho en milieu paysan au Mali* (43 pp.). Montpellier: Cirad.

Sindzingre, A. (2005, août 29–31). The multidimensionality of poverty: An institutionalist perspective. In *International conference the many dimensions of poverty*. Brasilia: International Poverty Centre, UNDP.

Sinzogan, A. A. C., Van Mele, P., & Vayssières, J.-F. (2008). Implications of on-farm research for local knowledge related to fruit flies and the weaver ant *Oecophylla longinoda* in mango production. *International Journal of Pest Management, 54*(3), 241–246.

Sissoko, S., Doumbia, S., Vaksmann, M., Hocdé, H., Bazile, D., Sogoba, B., Kouressy, M., Vom, B. K., Coulibaly, M., Touré, A., & Dicko, B. G. (2008). Prise en compte des savoirs paysans en matière de choix variétal dans un programme de sélection. *Cahiers Agricultures, 17*(2), 128–133.

Smektala, G., Peltier, R., Sibelet, N., Leroy, M., Manlay, R., Njiti, C. F., Ntoupka, M., Njiemoun, A., Palou, O., & Tapsou. (2005). Parcs agroforestiers sahéliens: De la conservation à l'aménagement. *VertigO, 6*(2). Montreal: Institut des sciences de l'environnement, Université du Québec. http://vertigo.revues.org/index4410.html. Retrieved 20 Mar 2014.

Smil, V. (1991). *General energetics: Energy in the biosphere and civilization*. New York: Wiley.

Smit, J., Ratta, A., & Nasr, J. (1996). *Urban agriculture: Food, jobs and sustainable cities*. New York: UNDP, Habitat II Series.

Somarriba, E., Deheuvels, O., & Cerda, R. (CATIE). (2013, January 15). Trading-off cacao yields, carbon stocks and biodiversity in cocoa agroforestry. In *Séminaire agroécologie*, Montpellier.

Sotamenou, J., & Parrot, L. (2013). Sustainable urban agriculture and the adoption of composts in Cameroon. *International Journal of Agricultural Sustainability, 11*, 282–295.

Soti, V., Chevalier, V., Maura, J., Bégué, A., Lelong, C., Lancelot, R., Thiongane, Y., & Tran, A. (2013). Identifying landscape features associated with Rift Valley fever virus transmission, Ferlo region, Senegal, using very high spatial resolution satellite imager. *International Journal of Health Geographics, 12*(10), 11 pp. 10.1186/1476-072X-12-10. Retrieved 20 Mar 2014.

Sourisseau, J.-M., Bosc, P.-M., Fréguin-Gresh, S., Bélières, J.-F., Bonnal, P., Le Coq, J.-F., Anseeuw, W., & Dury, S. (2012). Les modèles familiaux de production agricole en question. Quelle méthode pour analyser leur diversité ? *Autrepart, 3*(62), 159–181.

Stiglitz, J. E. (2006). *Un autre monde, contre le fanatisme du marché* (452 pp.). Paris: Fayard.

Stolcke, V. (1986). *Cafeicultura: Homens, mulheres e capital (1850–1980)*. São Paulo: Brasiliense.

Subedi, A., Chaudhary, P., Baniya, B. K., Rana, R. B., Tiwari, R. K., Rijal, D. K., Sthapit, B. R., & Jarvis, D. I. (2003). Who maintains crop genetic diversity and how? Implications for on-farm conservation and utilization. *Culture and Agriculture, 25*(2), 41–50.

Subervie, J., & Vagneron, I. (2013). A drop of water in the Indian Ocean? The impact of GlobalGap Certification on lychee farmers in Madagascar. *World Development, 50*, 57–73.

Swanson, B. E. (2006). The changing role of agricultural extension in a global economy. *Journal of International Agricultural and Extension Education, 13*, 5–17.

Tabashnik, B. E., Brévault, T., & Carrière, Y. (2013). Insect resistance to Bt crops: Lessons from the first billion acres. *Nature Biotechnology, 31*, 510–521.

Tatsidjodoung, P., Dabat, M. H., & Blin, J. (2012). Insights into biofuel development in Burkina Faso: Potential and strategies for sustainable energy policies. *Renewable and Sustainable Energy Reviews, 16*(7), 5319–5330.

Taylor, A. (2001). *American colonies. The settling of north America* (544 pp.). New York: Penguin Books.

Tchayanov, A. V. (1972). Pour une théorie des systèmes économiques non capitalistes. 1re édition 1924. *Analyse et prévision, 13*, 19–51.

Tchayanov, A. V. (1990). *L'organisation de l'économie paysanne* (1re éd. 1923, 344 pp.). Paris: Librairie du Regard.

Temple, L., Marquis, S., & Simon, S. (2008). Le maraîchage périurbain à Yaoundé est-il un système de production localisé innovant ? *Économies et sociétés, 30*, 2309–2328.

Temple, L., Bonin, M., Houdart, M., & Joubert, N. (2010). Déterminants institutionnels de la diminution de pesticides dans la bananeraie antillaise: Nécessité d'indicateurs d'évaluations partagés. In *Colloque SFER: La réduction des pesticides agricoles – enjeux, modalités et consé quences* (16 pp.), Lyon.

Temple, L., Boyer, J., Briend, A., & Dameus, A. (2013). Les conditions socio-économiques de l'innovation agro-écologique pour la sécurisation alimentaire dans les jardins agroforestiers en Haïti. *Facts*, accepted in press.

Thiaucourt, F., Aboubakar, Y., Wesonga, H., Manso-Silvan, L., Blanchard, A., Schudel, A., & Lombard, M. (2004). Contagious bovine pleuropneumonia vaccines and control strategies: Recent data, In *Control of infectious animal diseases by vaccination* (developments in biologicals) (pp. 99–111).

Thurlow, J. (2010). Economic wide effects of bioenergy development in bioenergy and food security: The BEFS Analysis for Tanzania. Environment and Natural Resources Management, Working Paper, 35. Rome: FAO.

Timmer, C. P. (2009). *A world without agriculture: The structural transformation in historical perspective* (83 pp.). Washington, DC: The American Enterprise Institute Press.

Timone, E. (2013). La culture du palmier à huile en région amazonienne: Entre acceptation, résignation et résistance. Analyses des dynamiques et conflits dans la microrégion de Tomé-Açú,

Pará, Brésil. AgroParisTech Master 2, Sciences et techniques du vivant et de l'environnement, spécialité Environnement, développement, territoires, sociétés, 2012–2013, 72 pp.

Todd, E. (2011). *L'origine des systèmes familiaux. 1. L'Eurasie* (768 pp.). Paris: coll. NRF Essais, Gallimard.

Ton, G. (2012). La inteligencia organizacional: La riqueza de las organizaciones de agricultores. *Leisa, Revista de agroecología, 28*(2), 20–21.

Torquebiau, E. F. (2000). A renewed perspective on agroforestry concepts and classification. *Comptes rendus de l'Académie des sciences, Série III Sciences de la vie, 323*, 1009–1017.

Torquebiau, E., Cholet, N., Ferguson, W., & Letourmy, P. (2013). Designing an index to reveal the potential of multipurpose landscapes in Southern Africa. *Land, 2*(4), 705–725.

Toulmin, C. (2009). Securing land and property rights in sub-Saharan Africa: The role of local institutions. *Land Use Policy, 26*, 10–19.

Touré, S., & Mortelmans, J. (1990). Impact de la trypanosomose animale africaine. *Bulletin des séances, Académie royale des sciences d'outre-mer, 36*(2), 239–257.

Triomphe, B., & Rajallahti, R. (2013). From concept to emerging practice: What does an innovation system perspective bring to agricultural and rural development? In E. Coudel, H. Devautour, C. Soulard, G. Faure, & B. Hubert (Eds.), *Renewing innovation systems in agriculture and food: How to go towards more sustainability?* (pp. 57–76). Wageningen: Wageningen Academic Publishers.

Tritsch, I., Gond, V., Oszwald, J., Davy, D., & Grenand, P. (2012). Dynamiques territoriales des amérindiens wayãpi et teko du moyen Oyapock, Camopi, Guyane française. *Bois et forêts des tropiques, 311*, 49–61.

UNCTAD. (2013). Wake up before it's too late. Make agriculture truly sustainable now for food security in a changing climate. United Nations, UN Conference on Trade and Development. *Trade and Environment Review 2013*, New York, 341 pp.

Unep. (2010). A brief for policymakers on the green economy and millennium development goals. United Nations Environment programme. http://www.unep.ch/etb/publications/Green%20Economy/Brief%20Policymakers%20MDGs%20Summit%20Sept%202010/GREENECO-MDGs%20Policymakers%20Brief.pdf. Retrieved 20 Mar 2014.

United Nations. (2006). *Human development report 2006* (422 pp.). New York: United Nations Development Programme.

United Nations. (2013). *Millenium development goals: 2013 report* (63 pp.). New York: United Nations. http://www.un.org/millenniumgoals/poverty.shtml. Retrieved 20 Mar 2014.

Vakulabharanam, V. (2013). Fighting poverty through good governance using randomized experiments. *Development and Change, 44*(4), 1027–1037.

Valeix, S. (2012). *La surveillance des maladies animales à l'échelle locale: études des facteurs liés aux logiques d'une communauté villageoises en Thaïlande* (43 pp.). Mémoire de master SAEPS, UM2-Cirad-INP Toulouse.

Valette, É., Chéry, J.-P., Debolini, M., Azodjilande, J., François, M., & El Amrani, M. (2013). Urbanisation en périphérie de Meknès (Maroc) et devenir des terres agricoles. L'exemple de la coopérative agraire Naïji. *Cahiers Agricultures*.

Vall, E., & Diallo, M. A. (2009). Savoirs techniques locaux et pratiques: La conduite des troupeaux aux pâturages (ouest du Burkina Faso). *Natures sciences sociétés, 17*(2), 122–135.

Vall, E., Blanchard, M., Diallo, M., Dongmo, A., & Bayala, I. (2009, April 20–23). Savoirs techniques locaux, sources d'innovations ? Production de savoirs actionnables dans une démarche de recherche action en partenariat. In B. L. Seiny, & P. Boumard (Eds.), *Actes du colloque Prasac-Ardesac, Savanes africaines en développement: Innover pour durer* (15 pp.). Garoua: Cameroun.

Vall, E., Blanchard, M., Koutou, M., Coulibaly, K., Diallo, M. A., Chia, E., Traoré, L., Tani, F., Andrieu, N., Ouattara, B., Dugué, P., & Autfray, P. (2012, May 14–17). Recherche action en partenariat et innovations face aux changements globaux de l'Afrique subsaharienne. In P. Sérémé, & H. Roy-Macauley (Eds.), *Empowering the rural poor to adapt to climate change and variability in west and central Africa. Proceedings of CORAF/WECARD 3rd agricultural science week* (pp. 76–81), Ndjaména, Chad.

Van der Ploeg, J. D. (2008). *The new peasantries: Struggle for autonomy and sustainability in an era of empire and globalization* (356 pp.). Sterling: Earthscan.

Van der Ploeg, J. D. (2013). *Peasant and the art of farming. Chayanovian manifesto* (157 pp.). Canada: Agrarian Change and Peasant Studies, Fernwood Publishing.

Van Ittersum, M. K., Cassman, K. G., Grassini, P., Wolf, J., Tittonell, P., & Hochman, Z. (2013). Yield gap analysis with local to global relevance. A review. *Field Crops Research, 143*, 4–17.

Van Mele, P., & Vayssières, J.-F. (2007). Weaver ants help farmers to capture organic markets. *Pesticide News, 75*, 9–11.

Vayssières, J. F., Cayol, J. P., Perrier, X., & Midgarden, D. (2007). Impact of methyl eugenol and malathion bait stations on non-target insect populations in French Guiana during an eradication program for *Bactrocera carambolae*. *Entomologia Experimentalis et Applicata, 125*(1), 55–62.

Vayssières, J., Thévenot, A., Vigne, M., Cano, M., Broc, A., Bellino, R., Diacono, E., De Laburthe, B., Bochu, J. L., Tillard, E., & Lecomte, P. (2012). Évaluation des inefficiences zootechnique et environnementale pour intensifier écologiquement les systèmes d'élevage tropicaux. *Revue d'élevage et de médecine vétérinaire des pays tropicaux, 64*(1–4), 73–79.

Vejpas, C., Bousquet, F., Naivinit, W., & Trébuil, G. (2004, October 15–17). Participatory modelling for managing rainfed lowland rice varieties and seed system in lower Northeast Thailand. In *Proceedings of Mekong rice conference* (p. 15). Ho Chi Minh City, Vietnam: IRRI Press.

Vermeulen, S., & Goad, N. (2006). Towards better practice in smallholder palm oil production. IIED Report, 55 pp.

Verspieren, M. -R., & Chia, E. (2012). Rôle d'une recherche-action sur la diffusion des savoirs et la modification du contexte social. In B. Bourrassa, & M. Boudjaoui (dir.), *Des recherches collaboratives en sciences humaines et sociales (SHS) – enjeux, modalités et limites* (pp. 47–76). Quebec: Presses universitaires Laval.

Vigouroux, Y., Barnaud, A., Scarcelli, N., & Thuillet, A. C. (2011). Biodiversity, evolution and adaptation of cultivated crops. *CR Biologies, 334*, 450–457.

Vivien, F. D. (2002). De Rio à Johannesburg les négociations autour de la diversité biologique. *Ecologie et politique, 26*, 35–53.

Von Maltitz, G., & Stafford, W. (2011). Assessing opportunities and constraints for biofuel development in sub-Saharan countries. CIFOR, Working paper, 66 pp.

Von Maydell, H. J. (1983). *Arbres et arbustes du Sahel* (532 pp.). Bonn: GTZ

Vorley, B., Fearne, A., & Ray, D. (Eds.). (2007). *Regoverning markets: A place for small scale producers in modern agrifood chains?* (258 pp.). Gower.

Wallerstein, I. M. (1974). *Capitalist agriculture and the origins of the European world-economy in the sixteenth century*. New York: Academic Press.

Wallerstein, I. (1989). *The modern world-system*. San Diego: Academic Press.

Wane, A., Ancey, V., Touré, I., Kâ, S. N., & Diao-Camara, A. (2010). L'économie pastorale face aux incertitudes. Le salariat au Ferlo (Sahel sénégalais). *Cahiers Agricultures, 19*(5), 359–365.

Wardell, D. A. (2003). Empire forestry in the margins of empire. Forest reservation in the northern territories of the Gold Coast Colony (pp. 75–114). SEREIN Occasional Paper no. 15, Institute of Geography, University of Copenhagen.

Waret-Szkuta, A., Ortiz-Pelaez, A., Pfeiffer, D. U., Roger, F., & Guitian, F. J. (2011). Herd contact structure based on shared use of water points and grazing points in the Highlands of Ethiopia. *Epidemiology and Infection, 139*(6), 875–885.

Weber, M. (1991). *Histoire économique: Esquisse d'une histoire universelle de l'économie et de la société*. Paris: Gallimard.

Wegner, G., & Pascual, U. (2011). Cost-benefit analysis in the context of ecosystem services for human well-being: A multidisciplinary critique. *Global Environmental Change: Human and Policy Dimensions, 21*, 492–504.

White, B. (2012). Agriculture and the generation problem: Rural youth, employment and the future of farming. *IDS Bulletin, 43*(6), 9–19.
WHO. (2005). Ecosystems and human well-being: Health synthesis. A report of the Millennium Ecosystem Assessment. Geneva: WHO. http://www.unep.org/maweb/en/Framework.aspx. Retrieved 20 Mar 2014.
WHO. (2013a). Initiative to estimate the global burden of foodborne diseases, The unknown Burden. Geneva: WHO. http://www.who.int/foodsafety/foodborne_disease/ferg/en/index1.html. Retrieved 20 Mar 2014.
WHO. (2013b). The 17 neglected tropical diseases. Geneva: WHO, http://www.who.int/neglected_diseases/diseases/en/. Retrieved 20 Mar 2014.
Wiese, M., & Wyss, K. (1998). *Les populations nomades et la santé humaine et animale en Afrique et notamment au Tchad*. APT, Institute of Physical Geography, University of Freiburg, Germany, Swiss Tropical Institute, Basle.
Wolf, E. R. (1966). *Peasants*. Chicago: Prentice-Hall.
Wood, D., & Lenne, J. M. (1997). The conservation of agrobiodiversity on farm: Questioning the emerging paradigm. *Biodiversity and Conservation, 6*, 109–129.
World Bank. (2006). *Enhancing agricultural innovation: How to go beyond the strengthening of research systems*. Washington, DC: World Bank.
World Bank. (2007). *World development report 2008* (386 pp.). Washington, DC: The International Bank for Reconstruction and Development.
World Bank. (2013). *Latin America and the Caribbean Poverty and Labor Brief, June 2013: Shifting gears to accelerate shared prosperity in Latin America and the Caribbean*. Washington, DC: World Bank.
Wrigley, E. A. (1988). *Continuity Chance and Change. The Characters of the Industrial Revolution in England*. Cambridge/New York: Cambridge University Press.
Yadouleton, A., Martin, T., Padonou, G., Chandre, F., Asidi, A., Djogbenou, L., Dabiré, R., Aïkpon, R., Boko, M., Glitho, I., & Akogbeto, M. (2011). Cotton pest management practices and the selection of pyrethroid resistance in Anopheles gambiae population in Northern Benin. *Parasites and Vectors, 4*, 60.
Yapi, A. M., & Debrah, S. K. (1998). Evaluation de l'impact des recherches variétales de sorgho et de mil en Afrique de l'Ouest et du Centre. In *Amélioration du sorgho et de sa culture en Afrique de l'Ouest et du Centre* (pp. 215–221). Actes de l'atelier de restitution du programme conjoint sur le sorgho ICRISAT-CIRAD. Bamako, Mali, 17–20 March 1997. Montpellier: CIRAD-CA.
Zhang, L. (Ed.). (2007). *The evolution of poverty reduction policies in China, 1949–2005* (280 pp.). Beijing: ICPRC.

Printed in the USA
CPSIA information can be obtained
at www.ICGtesting.com
CBHW060237071024
15472CB00003B/62